Cryopreservation of Plant Cells and Organs

Editor

K. K. Kartha, Ph.D.
Senior Research Officer
Plant Biotechnology Institute
National Research Council
Saskatoon, Saskatchewan
Canada

CRC Press, Inc.
Boca Raton, Florida

Library of Congress Cataloging in Publication Data
Main entry under title:

Cryopreservation of plant cells and organs.

  Bibliography: p.
  Includes index.
  1. Plant cells and tissues—Cryopreservation.
I. Kartha, K. K. (Kutty K.)
QK725.C78   1985   581'.07'24   84-7694
ISBN 0-8493-6102-8

This book represents information obtained from authentic and highly regarded sources. Reprinted material is quoted with permission, and sources are indicated. A wide variety of references are listed. Every reasonable effort has been made to give reliable data and information, but the author and the publisher cannot assume responsibility for the validity of all materials or for the consequences of their use.

Direct all inquiries to CRC Press, Inc., 2000 Corporate Blvd., N.W., Boca Raton, Florida, 33431.

International Standard Book Number 0-8493-6102-8

Library of Congress Card Number 84-7694
Printed in the United States

DEDICATED TO MY LATE BELOVED FATHER, DEVAN RAMAN KARTHA

PREFACE

In the rapidly emerging domain of biotechnology, plant tissue culture, from a modest start, has now attained a unique and respectable status, more so in the last decade. Since plant cells or organs are amenable, in most cases, to in vitro culture resulting in the regeneration of entire plants, tissue culture techniques have been extensively used not only in the rapid clonal multiplication and eradication of systemic pathogens, notably viruses, but also in addressing problems related to physiology, genetics, developmental biology, and biochemistry. Recently, it was also recognized that by an interdisciplinary approach involving tissue culture and cryobiology, it should be feasible to preserve plant organs and cell lines for extended periods of time for the purpose of long-term preservation of germplasm of crop species or experimental material in a genetically unaltered state. Although this area of research (cryopreservation of plant cells, tissues, and organs) is still in its infancy as compared to the developments made with mammalian systems, remarkable progress has been made during the last 5 years. The purpose of this volume, therefore, is to provide the reader with an in-depth look at the subject matter area, analyze various critical components involved in the methodology, and present the progress made so far in the cryopreservation of various kinds of plant material.

Since plant cells, tissues, or organs form the basic experimental material in the cryopreservation study, the book opens with a chapter on plant cell culture in order for the reader to gain an overall appreciation of this field of research. In my opinion, a clear understanding of the various events a cell is subjected to during freezing and thawing is imperative before cryopreservation is attempted. The following three chapters provide basic information concerning the principles, nature, and mechanism of freezing injury to plant cells at various levels and attempt to suggest various ways and means of overcoming them for devising successful cryopreservation protocols. In subsequent chapters the role of cryoprotectants in the viable freezing of plant cells and the cryopreservation studies carried out with several types of plant material are examined in detail.

The authors of various chapters are internationally known scientists in the area of their expertise. I am extremely grateful to the contributors without whose cooperation and help this assignment could not have been completed. If the information presented in this book generates added interest and research activity among people considering cryopreservation of plant cells, tissues, and organs, the very purpose of the book, in my opinion, is served.

K. K. Kartha

THE EDITOR

Dr. K. K. Kartha is a Senior Research Officer at the Plant Biotechnology Institute of the National Research Council of Canada, Saskatoon. He obtained his M.Sc. (Agriculture) and Ph.D. degrees in India. His Doctorate in Mycology and Plant Pathology was obtained from the Indian Agricultural Research Institute, New Delhi, in 1969. From 1970 to 1972, he did his postdoctoral research at the Institut National de la Recherche Agronomique, Versailles, France, in the laboratories of Dr. M. T. Cousin and the late Professor Georges Morel. It was Professor Morel who introduced him to plant tissue culture research.

He joined on staff of the Plant Biotechnology Institute in 1973. He has published over 50 scientific papers in the area of plant pathology, plant tissue culture and cryopreservation. He has also contributed several chapters in various books. In 1981 he was a recipient of the George M. Darrow Award for excellence in research from the American Society for Horticultural Science for his work on the cryopreservation of strawberry meristems and mass propagation of plantlets. He was an invited symposium speaker at several international symposia including the Rockefeller Foundation Conference on Genetic Engineering for Crop Improvement held in New York in 1980.

He is a member of the Society for Cryobiology, Canadian Society of Plant Physiologists, Canadian Phytopathological Society, International Association for Plant Tissue Culture, and a life member of the Indian Phytopathological Society. Currently he is the elected National Correspondent for Canada for the International Association for Plant Tissue Culture.

His research interests include tissue culture, virus elimination, cryopreservation, cloning agricultural plants through in vitro techniques, and somatic cell hybridization by fusion of plant protoplasts.

TABLE OF CONTENTS

Chapter 1

PLANT CELL CULTURE*

Friedrich Constabel

TABLE OF CONTENTS

* NCCC #20993.

I. INTRODUCTION

Over 40 years ago White and Gautheret initiated plant cell culture as a method of experimental morphology.[1,2] Early objectives were to parallel animal cell culture and demonstrate theoretically indefinite longevity of meristems. Since then the cell culture method has permitted substantial advances in understanding growth and differentiation of plants. A cornerstone was the realization of totipotency of nonzygotic cells by Steward and Reinert.[3,4] More recently, plant cell culture has become a tool of geneticists and is considered ancillary to analyzing and manipulating the plant genome and to improving crops. Cryopreservation of cells and tissues is an integral part of this effort.

II. DEFINITION

The term *plant cell culture* refers to a variety of techniques which permit the growth and development, in vitro, of protoplasts, single cells, tissues, and organs derived from seed plants and ferns in a well-defined environment. As with animal cell culture, plant cell culture is not to suggest a suspension of uniformly single cells as can be obtained with microorganisms. The term is justified, however, when viewing the general objective to be attained by manipulating plant material at the level of cells.

III. METHOD

Plant cell culture requires the excision of a piece of tissue from a given plant and transfer of this tissue, the explant, to a nutrient medium formulated to induce and sustain cell multiplication and formation of new growth, callus. Step-by-step descriptions of procedures are presented in various laboratory manuals.[5,6] Isolation of callus from the explant and serial subculture in fresh media leads to the establishment of a callus strain or cell line. Employing liquid media and agitation will result in suspension cultures of smaller and smaller callus pieces, the formation of a cell suspension. Similarly, callus and cell suspensions can be obtained by starting cultures with cells freed from their walls, i.e., protoplasts.

The problem of not being able to create permanent, true cell suspensions of seed plants has often been attacked.[7] Sieves, wall-digesting enzymes, and a variety of hormones have been employed, but to no avail. The formation of a phragmoplast and primary wall between daughter cells appears to be an integral part of cell division and, so far, not accessible to biochemical or mutational inhibition. Callus, then, remains the predominant form of cell material open to in vitro culture.

IV. CALLUS

Coined for wound tissue, the term callus here refers to a more or less unorganized mass of parenchyma cells sometimes including sklerenchyma, i.e., tracheids. Callus originates from parenchyma cells of the explant by way of cell division. Thus, the size of the initial callus can be a function of the amount of parenchyma in the explant. Callus growth is sustained by the activity of meristematic cells which may remain diffuse or be grouped in islands and zones not unlike cambia. In time callus may attain a diameter of several centimeters. As a consequence, physiological gradients will establish themselves. This development would clearly contravene the purpose of plant cell culture, i.e., the study of growth and differentiation of cells outside gradients as characterize cells *in situ*. Subculturing calli is easily accomplished. The heterogeneity of this material, however, poses the problem of selecting those pieces as inocula for transfer which best suit subsequent experimentation. For example, selecting the fastest growing portions may result in fine-cell suspensions and simultaneously in loss of morphogenetic capability. More than any other aspect of plant cell culture, growing callus can, therefore, truly be called an art.

V. NUTRIENT MEDIA

Growth of callus and cell suspensions is sustained by nutrient media. The most common media have been formulated by Murashige and Skoog for tobacco-pith tissue and by Gamborg et al. for soybean cell suspensions.[8,9] The merits of these two media have been discussed by Gamborg et al.[10] Besides mineral salts, carbohydrates, and vitamins which constitute such media, it is their fourth component, the hormones or growth regulators, which affects callus growth. While plant cell culture was thought to be primarily a method to detect and assay plant hormones and to analyze their molecular mode of action, it has been the response of plant cells to hormone treatment which has filled plant cell culture libraries. Only the detection of 6-furfurylaminopurine was related to growing tissues in vitro.[11] High-pressure liquid chromatographic and immunological analytical techniques have replaced bioassays today.[12]

In attempts to elucidate tne reaction of cells to hormones, the attack is directed at the receptor site and at the reaction product. Libbenga et al., using tobacco cells and protoplasts, identified soluble in addition to insoluble, i.e., membrane-bound receptors for auxin.[13,14] These authors demonstrated binding of auxin to a receptor protein. The hormone-receptor protein complex formed specifically binds to certain nonhistone chromatin proteins called acceptors. This reaction, then, brings about an increase in template availability. The first response of hormone to follow is probably the synthesis of new mRNAs which are translated into new proteins, regulatory proteins which in turn derepress additional genes. For cytokinins there is similar strong evidence in favor of a translational control of growth.

The attack aimed at the reaction products to analyze the mode of action of hormones appears impeded by the difficulty of identifying such a product. Sung and Okimoto identified proteins which occur in carrot cell cultures upon elimination of auxins from the media and onset of embryogenesis.[15] Sussex described marker proteins which appear during a period of extensive growth of embryos.[16]

Media after Murashige and Skoog, and after Gamborg et al., have proven to be well-balanced nutrients and, therefore, have found world-wide acceptance. In this way the two media have greatly contributed to some standardization of callus and cell-suspension culture. Advances in methodology require, however, that new media are formulated to provide for growth of novel materials or growth in novel environmental conditions, i.e., for growth of anthers and microspores, immature embryos, for protoplasts, hybrid cells, mutants, and cells cultured at low density.[17-23] Most of these media are characterized by a numerical increase of components, carbohydrates, vitamins, C3 acids, complex substances, spent or conditioned media, or substances emanating from feeder layers of immobilized, X-rayed or actively growing cells.[24]

Callus and cell suspensions, even when green and exposed to light and good aeration, will not grow without an adequate supply of digestible carbohydrates. Attempts to increase the concentration of chlorophyll and differentiation of chloroplasts in cultured cells, thereby aiming at photoautotrophy and reduction of the nutrient media to mineral salts have been numerous. Adequate callus has been selected for green pigmentation with tobacco tissue over many subcultures, using media with 3% sucrose, 2 μM NAA, and 6 to 14,000 lx, or with *Chenopodium rubrum*, using media with 3 to 0.5% glucose and 0.01 μM 2,4-dichlorophenoxyacetic acid and 4 to 8000 lx.[25,26] Today these cultures grow as fine, deeply green cell suspensions in the presence of only mineral salts, 1% CO_2, and light. This condition is thought to be most desirable for plant cells. One may expect, therefore, that more species will be brought into such an environment in the future. Photoautotrophy would also permit the closest approach of standardized plant cell culture.

VI. GROWTH

Growth by cell division is the result of one or a sequence of cell cycles. The control of cell cycles would appear to be a problem studied most suitably with synchronized cells cultured in vitro. Cell suspensions, indeed, have been amenable to synchronization, i.e., with an amplification factor which would allow biochemical analysis of phased nuclei and cells. Still, the perfection in synchronization as achieved with mammalian cells due to unique behavior of mitotic cells has not been paralleled with plant cells. Experimentation with hydroxyurea permitted 34% of cells to enter mitosis in synchrony.[27] A combination of FUdR (1 μg/mℓ) and uridine (0.5 μg/mℓ) induced 75% of cells to simultaneously enter mitosis.[28] Difficulties in synchronizing cultured plant cells are twofold: (1) lack of uniformity of starting material and (2) relative slowness in response of cells to inducing factors. Good results have been obtained with parenchyma cells of explants and with fine suspensions of rapidly growing cells of sycamore.[29,30] The problem involved in applying antimetabolites to induce synchrony is the necessity to subject the cells to several washes with fresh media, thereby removing conditioning factors required for rapid onset of growth. Constabel and Kurz employed nitrogen, and later, ethylene and CO_2 for temporary arrest of cells without disturbing media conditions.[31,32] Results obtained with synchronized material allowed calculation of the duration of individual phases of the cell cycle for various species. For example, the mean cycle time for sycamore cells was found to be 22.3 hr.[33] Synchronization also permitted demonstration that DNA replication in plant cells follows a pattern found earlier in mammalian cells, i.e., small intermediates of DNA were synthesized and then joined together, eventually yielding replication units which varied in length from 5 to 50 μm and in turn formed clusters of about 100 genes.[34]

Growth kinetics have been established for many callus and cell suspension cultures. They usually are the basis for studies directed towards cell development, specialization, and regenerative processes. Truly sigmoid growth curves resulted in a few instances only, i.e., when using most rapidly growing cell suspensions. Still, in most cultures a lag phase, a phase of accelerated ("logarithmic") and of decelerated and stationary growth can be identified. Of increasing interest are the growth kinetics of photoautotrophic cell cultures. Using material derived from *Chenopodium rubrum*, Hüsemann observed that during a 14-day growth period the increase in cell density and fresh weight followed a lag period of about 1 to 2 days, a 4-day period of exponential growth (60 to 80% increase in cell number in 2 days) with doubling times of about 55 hr, 4 days in which the rate of cell division declined to a 20% increase in cell numbers in 2 days, and stationary growth to day 14.[35] The time course of increase in chlorophyll of photoautotrophic cell cultures was linear except for an initial lag phase of about 4 days. Remarkably, these cells can be maintained in exponential growth without loss of photosynthetic activity. This behavior would indicate simultaneity of growth and differentiation.

VII. DIFFERENTIATION

Genetic, physiological, and morphological processes which bring about the specialization of cells are referred to as differentiation. Cells which have completed their development and have acquired a special function within the plant organism are differentiated cells.[36] The analysis of factors which control differentiation is often called morphogenesis. In the context of plant cell culture hormones, nutrients, light, and temperature, all have been identified as morphogenetic factors.

The focal point of interest in differentiation of plant cells grown in vitro is the observation that such cells are able to realize totipotency, i.e., to differentiate and grow into entire, functional plants theoretically identical to the original plants. The number and variety of species which have successfully been made to develop from a cell or even protoplast to plants is remarkably great and still growing.[37-39] While immensely fertile as a concept and profitable as it permits industrial plant propagation, totipotency today is subject to sober scrutiny because of mounting evidence against its general applicability. First, cells grown in vitro may undergo genetic changes which often result in loss of totipotency. Secondly, a variety of materials will not respond with any differentiation and plant regeneration even under permissive conditions. Thirdly, plants regenerated from cells cultured in vitro do not necessarily perform as well as the original plants. Still, totipotency is fundamental to experimentation directed to better understanding and exploitation of differentiation, specifically the formation of plant products, of roots, shoots, and embryos.

A. Phytoproducts

Since their inception, cells cultured in vitro have been observed to synthesize and accumulate storage proteins, fats, carbohydrates, a variety of antibiotics, and secondary products. The latter have received most attention because they are the simplest products of morphogenetic processes, can be used as markers in somatic-cell genetics, and are industrial commodities. Typically, they are species specific and structurally occur in an almost infinite variety. Terpenoids, alkaloids, flavonoids, and glycosides are the most prominent groups of secondary metabolites. In plants, their formation is controlled by genomic, physiological, and morphological conditions. In *Sinapis alba*, for instance, cyanidin glycosides are formed as a response to photomorphogenetic factors in the lower epidermis of cotyledons and subepidermis of hypocotyls.[41] The challenge with plant cells cultured in vitro would be to devise genomic and physiological conditions, and to compensate for morphological conditions to permit product formation. Cloning cells selected for high yields and varying the nutritional and physical conditions of such cells would be the arsenal of tools to accomplish this.

Cells which undergo division may contain secondary metabolites.[42] In general, however, product formation occurs after cells have ceased to divide and entered what has been referred to as the idiophase. In praxi, this condition is attained when cells have exhausted nutrients of the media or upon transfer to media with reduced levels of hormones and nutrients. A number of media which promote product formation have been defined; the most widely accepted is the alkaloid production medium of Zenk et al.[43] In comparison to nutrient media mentioned earlier designed to stimulate growth, Zenk's medium features a low concentration of nitrate, phosphate, and auxin, and sucrose at a level of 5%. The carbohydrate is added not only as a source of energy, but also as material to be converted to secondary products. Light source and photoperiod as well as temperature, aeration, agitation, and pH, may be morphogenetic factors equal in importance as are media components.

Since publication of results by Zenk et al.,[43] it is generally agreed that, on the average, callus and cell suspensions derived from high-yielding plants are higher yielding than those from low-yielding plants. The observation of aberrant, low-yielding or nonproducing cultures can be explained by the heterogeneity of the explanted tissues as well as the mutations which may occur upon in vitro culture, specifically upon exposure to high concentrations of auxins.

Some secondary products are preferentially accumulated in cells with special structural features, i.e., in glands, ducts, laticifers. Callus and cell suspensions do not attain this stage of differentiation. It therefore appeared doubtful whether material cultured in vitro would accumulate respective products, i.e., mint oil terpenoids or latex alkaloids, for instance. More recently, however, menthone and menthol, components of mint oil, and codeine, a component of poppy latex, have been detected in cultured cells.[44,45] The biosynthetic capacity of plant cells appears not genetically linked to simultaneous structural differentiation.[46]

The significance of plant cell culture for fundamental research of secondary metabolism may be highlighted by showing inroads made by Hahlbrock et al.[48] into regulatory mech-

B. Habituation

The induction of callus formation with explants and maintenance of subcultures of such callus generally require the presence of hormones in the nutrient medium. Sometimes, however, subcultures lose their requirement for hormones and can be maintained without external supply of hormones; material cultured has acquired the ability to produce hormones in amounts sufficient to promote growth. The phenomenon was first observed by Gautheret and referred to as habituation. Binns and Meins discovered habituation for auxin and for cytokinin.[65] Upon analysis, habituation was described as stable at the cellular level because in cell clones this trait was inherited by the progeny. Habituation also proved to be reversible, because callus from regenerated plants of habituated cell clones reverted to hormone requirement. Therefore, habituation was demonstrated to be epigenetic in nature.

Results would highlight the need for rigorous experimental analysis of variation in cell cultures, including cloning, subcloning, and plant regeneration before an opinion on the genetic or epigenetic nature of such variation is given.

C. Mutation

Mutant cell lines are desirable for two reasons: they may show improved yields of phytoproducts, alkaloids for example, and they may provide a marker system to be used in subsequent genetic and biochemical experimentation. Mutagenic agents, physical and chemical, and their effects on plant cells cultured in vitro have been well defined. It may be added that dose-response curves differ for each cell line. The problem with mutant cell lines is their leakiness. It results from defects being associated with recessive genes, the presence of multiallelic genes, or contamination of the mutant by wild-type cells. The mutagenic system would be improved by aiming at mutations or dominant genes of traits coded for by only one pair of allels and by stringent selection procedures.

Haploid cells grown by way of anther and microspore culture showed experimentally induced mutation. Subsequent diploidization can be achieved by colchicine treatment and cell fusion. As a result, mutations have been obtained in a homozygous state in doubled amphidiploid cell lines. Regeneration led to respective mutant plants, with *Nicotiana tabacum*, for instance.[66]

D. Cell Fusion and Hybridization

The establishment of techniques to isolate and culture plant cells without walls, protoplasts, opened the way to cell fusion. The discovery of polyethylene glycol and a few other chemicals as powerful fusogenic agents has made the generation of somatic-cell hybrids reality.[67] Several cell hybrids have been grown to plants showing more or less intermediate character.[59] Critical in all cases was the application of successful hybrid selection systems. Cocking lists those methods which have proven useful in selecting hybrid callus and plants.[68]

1. Use of selective media for the selective growth of somatic hybrids (*Nicotiana glauca* + *N. langsdorffii*).[69]
2. Use of light-sensitive, chlorophyll-deficient mutants for complementation selection (*N. tabacum* + *N. tabacum*).[70]
3. Use of complementation selection, coupled with differential media growth, by fusing wild type and albino protoplasts (*Petunia parodii* + *P. hybrida*).[71]
4. Use of biochemical mutants for complementation selection of somatic hybrids (*N. tabacum* + *N. knightiana*).[72]
5. Use of two nonallelic albino mutants for complementation selection of somatic hybrids (*Datura innoxia* + *D. innoxia*).[73]
6. Heterokaryon selection by mechanical isolation (*Glycine max* + *N. glauca*).[74]

Once generated, hybrid cells may be employed in answering a number of questions on speciation and incompatibility of species, on chromosome segregation and gene mapping, and on the inheritance of cytoplasmic traits.

Somatic cell hybridization has been used for the transfer of male sterility. Through cell fusion, cytoplasm was transferred from irradiated *N. tabacum* to *N. plumbaginifolia* cells with streptomycin resistance as the marker. By selection for streptomycin resistance, cytoplasmic male sterility, a mitochondrial trait, could be transferred.[75]

E. Gene Transfer

Several avenues are being explored to arrive at gene transfer. One way has been found in generating fusion products, using as partners protoplasts with nuclei in metaphase. Here, chromosome segregation is achieved not during long-term hybrid cell culture, but with the initial step, i.e., fusion.[76] Individual DNA segments may be packaged with membranes, i.e., enclosed by lipid vesicles and made to fuse with host cells.[77] Finally, and with remarkable success, plasmids may be used as vectors for gene transfer.[78]

Ti-plasmids, carried by *Agrobacterium tumefaciens*, have been shown to be responsible for crown gall formation in plants. Ti-plasmids are natural gene vectors with which *Agrobacterium* achieve the transfer and stable maintenance of a defined DNA segment (called T region) into the nucleus of transformed plant cells. Using site-specific mutagenesis, it was possible to introduce mutations in different parts of the T region. The transcription of the T-DNA in wild-type and mutant crown galls was compared and it was found that the induction of specific developmental patterns could be correlated with the presence or absence of specific T-DNA transcripts. Double mutants were obtained in which the expression of all the genes for tumor formation was abolished. Tobacco, potato, and petunia plant cells harboring the inactivated T-DNAs were shown to regenerate normal, fertile plants which transmit the T-DNA segment as a single Mendelian locus.[79]

IX. CONCLUSION

Cell cultures have proven to be an extremely useful tool, because (1) they reduce biochemical processes that occur in whole plants to cells or groups of cells which can be kept under strictly controlled environmental conditions, (2) they can be obtained as haploids, (3) they can be genetically modified through a variety of mutations and hybridization, and (4) they can be made to regenerate plants and thus be employed in plant propagation and plant breeding. Cells in vitro, however, are also prone to spontaneous changes. While this variability may be desirable in some instances, generally it is undesirable and should be prevented as far as possible. Plant cell culture, therefore, has to be supplemented by cryogenic methodology.

REFERENCES

1. **White, P. R.,** Potentially unlimited growth of excised plant callus in an artificial nutrient, *Am. J. Bot.,* 26, 59, 1939.
2. **Gautheret, R. J.,** Sur la possibilité de réaliser la culture indéfinie des tissus de tubercules de carotte, *C. R. Acad. Sci.,* 208, 118, 1939.
3. **Steward, F. C.,** Growth and development of cultivated cells. III. Interpretations of the growth from free cell to carrot plant, *Am. J. Bot.,* 45, 709, 1958.
4. **Reinert, J.,** Untersuchungen über die Morphogenese an Gewebekulturen, *Ber. Dtsch. Bot. Ges.,* 71, 15, 1958.
5. **Thomas, E. and Davey, M. R.,** From single cells to plants, in *The Wykeham Science Series,* Wykeham, London, 1975.

6. **Wetter, L. R. and Constabel, F., Eds.,** *Plant Tissue Culture Methods,* 2nd ed., NRCC #19876, Nat. Res. Counc. Can., Ottawa, 1982.

7. **King, P. J., Mansfield, K. J., and Street, H. E.,** Control of growth and cell division in plant cell suspension cultures, *Can. J. Bot.,* 51, 1807, 1973.

8. **Murashige, T. and Skoog, F.,** A revised medium for rapid growth and bioassays with tobacco tissue cultures, *Physiol. Plant.,* 15, 473, 1962.

9. **Gamborg, O. L., Miller, R. A., and Ojima, K.,** Nutrient requirements of suspension cultures of soybean root cells, *Exp. Cell Res.,* 50, 151, 1968.

10. **Gamborg, O. L., Murashige, T., Thorpe, T. A., and Vasil, I. K.,** Plant tissue culture media, *In Vitro,* 12, 473, 1976.

11. **Miller, C. O., Skoog, F., Okumura, F. S., von Salza, M. H., and Strong, F. M.,** Structure and synthesis of kinetin, *J. Am. Chem. Soc.,* 77, 2662, 1955.

12. **Horgan, R.,** Modern methods in plant hormone analysis, in *Progress in Phytochemistry,* Vol. 7, Reinhold, L., Harborne, J. B., and Swain, T., Eds., Pergamon Press, Oxford, 1981, 137.

13. **Libbenga, K. R.,** Hormone receptors in plants, in *Frontiers of Plant Tissue Culture 1978,* Thorpe, T. A., Ed., Int. Assoc. Plant Tissue Culture, University of Calgary, Alberta, 1978, 325.

14. **Vreugdenhil, D., Burgers, A., Harkes, P. A. A., and Libbenga, K. R.,** Modulation of the number of membrane-bound auxin-binding sites during the growth of batch cultured tobacco cells, *Planta,* 152, 415, 1981.

15. **Sung, Z. R. and Okimoto, R.,** Embryonic proteins in somatic embryos of carrot, *Proc. Nat. Acad. Sci. U.S.A.,* 78, 3683, 1981.

16. **Sussex, I. M.,** personal communication.

17. **Nitzsche, W. and Wenzel, G.,** *Haploids in Plant Breeding,* Verlag Paul Parey, Berlin, 1977.

18. **Stewart, J. McD.,** In vitro fertilization and embryo rescue, *Exp. Bot.,* 21, 301, 1981.

19. **Constabel, F.,** Isolation and culture of plant protoplasts, in *Plant Tissue Culture Methods,* 2nd ed., NRCC #19876, Wetter, L. R. and Constabel, F., Eds., Nat. Res. Counc. Can., Ottawa, 1982, 38.

20. **Kao, K. N.,** Plant protoplast fusion and isolation of heterokaryocytes, in *Plant Tissue Culture Methods,* 2nd ed., NRCC #19876, Wetter, L. R. and Constabel, F., Eds., Nat. Res. Counc. Can., Ottawa, 1982, 49.

21. **Weber, G. and Lark, K. G.,** An efficient plating system for rapid isolation of mutants from plant cell suspensions, *Theor. Appl. Genet.,* 55, 81, 1979.

22. **Kao, K. N. and Michayluk, M. R.,** Nutritional requirements for growth of *Vicia hajastana* cells and protoplasts at a very low population density in liquid media, *Planta,* 126, 105, 1975.

23. **Caboche, M.,** Nutritional requirements of protoplast derived haploid tobacco cells grown at low cell densities in liquid medium, *Planta,* 149, 7, 1980.

24. **Raveh, D., Huberman, E., and Galun, E.,** *In vitro* culture of tobacco protoplasts: use of feeder layer techniques to support division of cells plated at low densities, *In Vitro,* 9, 216, 1973.

25. **Yamada, Y., Sato, F., and Hagimori, M.,** Photoautotrophism in green cultured cells, in *Frontiers of Plant Tissue Culture 1978,* Thorpe, T. A., Ed., Int. Assoc. Plant Tissue Culture, University of Calgary, Alberta, 1978, 453.

26. **Hüsemann, W. and Barz, W.,** Photoautotrophic growth and photosynthesis in cell suspension cultures of *Chenopodium rubrum, Physiol. Plant.,* 40, 77, 1977.

27. **Eriksson, T.,** Partial synchronization of cell division in suspension cultures of *Haplopappus gracilis, Physiol. Plant.,* 19, 900, 1966.

28. **Chu, Y. E. and Lark, K. G.,** Cell-cycle parameters of soybean (*Glycine max* L.) cells growing in suspension culture: suitability of the system for genetic studies, *Planta,* 132, 259, 1976.

29. **Yoeman, M. M. and Evans, P. K.,** Growth and differentiation of plant tissue cultures. II. Synchronous cell divisions in developing callus cultures, *Ann. Bot. (London),* 31, 323, 1967.

30. **King, P. J., Mansfield, K. J., and Street, H. E.,** Control of growth and metabolism of cultured plant cells, *Can. J. Bot.,* 51, 1807, 1973.

31. **Constabel, F., Kurz, W. G. W., Chatson, B., and Gamborg, O. L.,** Induction of partial synchrony in soybean cell cultures, *Exp. Cell Res.,* 85, 105, 1974.

32. **Constabel, F., Kurz, W. G. W., Chatson, K. B., and Kirkpatrick, J. W.,** Partial synchrony in soybean cell suspension cultures induced by ethylene, *Exp. Cell Res.,* 105, 263, 1977.

33. **Gould, A. R. and Street, H. E.,** Kinetic aspects of synchrony in suspension cultures of *Acer pseudoplatanus* L., *J. Cell Sci.,* 17, 337, 1975.

34. **Cress, D. E., Jackson, P. J., Kadouri, A., Chu, Y. E., and Lark, K. G.,** DNA replication in soybean protoplasts and suspension-cultured cells: comparison of exponential and fluorodeoxyuridine synchronized cultures, *Planta,* 143, 241, 1978.

35. **Hüsemann, W.,** Growth characteristics of hormone and vitamin independent photoautotrophic cell suspension cultures from *Chenopodium rubrum, Protoplasma,* 109, 415, 1981.

36. **Esau, K.,** *Plant Anatomy,* 2nd ed., John Wiley & Sons, New York, 1965.

37. **Murashige, T.,** The impact of plant tissue culture on agriculture, in *Frontiers of Plant Tissue Culture 1978,* Thorpe, T. A., Ed., Int. Assoc. Plant Tissue Culture, University of Calgary, Alberta, 1978, 15.

38. **Harney, P. M.,** Tissue culture propagation of some herbaceous horticultural plants, in *Application of Plant Cell and Tissue Culture to Agriculture and Industry,* Tomes, D. T., Ellis, B. E., Harney, P. M., Kasha, K. J., and Peterson, R. L., Eds., University of Guelph, Ontario, Canada, 1982, 187.

39. **Thorpe, T. A.,** Callus organization and de novo formation of shoots, roots and embryos *in vitro,* in *Application of Plant Cell and Tissue Culture to Agriculture and Industry,* Tomes, D. T., Ellis, B. E., Harney, P. M., Kasha, K. J., and Peterson, R. L., Eds., University of Guelph, Ontario, Canada, 1982, 115.

40. **Arcia, M. A., Wernsman, E. A., and Burk, L.,** Performance of anther-derived dihaploids and their conventionally inbred parents as lines, in F_1 hybrids, and in F_2 generations, *Crop Sci.,* 17, 413, 1978.

41. **Mohr, H.,** *Lehrbuch der Pflanzenphysiologie,* Springer-Verlag, Berlin, 1969.

42. **Constabel, F., Shyluk, J. P., and Gamborg, O. L.,** The effect of hormones on anthocyanin accumulation in cell cultures of *Haplopappus gracilis, Planta,* 96, 306, 1971.

43. **Zenk, M. H., El-Shagi, H., Arens, H., Stöckigt, J., Weiler, E. W., and Deus, B.,** Formation of the indole alkaloids serpentine and ajmalicine in cell suspension cultures of *Catharanthus roseus,* in *Plant Tissue Culture and its Bio-technological Application,* Barz, W., Reinhard, E., and Zenk, M. H., Eds., Springer-Verlag, Berlin, 1977, 27.

44. **Hudson, M. J. and Goldsworthy, A.,** Microdetermination of essential oils in plant tissues, *Plant Physiol.,* 67 (Suppl.), 143, 1981.

45. **Tam, W. H. J., Constabel, F., and Kurz, W. G. W.,** Codeine from cell suspension cultures of *Papaver somniferum, Phytochemistry,* 19, 486, 1980.

46. **Dougall, D. K.,** Factors affecting the yields of secondary products in plant tissue cultures, in *Plant Cell and Tissue Culture, Principles and Applications,* Sharp, W. R., Larsen, P. O., Paddock, E. F., and Raghavan, V., Eds., Ohio State University Press, Columbus, 1977, 727.

47. **Schröder, J.,** Light induced increase of m-RNA for phenylalanine-ammonia lyase in cell suspension cultures of *Petroselinum hortense, Arch. Biochem. Biophys.,* 182, 488, 1977.

48. **Schröder, J., Kreuzaler, F., Schäfer, E., and Hahlbrock, K.,** Concomittant induction of phenylalanine-ammonia lyase and flavanone synthase m-RNA in irradiated plant cells, *J. Biol. Chem.,* 254, 57, 1979.

49. **Berlin, J. and Widholm, J. M.,** Metabolism of phenylalanine and tyrosine in tobacco cell lines resistant and sensitive to p-fluorophenylalanine, *Phytochemistry,* 17, 65, 1978.

50. **Tisserat, B., Esan, E. B., and Murashige, T.,** Somatic embryogenesis in angiosperms, in *Horticultural Reviews,* Vol. 1, Janick, J., Ed., AVI Publ., Westport, Conn., 1979, 1.

51. **Haccius, B.,** Question of unicellular origin of non-zygotic embryos in callus cultures, *Phytomorphology,* 28, 74, 1978.

52. **Fujimura, F. and Komamine, A.,** Synchronization of somatic embryogenesis in a carrot cell suspension culture, *Plant Physiol.,* 64, 162, 1979.

53. **Green, C. E. and Phillips, R. L.,** Plant regeneration from tissue cultures of maize, *Crop Sci.,* 15, 417, 1975.

54. **Lu, C. Y. and Vasil, I. K.,** Somatic embryogenesis and plant regeneration in tissue cultures of *Panicum maximum, Am. J. Bot.,* 69, 77, 1982.

55. **Smith, S. M. and Street, H. E.,** The decline of embryogenic potential as callus and suspension cultures of carrot (*Daucus carota* L.) and serially subcultured, *Ann. Bot.,* 38, 223, 1974.

56. **Vasil, I. K.,** Androgenetic haploids, *Int. Rev. Cytol.,* 11A, 195, 1980.

57. **Thomas, E., Hoffmann, F., Potrykus, I., and Wenzel, F.,** Protoplast regeneration and stem embryogenesis of haploid androgenetic rape, *Mol. Gen. Genet.,* 145, 245, 1976.

58. **Gengenbach, B. G.,** Development of maize caryopses resulting from in vitro pollination, *Planta,* 134, 91, 1977.

59. **Melchers, G.,** The first decennium of somatic hybridization by fusion of protoplasts, in Proc. 5th Int. Conf. Plant Tissue Culture, Fujiwara, A., Ed., Tokyo, 1982, 13.

60. **Ashmore, S. E. and Gould, A. R.,** Karyotype evolution in a tumor derived plant tissue culture analyzed by Giemsa C-banding, *Protoplasma,* 106, 297, 1981.

61. **Bayliss, M. W.,** Chromosomal variation in plant tissues in culture, *Int. Rev. Cytol.,* 11A, 113, 1980.

62. **Shepard, J. F., Bidney, D., and Shahin, E.,** Potato protoplasts in crop improvement, *Science,* 208, 17, 1980.

63. **Constabel, F., Kurz, W. G. W., and Kutney, J. P.,** Variation in cell cultures of periwinkle, *Catharanthus roseus* (L.) G. Don., in Proc. 5th Int. Conf. Plant Tissue Culture, Fujiwara, A., Ed., Tokyo, 1982, 301.

64. **Filner, P.,** Heritable increases in urease during urease-limited growth of cultured tobacco cells, in Proc. 5th Int.Conf. Plant Tissue Culture, Fujiwara, A., Ed., Tokyo, 1982, 235.

65. **Binns, A. N. and Meins, F., Jr.,** Habituation of tobacco pith cells for factors promoting cell division is heritable and potentially reversible, *Proc. Natl. Acad. Sci. U.S.A.,* 70, 2660, 1979.

66. **Müller, A. J. and Mendel, R. R.,** Nitrate reductase deficient tobacco mutants and the regulation of nitrate assimilation, in Proc. 5th Int. Conf. Plant Tissue Culture, Fujiwara, A., Ed., Tokyo, 1982, 233.

67. **Kao, K. N. and Michayluk, M. R.,** A method for high-frequency intergeneric fusion of plant protoplasts, *Planta,* 115, 335, 1974.
68. **Cocking, E. C. and Riley, R.,** Application of tissue culture and somatic hybridization to plant improvement, in *Plant Breeding,* Vol. 11, Frey, K. J., Ed., Iowa State University Press, Ames, 1979, 85.
69. **Smith, H. H., Kao, K. N., and Combatti, N. C.,** Interspecific hybridization by protoplast fusion in *Nicotiana:* confirmation and extension, *J. Hered.,* 67, 123, 1976.
70. **Melchers, G. and Labib, G.,** Somatic hybridization of plants by fusion of protoplasts. I. selection of light resistant hybrids of haploid light sensitive varieties of tobacco, *Mol. Gen. Genet.,* 135, 277, 1974.
71. **Cocking, E. C., George, D., Price-Jones, J. J., and Power, J. B.,** Selection procedures for the production of inter-species somatic hybrids of *Petunia hybrida* and *Petunia parodii.* II. Albino complementation selection, *Plant Sci. Lett.,* 10, 7, 1977.
72. **Maliga, P., Lazar, G., Joo, F., Nagy, A. H., and Menczel, L.,** Restoration of morphogenic potential in *Nicotiana* by somatic hybridization, *Mol. Gen. Genet.,* 157, 291, 1977.
73. **Schieder, O.,** Hybridization experiments with protoplasts from chlorophyll-deficient mutants of some Solanaceous species, *Planta,* 137, 253, 1977.
74. **Kao, K. N.,** Chromosomal behavior in somatic hybrids of soybean *Nicotiana glauca, Mol. Gen. Genet.,* 150, 225, 1977.
75. **Belliard, G., Pelletier, G., Vedel, F., and Quetier, F.,** Morphological and chloroplast DNA distribution in different cytoplasmic parasexual hybrids of *Nicotiana tabacum, Mol. Gen. Genet.,* 165, 231, 1978.
76. **Dudits, D., Koncz, Cs., Bajszar, G., Hodlaczky, G., Lazar, G., and Horvath, G.,** Intergeneric transfer of nuclear markers through fusion between dividing and mitotically inactive plant protoplasts, in *Advances in Protoplast Research,* Ferenczy, L. and Farkas, G. L., Eds., Pergamon Press, Oxford, 1980.
77. **Nagata, T., Okada, K., Takebe, I., and Matsui, C.,** Delivery of tobacco mosaic virus RNA into plant protoplasts mediated by reverse-phase evaporation vesicles (Liposomes), *Mol. Gen. Genet.,* 184, 161, 1981.
78. **Wullems, G. J., Krens, F. A., Peerbolte, R., and Schilperoort, R. A.,** Transformed tobacco plants regenerated after single cell transformation, in Proc. 5th Int. Conf. Plant Tissue Culture, Fujiwara, A., Ed., Tokyo, 1982, 505.
79. **Schell, J., van Montagu, M., Holsters, M., Leemans, J., de Greve, H., Hernalsteens, J. P., Willmitzer, L., Otten, L., and Schröder, J.,** Plant cell transformation and genetic engineering, in Proc. 13th *Int. Bot. Cong.,* Carr, I. J., Ed., Sydney, 1981, Abstr. #120306.

Chapter 2

BASIC PRINCIPLES OF FREEZING INJURY TO PLANT CELLS; NATURAL TOLERANCE AND APPROACHES TO CRYOPRESERVATION*

Harold T. Meryman and Robert J. Williams

TABLE OF CONTENTS

* Contribution No. 597 from the American Red Cross Blood Services. Supported in part by NIH Grant GM 17959 and BRSG NO. 2 S07 RR05737.

I. INTRODUCTION

Freezing injury to plants and the mechanisms that they have devised to avoid or tolerate freezing are, of course, subject to fundamental physical and biological principles, some long known but others only recently being understood as the science of cryobiology develops. This chapter will attempt to outline those principles and will also explore ways by which living organisms, particularly plants, have exploited them to acquire natural tolerance to subfreezing temperatures. This chapter also discusses the principles and practices of the cryopreservation of cells and tissues not possessing natural tolerance to freezing.

II. FREEZING INJURY

Cryobiology is a relatively young science, surprisingly so in view of the fact that freezing and its effects have always been an important part of our environment. Although there are records of experimental work dating from the mid-19th century, serious investigations into the mechanisms of freezing injury in biological systems did not really begin until the 2nd quarter of the 20th century, principally through the efforts of Luyet and colleagues; Luyet and Gehenio's book, *Life and Death at Low Temperatures*,[1] is a classic review of a scientific discipline in its infancy. These early studies were dominated by the belief that physical damage by ice crystals was the principal cause of freezing injury. Although the idea that cell dehydration might somehow be associated with injury was by no means new,[2] it was probably Lovelock[3,4] who first succeeded in diverting attention from the direct effects of ice crystals and demonstrated a clear correlation between the concentration of extracellular solutes and cell death. Although confusion continued to exist between the direct and indirect effects of ice, this was largely dispelled by Mazur's[5] formulation of a two-factor hypothesis for freezing injury which proposed that cell injury was the result of either the concentration of solutes by extracellular ice, producing what he referred to as "solution effects", or resulted from the presence of intracellular ice which presumably caused mechanical injury. Although these principles were largely developed using animal cells and microorganisms, current understanding of freezing injury and freezing tolerance in both plant and animal cells suggests that they have general application.

A. Ice Nucleation and Growth

Although water is popularly thought of as freezing at 0°C, in fact it rarely does so and can be supercooled to far below that temperature. It is generally believed that to initiate ice a crystal nucleus must be present. Nuclei are presumed to be regions on nonaqueous surfaces: container walls, dust particles, or macromolecules that act as templates for crystal growth, referred to as heterogeneous nuclei. The smaller the nucleus, the lower the temperature must be before it will support ice growth. As a result, water will supercool until the temperature is low enough for the largest nucleus present to initiate crystallization. In the absence of

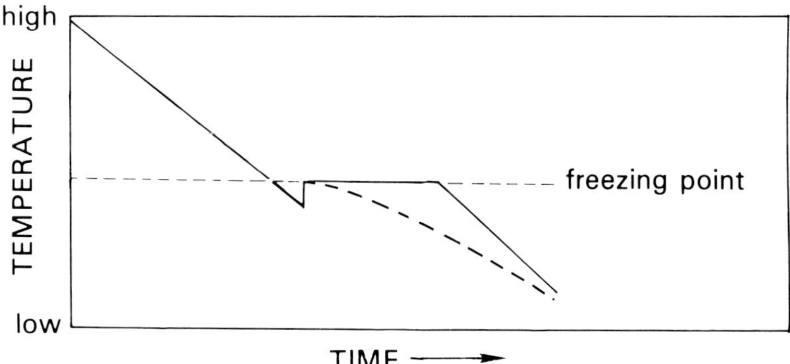

FIGURE 1.　Schematic for the time-temperature course of freezing of an aqueous solution. The solid line is for pure water, the dashed line is for a solution.

heterogeneous nuclei, water can be supercooled to approximately $-40°C$, at which temperature random aggregations of water molecules can nucleate ice.

When ice forms from water, the transition from disorder to order results in the release of energy which appears as the latent heat of fusion. The amount of heat to be removed in the water-ice transition is substantial; the amount required to convert 1 g of water at $0°$ to ice at $0°C$ is 80 times that required to change the temperature of 1 g of water by $1°$. When water is supercooled, the heat deficit generated by cooling the water below its nominal freezing point can absorb some of the heat of fusion so that when ice is seeded it propagates rapidly throughout the specimen and the temperature will rise to the nominal freezing (melting) temperature.

The presence of solutes dissolved in water will lower the freezing point in approximate proportion to the molal concentration of the solution. A 1 m solution of an ideal solute, one that does not interact in any way with water, will lower the freezing point by $1.86°C$. This means that when freezing takes place in an aqueous solution, as water is frozen out, the solution becomes more concentrated and its freezing point is accordingly reduced.

The interplay of these factors is illustrated in Figure 1. As heat is withdrawn from the specimen at a uniform rate, the temperature falls in linear fashion, passing below $0°C$ and continuing to fall until the temperature is low enough for the largest nucleus present to support crystal growth. At this point, ice propagates rapidly as a portion of the latent heat is absorbed by the heat deficit of the supercooled water. When the mixture of water and ice has been warmed to the melting point by the release of latent heat, further ice formation must await further heat removal from the specimen. If the specimen is pure water, illustrated by the solid line, the temperature will remain at $0°C$ until all of the water is frozen (assuming equilibrium throughout the specimen). If the specimen is a solution, the melting point will be below $0°C$ and, as freezing progresses, the solution becomes more concentrated and the freezing point progressively falls, as illustrated by the dashed line in Figure 1.

There are two other parameters of importance to an understanding of the freezing process. As stated above, in the absence of a heterogeneous nucleus, water can be supercooled to about $-40°C$. Below this temperature, water becomes selfnucleating, a process known as homogeneous nucleation. The homogeneous nucleation temperature (T_H) is also depressed by the addition of solutes, although the relationship between concentration and T_H is more complex than with freezing-point depression (see Figure 2). The final parameter of importance is the glass transformation temperature (T_G). This is the temperature below which an unfrozen liquid is too viscous to permit crystal growth. For example, water vapor allowed to condense molecule-by-molecule on a surface cooled by liquid nitrogen ($-198°C$) will form a glass. The glassy state will persist as the temperature is raised until T_G is reached,

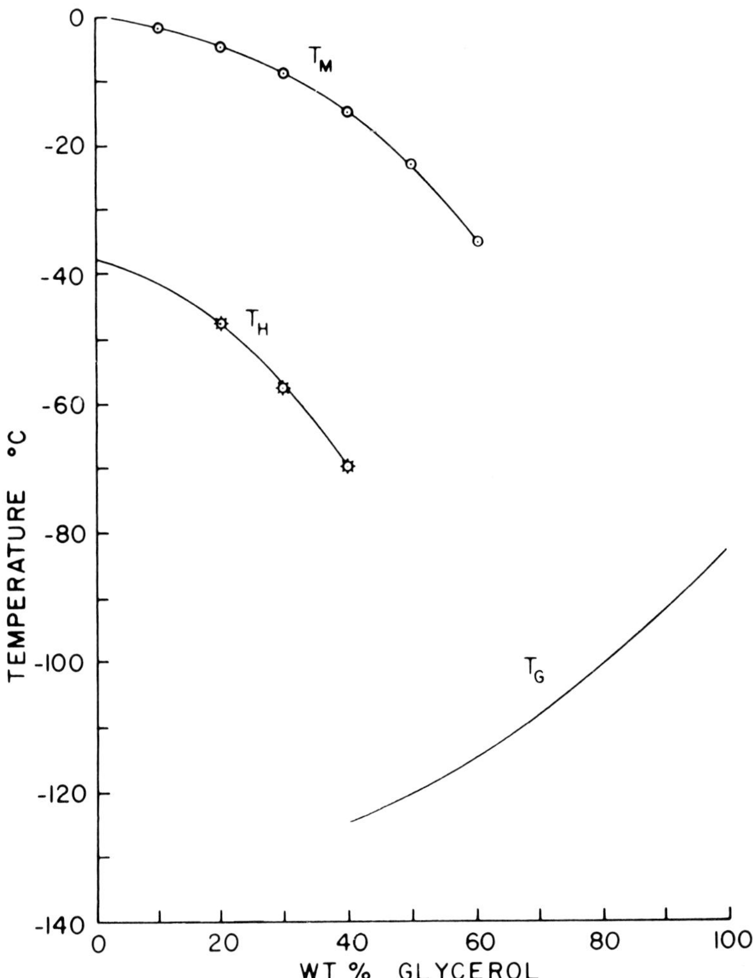

FIGURE 2.   This phase diagram for glycerol is qualitatively applicable to any aqueous solution. As the solute concentration increases, both the melting point (T_M) and the homogeneous nucleation temperature (T_H) fall while the glass transformation temperature (T_G) rises.[6]

at which temperature nucleation and crystal growth will take place. Above T_G, a process called recrystallization can also take place. Since smaller crystals have a higher vapor pressure than large, the large crystals will tend to grow at the expense of small with the result that many small crystals, which in a biological setting might be innocuous, can recrystallize into a few large, destructive crystals (Figure 3).

B. Freezing in Biological Systems

When freezing is slow, as is generally the case in natural freezing, with rare exceptions ice forms in the extracellular spaces by heterogeneous nucleation. There is evidence that cells rarely contain nuclei. Even if they did, slow freezing would still produce extracellular ice since, if the first nucleus to become critical were to be intracellular, it would promptly grow, burst the cell, and become extracellular.

In a cellular system, extracellular ice formed during slow cooling probably is the result of one or a very few nucleation events. The initial supercooling is evidence that there are no nuclei present capable of initiating freezing near the melting temperature. When nucleation

FIGURE 3. Recrystallization of ice. Water vapor was allowed to condense on a liquid-nitrogen-cooled surface in a vacuum to produce a glass. Samples were then permitted to warm briefly to permit recrystallization. The surface was then recooled to liquid-nitrogen temperature and replicated by silicon monoxide evaporation for electron microscopy.[7]

occurs, the temperature will abruptly rise as latent heat is released. Since no nuclei exist that are large enough to nucleate at a temperature higher than that at which initial freezing took place, crystal growth can only progress from the initial crystal. Only if the growth of the first crystal cannot supply the heat being removed will the specimen temperature fall below the initial nucleating temperature so that other nuclei may become critical.

Intact cell membranes appear to be impermeable to ice. Mazur[8] has suggested that at temperatures 10 to 20°C below zero the radius of curvature of a growing ice front may approximate the dimensions of membrane pores and permit the entry of ice into the cell. It is more probable that when ice develops within cells at temperatures above $-40°C$ it is because freezing injury has created membrane defects through which extracellular ice can seed the interior. Recent studies by Steponkus[9] have demonstrated cell rupture prior to flashing in naked protoplasts.

In any event, slow cooling generally produces extracellular ice. In the case of plant cells, the ice invades the cell wall but not the plasmalemma. As the ice grows, the extracellular solution is concentrated. The resulting increase in osmotic pressure causes water to leave the cell, reducing cell volume and concentrating the cell contents. The precise mechanism by which cell dehydration causes injury will be examined in subsequent paragraphs, but it

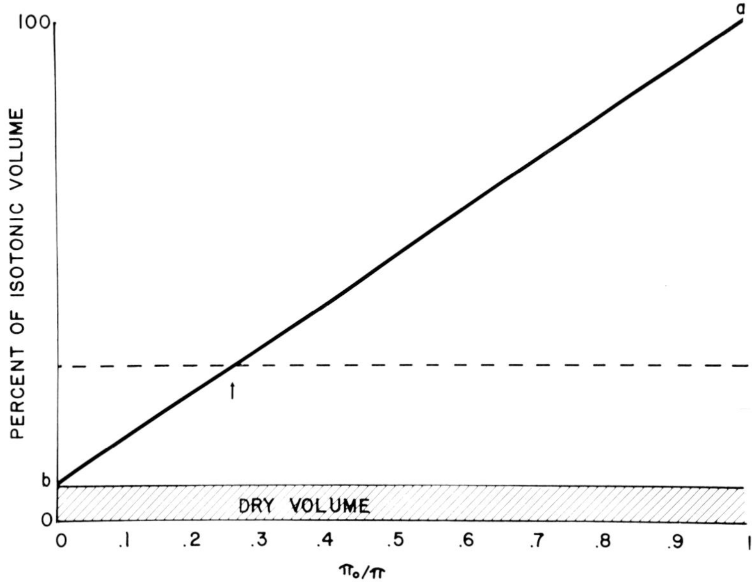

FIGURE 6.   Boyle-van't Hoff plot of a typical cell: π_o is defined as isotonic osmolality; π is the suspending osmolality. The cell is therefore at isotonic volume at $\pi_o/\pi = 1$ at the right edge of the graph. Suspending osmolality increases from right to left and, at $\pi_o/\pi = 0$, infinite osmolality has been reached. A cell behaving as an ideal osmometer, shrinking in response to increasing osmolality, neither losing nor gaining solute, will describe a straight line on this graph. If that line is projected to $\pi_o/\pi = 0$, it will intersect the ordinate at a volume b equal to the dry volume with all water osmotically removed. A real cell, of course, would lose its integrity before reaching that point, presumably after its volume was reduced to less than the critical minimum volume. Minimum volume is indicated in the figure by the dashed line. (Reproduced with permission, from the *Annual Review of Biophysics and Bioengineering,* 3, 350, ©1974 by Annual Reviews Inc.)

is plotted against the reciprocal of extracellular osmolality, a cell behaving as an ideal osmometer will produce a straight line. When this line is projected to the ordinate, the intercept, b, will define the volume of the osmotically inactive constituents of the cell (Figure 6).

With modest changes in osmolality, most cells appear to behave as osmometers, at least within the precision of available measurement. As Höfler[19] has shown, plant-cell volume can be measured with considerable precision provided one is judicious in the selection of cell types to be examined. In our laboratory,[20] we have shown that the volume of selected individual plant cells can be measured relative to their normal volume with a precision of 3%.

Three distinct types of volumetric behavior may be seen when cells are suspended in solutions made increasingly hypertonic with nonpenetrating solutes. First, the cell may behave as an apparently ideal osmometer displaying straight-line behavior on the Boyle-van't Hoff plot with extrapolation of the line to the ordinate indicating a volume approximately equal to the nonaqueous components of the cell. Examples of such behavior are the wheats illustrated in Figure 5.

In the second type of behavior, cell volume also displays a straight line on the Boyle-van't Hoff plot, but the extrapolation of the curve intercepts the ordinate at a volume substantially greater than can be attributed to nonaqueous constituents. Presuming that the minimum volume where cell injury takes place is unchanged, raising the intercept serves to shift the point at which minimum volume is achieved to a higher osmolality, affording a degree of cryoprotection.

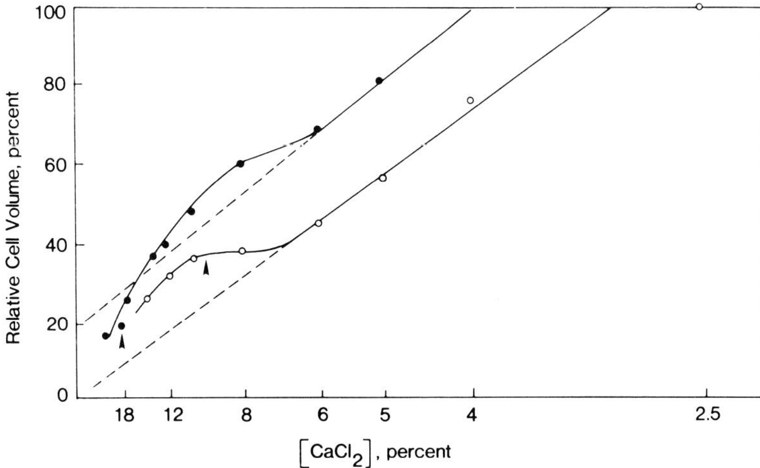

FIGURE 7. Volumetric data for hardy (●) and nonhardy (○) Kharkov cells, plasmolyzed by suspension in hyperosmotic $CaCl_2$ solutions, illustrating that, like many other plants, intracellular osmolality is increased with hardening and the intercept b on the ordinate is raised. However, unlike most other hardy cells, the plateau is much reduced and the killing point occurs at a substantially lesser volume than in the nonhardy cell.

In the third form of osmotic behavior, cell volume may at first describe a straight line on the Boyle-van't Hoff plot, but then either abruptly or progressively stops responding to increased osmolality, displaying a volume plateau which may be extensive. This form of nonideality can serve to prevent cell volume from reaching minimum volume and can be a particularly effective cryoprotective mechanism.

The hypothesis that some form of membrane stress is imposed by volume reduction implies that a portion of the work done in reducing cell volume must be stored in the membrane as potential energy and the cell should fail to respond as a perfect osmometer and should remain larger than predicted. The discussion of the next several pages will present the principle pieces of experimental evidence that we have collected over the past decade which bear on the relationship between cell-volume reduction and cell injury.

E. Cell-Volume Reduction and the Loss of Membrane Material

Possibly our most provocative study has been that of the winter wheat, *Triticum aestivum* L., cv. Kharkov, which, when hardened, is able to withstand freezing to at least $-25°C$. Winter wheat cells in general behave as good osmometers. They show little departure from straight-line behavior, even when reaching and passing the killing point. These cells continue to plasmolyze far beyond the point of no return. Initially, deplasmolysis also appears to be normal. However, shortly before re-establishing initial volume, cells that have been plasmolyzed beyond the killing point will burst as though the membrane no longer had sufficient area to permit return to the original volume.[21] This phenomenon has been amply recorded also by Wiest and Steponkus,[22] by Siminovitch,[23] and by Singh.[24]

The behavior of unhardened cells of Kharkov is shown in the lower curve of Figure 7. The projection of the Boyle-van't Hoff curve intercepts the abscissa at approximately zero volume, implying that all cell water is osmotically active. There is a small plateau and, as is true of most plant cells we have studied that display volume plateaus, the killing point and the collapse of the plateau are simultaneous. As was true with the wheats in Figure 5, winter hardening involves an increase in intracellular osmolality so that plasmolysis begins at a higher extracellular concentration. The projected intercept with the abscissa is raised, implying, according to classical interpretation, the "binding" of some 20% of cell water.

The volume plateau in hardened cells is substantially diminished and the killing point is not associated with the point of collapse of the plateau, as occurs in tender Kharkov cells, but occurs at a far higher osmolality.

Of particular significance is the observation that as the volume of hardy Kharkov cells approaches the plateau phase, refractile droplets begin to appear in the cytoplasm.[21] As plasmolysis continues, these lipid particles grow to about 5 μm in diameter. During deplasmolysis they vanish as the cell approaches normal volume with the clear implication that these are lipid depots, storing membrane lipid during plasmolysis and returning it to the membrane with deplasmolysis. The observations that conventional cells burst on deplasmolysis and that hardy Kharkov has evolved a mechanism to recapture lost membrane lipid imply that loss of material from the membrane may be one cause of plant cell injury from plasmolysis.

Wiest and Steponkus,[22] using naked protoplasts of spinach cells, have also reported experiments implying the loss of membrane material following osmotic volume reduction. More recently, in an elegant series of videotaped microscopic observations and by scanning electron microscopy, Steponkus and Gordan-Kamm[9,25] have shown that in a hardy variety of rye the membrane material is extruded from the membrane into droplets attached by stalks to the membrane exterior surface whence it can be returned to the membrane on deplasmolysis of the protoplast, an adaptation differing from that of Kharkov but serving the same purpose.

Analogous behavior has been demonstrated with artificial lecithin vesicles by Boroske and Elwenspoek.[26] Vesicles transferred to solutions of elevated osmolality lost volume but periodically passed through phases of instability during which the vesicles became irregular, small droplets budded off, and the vesicles once more became spheroidal.

Additional evidence in support of the concept of osmotically induced membrane strain can be obtained by means of the Langmuir trough.[27,28] The apparatus consists simply of a rectangular tray of water. Spanning one dimension of the rectangle is a bar that can be moved from one end to the other to reduce the surface area of the trough. If a few nmol of lipid in a volatile solvent are added to the water, the lipid will form a monolayer on the surface with the hydrophobic tails extending into the air and the hydrophilic head groups into the water. When the bar is moved, compressing the monolayer, the pressure required increases as the area is decreased.

Figure 8 illustrates the classical behavior of a monolayer of stearic acid. Reduction in surface area results in a proportionate increase in pressure to a limiting value, P*. At this point, the pressure exerted on the monolayer has exceeded its capacity to resist, a small amount of lipid is lost from the interface, lowering the pressure to a stable value, P, which can be tolerated by the monolayer. This event probably corresponds to the zone of instability seen by Boroske and Elwenspoek[26] in lipid vesicles subjected to hyperosmolality, and may correspond to the collapse of the volume plateau observed in plant cells. As the monolayer area is further reduced, P remains constant as monolayer material is progressively excluded from the interface. The fact that the pressure does not return to P* implies that the initial collapse at P* constitutes a nucleating event which subsequently facilitates further lipid loss. We interpret this as an activation energy. During decompression, the pressure drops immediately, indicating that little if any of the excluded lipid is able to re-enter the interface.

In Figure 9, top, phospholipids from nonhardy Kharkov demonstrate some solubility and, although there is some return of lipids to the monolayer, losses from the interface are largely irreversible.

Phospholipids from hardened Kharkov, on the other hand (Figure 9, center), leave the monolayer during compression more readily and re-enter the interface to a greater degree. When water-soluble cell extract was added to the trough (Figure 9, bottom), the loss of monolayer material was almost continuous with decrease in area and was completely reversible. This implies the presence in the cytoplasm of a carrier mechanism facilitating the transport of lipid from the membrane to the storage vesicles and back.

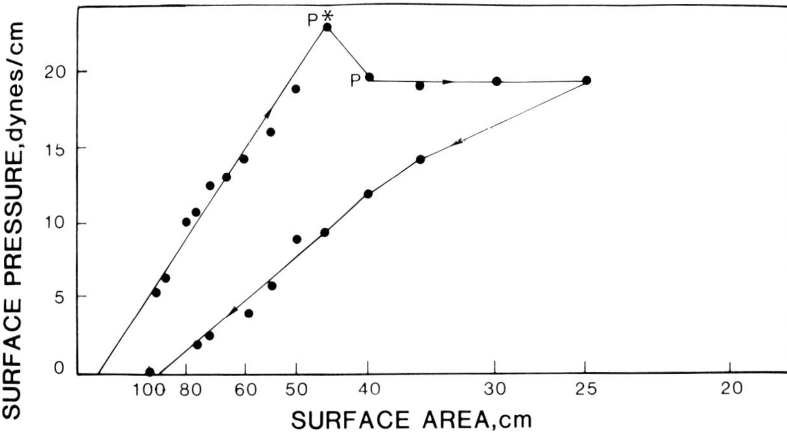

FIGURE 8. This figure illustrates a typical force-area diagram for a monolayer on the Langmuir trough — in this example, for stearic acid. Following the solid compression line from its intersection with the abscissa, representing zero pressure on the monolayer, a decrease in monolayer area requires an increasing force. In the range of 25 dyn/cm, the film has been compressed to nearly one fourth of its resting area and the stearic acid molecules are tightly packed in a quasi-crystalline array. Once collapse has occurred at P*, continued reduction in area leads to a continuous loss of membrane material. When the force is relaxed the monolayer expands, but at resting area (P = 0) the area is less than the initial area because of the loss of material.

The extent to which monolayer data can be compared to bilayer behavior is still open to debate (for a recent review see Gruen and Wolfe[30]). However, the analogy between the behavior of the lipid monolayer and of the intact cells is inescapable. In the unhardy cell, lipid is lost from the membrane or the monolayer as a result of volume reduction of the cell or area reduction of the cell and the loss is irreversible. Following hardening, lipid behavior is different and in both cell and monolayer, the lipid can be recaptured on deplasmolysis of the cell or re-expansion of the monolayer. The ultimate killing point of the cell may coincide with the exhaustion of membrane lipid suitable for storage and return.

Thus, there appears at this point to be little question but that one form of injury to plant cells from excessive plasmolysis is irreversible loss of membrane material, preventing the cell from refilling the cell wall without lysis. What remains a subject for controversy is whether cell-volume reduction is a passive event resulting in loss of surface tension leading to the loss of material in order to restore surface tension, analogous to reducing the volume of a soap bubble, or whether one should pursue the analogy of the monolayer where a reduction in area involves not only a reduction in tension but also the doing of work which is stored in the monolayer as a pressure.

F. Major Volumetric Anomalies

Although Kharkov does display some modest departure from ideal volumetric behavior, these departures can be remarkable in other plants. *Cornus florida*, for example, is an extreme case, displaying both a substantial rise in intercept b with hardening and the development of an extensive volume plateau.

Our data,[20] reproduced in Figure 10, show that bark cells from summer *C. florida* appear to be good osmometers and describe an essentially straight line on the Boyle-van't Hoff plot with plasmolysis commencing at 472 mOsm. However, this line extrapolates to 33% of isotonic volume, implying that nearly a third of the cell content is osmotically inactive. Even in summer, the killing temperature for *C. florida* cortical cells has been reported to be in the vicinity of −15°C.

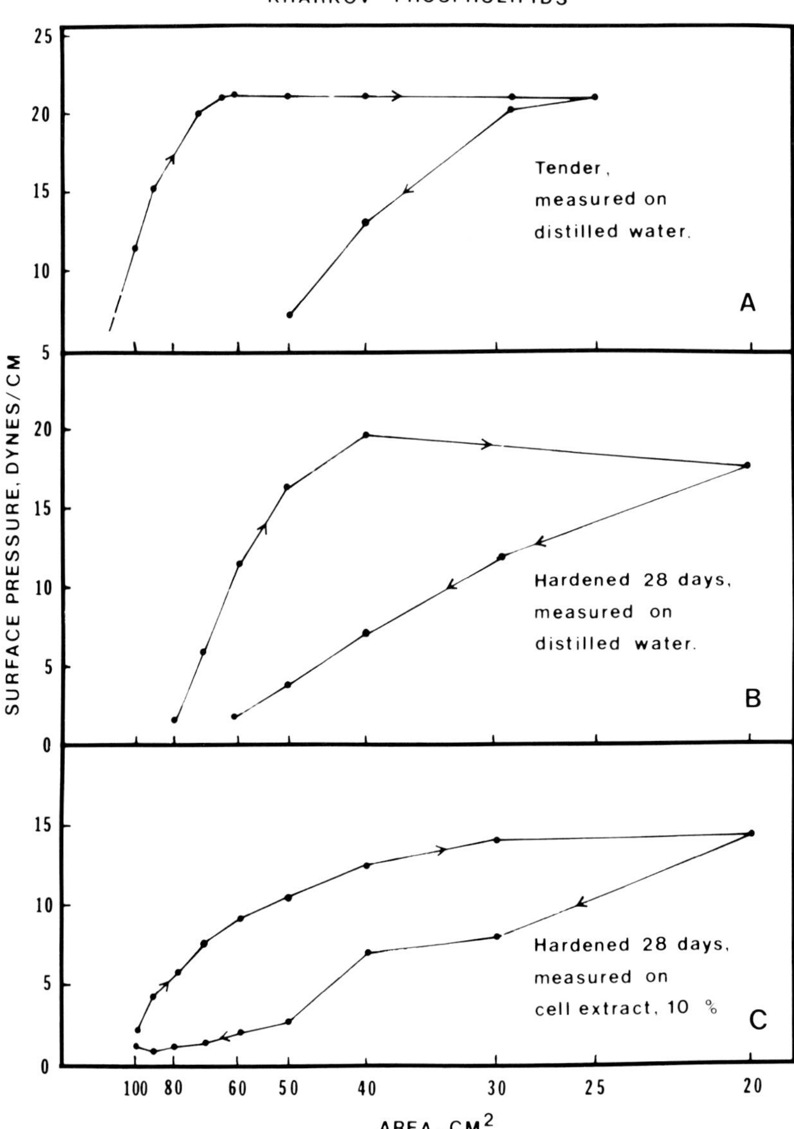

FIGURE 9.   Force-area diagram of purified phospholipid from Kharkov wheat cells. The behavior of nonhardened lipids is not particularly atypical. However, after hardening, the special-purpose lipids elaborated during hardening are more easily lost from the monolayer. These lipids, particularly in the presence of dilute cell sap, return readily to the monolayer on relaxation of pressure. If the monolayer is compressed much beyond that shown in the figure, the original area is not regained, suggesting that the retrievable lipid is especially designed for this purpose and is limited in quantity. (From Meryman, H. T. and Williams, R. J., *Crop Genetic Resources: Conservation of Difficult Material*, Withers, L. A. and Williams, J. T., Eds., I.U.B.S., Paris, 1980. With permission.)

Following cold hardening, there is a dramatic change in the volumetric behavior of *C. florida* bark cells. The elaboration of additional intracellular solutes, primarily peptidoglycans,[31] has increased intracellular osmolality to 814 mOsm, nearly twice the summer value. The point of incipient plasmolysis is thereby shifted to the left in the Boyle-van't Hoff plot. As plasmolysis commences, cell volume again initially demonstrates an essentially straight-

As the (
surface te
monolayer
plant-cell i
liquid pha
intracellul.
area assoc
be done ii
between ii
compartm
may be w

Depenc
which the
ideal osm
In some (
and is ex
and norm.
may be c
the Boyle
to which
structures
reduction
resists co
and, at sc
results. I
at either
repairing

In the
general 1
volume

A. Supe

Like
perature
the press
initiated
water n
by faili
nuclei f
bacteria
water.[44]

A nu
the pres
ray par
dition u
extrace
persist,
form o

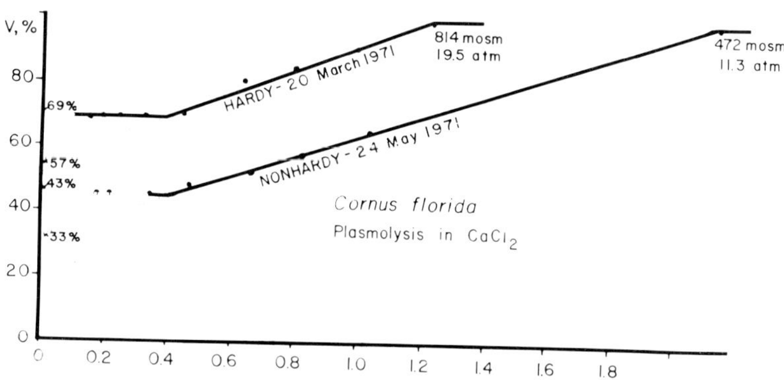

FIGURE 10. Cells from *C. florida* provide particularly striking examples of departure from normal isotonic behavior when plasmolyzed in hyperosmotic calcium chloride solutions. The end of the solid line at the left of the graph indicates the killing point. (From Meryman, H. T. and Williams, R. J., *Crop Genetic Resources: Conservation of Difficult Material*, Withers, L. A. and Williams, J. T., Eds., I.U.B.S., Paris, 1980. With permission.)

line relationship, but projects to a volume of 57% of normal, implying more than a 50% increase in the amount of osmotically inactive material. At approximately −4°C, at roughly 2 Osm, the cells cease to behave as osmometers. The volume remains unchanged as the extracellular osmolality continues to increase. The killing temperature is approximately −25°C at an osmolality of roughly 13.5 Osm.

When lipids from unhardened *C. florida* are examined on the Langmuir trough (Figure 11a) there is some bending of the curve at high pressure, implying a degree of solubility in water. There is also some evidence of lipid return to the monolayer with decompression. In Figure 11b, the nearly vertical initial rise implies that lipids from a hardened dogwood are relatively incompressible until collapse and extrusion from the monolayer occurs.

The first problem encountered in interpreting these data is in accounting for the very large component of osmotically inactive material implied by extrapolation of the curves to the intercept, b, on the ordinate. Attempts to demonstrate unusual osmotic behavior of cell sap extracted from cells have so far been unsuccessful, implying that if "bound" water accounts for a significant portion of the osmotically inactive material, it does so only in the living cell.

Evidence for the existence of large quantities of osmotically inactive water is scarce. Unfreezeable water attributed to monolayers on protein generally ranges from 0.2 to 0.5 g/g dry weight, occasionally as high as 1.0 g/g, quite insufficient to account for the 57% osmotically inactive volume implied by the Boyle-van't Hoff curve for *C. florida*. Some polymers, such as hemoglobin, appear to sequester water within the molecule so that the osmolality rises far more rapidly with concentration than predicted on the basis of molecular weight.[32] However, even starting with a hemoglobin concentration of 30% in the isotonic red cell, intercept b is raised sufficiently to simulate the "binding" of only 25% of cell water.[33] To our knowledge, intracellular macromolecular concentrations of this magnitude have not been reported for plant cells. The existence of osmotically inactive water to the extent implied by the volumetric measurements of hardy *C. florida* has yet to be directly demonstrated.

G. The Plateau

Explaining the plateau region is still more difficult. Either "bound" water or a changing osmotic coefficient of cell contents are equally unsatisfactory explanations. The usual models for "bound" water, entrapment within macromolecules or the reduction of entropy at surfaces, are not adequate to explain the abrupt change in behavior. Solutes such as hemoglobin

CRYOPROTECTION: EQUILIBRIUM (COLLIGATIVE) MODE

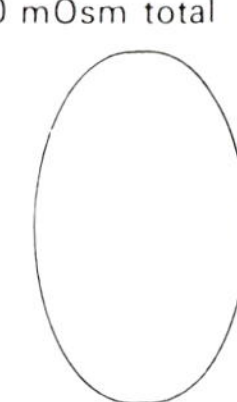

FIGURE 13.   Colligative cryoprotection. This schematic diagram illustrates how the presence of a penetrating agent in high concentration reduces the non-penetrating solute (NPS) concentration attained following freezing: a pure antifreeze effect. Since the glycerol is equally distributed inside and outside the cell, it will have no effect on the cell volume. As extracellular ice forms, water will be equally lost by both intracellular and extracellular solutions. The high solute concentration reduces the total amount of ice formed, thereby reducing the extent of volume lost by the cell. (From Meryman, H. T. and Williams, R. J., *Crop Genetic Resources: Conservation of Difficult Material*, Withers, L. A. and Williams, J. T., Eds., I.U.B.S., Paris, 1980. With permission.)

produce the very injury it is designed to prevent; second, it must be nontoxic in concentrations sufficiently high to produce a useful amount of freezing-point depression. That these simple criteria are sufficient is supported by the demonstration that penetrating solutes such as ethanol, methanol, a variety of glycols, and DMSO are all equally effective on a molal basis provided they are used at less than toxic concentrations.[18] Even ammonium acetate, which penetrates the human red cell, functions as a cryoprotectant, providing further evidence that electrolyte concentration per se is not a cause of freezing injury.[47]

In Figure 13, the principle of colligative cryoprotection is illustrated. Both the extra- and intracellular solutions have been supplemented by the addition of 7.0 Osm of glycerol. At −40°C, the vapor pressure at the surface of ice is such that it will be at equilibrium with a solution having an osmolality of approximately 22 Osm. If the extracellular solution had initially consisted of only sodium chloride, the salt concentration produced at −40°C would be more than 70 times isotonic, many times more than sufficient to produce 100% cell destruction. However, because of the elevation in initial osmolality contributed by the glycerol, only two thirds of the water (equal to one half the solution volume) need be frozen out to achieve 22 Osm, producing, even at −40°C, an extent of cell dehydration that will be easily tolerated.

Not surprisingly, valuable lessons regarding cryoprotectants have been learned from the study of organisms with acquired tolerance to cell dehydration from freezing, from drying, or from exposure to hyperosmotic environments. The elaboration of glycerol to forestall cell dehydration is not uncommon. Certain Alaskan insects have been shown to achieve whole-body concentrations of the order of 4.5 *M* (25%) in response to subfreezing temperatures.[48] *Dunaliella*, an alga found in the Dead Sea, can develop concentrations of the order of 4 *M*, as defense against a saline environment.[49] The carbohydrates sorbitol and trehalose have been shown to supplement glycerol as natural cryoprotectants in insects.[50]

Among organisms tolerant to drying, a number of protective compounds have been reported by Yancey et al.[49] Prominent among these are betaine, sarcosine, urea, sugars and sugar alcohols, various amino acids, polyols, and trimethylamine oxide. Of particular interest is the observation that this latter compound, a strong macromolecular stabilizer, is generally accompanied in tolerant organisms by urea, a destabilizer. It seems reasonable to interpret this observation as implying that the stabilizing/destabilizing balance of the normal aqueous environment must be maintained. The presence of a high concentration of stabilizer must, therefore, be offset by an associated destabilizer, possibly an important aspect of cryoprotectant toxicity.

It should be emphasized that natural cryoprotectants are not restricted to those capable of penetrating the cell membrane since they can be elaborated within the cells to begin with or, if they are not, adequate time is available during the hardening process for even very slowly permeating compounds to maintain equilibrium across the cell membrane. For artificial cryoprotection, permeation must be relatively rapid to be feasible, ruling out such naturally occurring solutes as trimethylamine oxide and most of the sugars and polysaccharides.

Among the various compounds available for artificial cryoprotection, glycerol appears to occupy a privileged position because of its lack of toxicity at high concentration. Of all the solutes that have been proposed as cryoprotectants, glycerol is the most waterlike in its ability to maintain the hydrophobic forces that are essential to the tertiary and quarternary conformation of macromolecules and the stability of membrane bilayers.[51] Glycerol is only slightly more stabilizing than water even at high concentrations. The surface energy of glycerol solutions is also comparable to that of water, unlike most other solutes which markedly alter solution surface energies and, thereby, the interfacial forces that are important for membrane integrity. An additional asset of glycerol is its relatively large molar volume so that during freezing it not only retains intracellular water colligatively but also contributes its own mass, thus further forestalling cell volume reduction.

From the standpoint of artificial cryoprotection, the principle and perhaps only shortcoming of glycerol is its relatively slow rate of passage across the cell membrane. Much of the success in the use of glycerol in red-cell freezing can be attributed to the fact that red cells possess a facilitated transport mechanism for glycerol with a half-time for glycerol penetration at 37°C of approximately 15 sec.[52] Even so, the rate of water flux is several orders more rapid than that of glycerol. This creates the potential for damaging osmotic gradients during the introduction and removal of the cryoprotectant. If a cell is transferred abruptly from a glycerol-free medium into one containing a high concentration of glycerol, water will rapidly leave the cell in response to the osmotic gradient before glycerol can enter and hyperosmotic damage may be done. In animal cells, the situation is even more serious during the removal of glycerol since few cells can tolerate an increase in volume by much more than a factor of two. This, of course, does not apply to cells with cell walls which protect the cells against hypotonic lysis.

Most cells do not possess a facilitated transport mechanism, and for them the transmembrane diffusion of glycerol is relatively slow with a half-time of the order of 3 to 4 min or more. Because of the care that must be taken in the introduction and removal of glycerol, the times required can be of the order of 30 to 60 min to achieve a concentration sufficiently high to result in good protection from freezing injury. Glycerol penetrates plant cells poorly and has not proven to be particularly useful in this application. The success of plant-cell cryoprotection using amino acids[53] is probably largely due to their relatively rapid permeation.

B. Colligative Cryoprotection in Practice

Since every cell has both a maximum and minimum tolerated volume, the introduction and removal of high concentrations of cryoprotectant must be carefully programmed to prevent exceeding the tolerance of the cell to osmotically induced volume changes. The procedure for designing a program for freezing cells in the equilibrium, or colligative, mode can, therefore, be outlined.

1. Determining the Osmotic Limits of the Cell

The response of the cell to hyper- and hypotonic solutions of nonpenetrating solutes must be determined in order to know the limits of cell volume within which the introduction and removal of cryoprotectants must be conducted. Hyperosmotic tolerance is implied by the killing temperature. For each degree of freezing, the osmolality of a solution in equilibrium with ice increases by 538 mOsm. Most cells appear to be able to tolerate hyperosmotic exposures of at least four times osmotic and, in special cases, osmolalities as high as eight times (lymphocytes and stem cells)[54] or even ten times isotonic (Chinese hamster ovary cells).[55] Human granulocytes, on the other hand, are instantly functionally inactivated by exposure to osmolalities above 750 mOsm.[56] For such cells, knowledge of their tolerance to osmotically induced volume changes is essential to the design of a glycerolizing and deglycerolizing protocol.

2. Determining the Kinetics of Permeation Into and Out of the Cell

The permeation rates can be influenced by temperature and by concentration. Rates can be determined using ^{14}C tagging[53] or, more easily, by following volume changes either by direct observation or by using electronic sizing with a device such as the Coulter Channelizer.* Knowledge of the kinetics of cryoprotectant permeation is essential to determining the equilibration times necessary during introduction and removal.

3. Introducing the Cryoprotectant

The limit of hyperosmotic tolerance established above determines the maximum extracellular concentration to which the cell can initially be exposed. If the maximum tolerance, for example, is to a 4 times isotonic (1200 mOsm) medium, the cells should not be abruptly transferred to a solution containing more than 900 mOsm of cryoprotectant plus isotonic (300 mOsm) saline or serum. By the same reasoning, a cell already equilibrated with a 1200 mOsm solution could safely be transferred to a 4800 mOsm solution (4500 mOsm cryoprotectant plus 300 mOsm saline or serum). For cells with cell walls, the presence of the saline is unnecessary, since the cell wall will prevent cell swelling.

4. Removing the Cryoprotectant

Cryoprotectant removal can be destructive for cells lacking cell walls. Some cells, such as red cells, will tolerate suspension in a one half isotonic solution. If such a cell were equilibrated with a 4800 mOsm glycerol-saline solution, it could be transferred to no less than a 2400 mOsm solution and allowed to equilibrate and return to normal volume. Even with this degree of tolerance to hypo-osmolality, it is evident that a sequence of transfers, from 4800 to 2400 to 1200 to 600 mOsm, and finally to isotonicity will be required. This multiple stepping can be avoided by the use of hyperosmotic solutions of nonpenetrating solutes, a procedure well established in the deglycerolizing of red cells.[56,57]

This technique can be used even should the cell have no tolerance to osmotic volume increase whatsoever provided it is tolerant to hyperosmolality. For example, a cell at 4800 mOsm can be transferred to a solution containing 3600 mOsm of glycerol and 1200 mOsm of salt. Initially, there is no change in total osmolality and, therefore, no change in volume. However, as the glycerol equilibrates across the membrane, the cell shrinks to a volume equivalent to that in a four-times isotonic saline solution because of the presence of the 1200 mOsm salt. The cell can then be transferred safely to a solution of one fourth the osmolality,

---

* Coulter Electronics, Hialeah, Florida.

in this case, to a solution containing only 1200 mOsm of saline. The cell will transiently swell to isotonic volume, then shrink again to the four-times volume as the glycerol leaves the cell. Safe transfer to an isotonic medium is then possible. The duration of the equilibration steps must be determined by the measurement of glycerol flux. As noted previously, cells with cell walls will be protected against hypo-osmotic swelling, greatly simplifying the process of cryoprotectant removal.

This stepwise sequence of suspensions can, of course, be replaced by a continuous change in the osmolality of both penetrating and nonpenetrating solutes. Formulations for such continuous loading and unloading protocols have been reported by several authors, particularly Levin and Miller.[59] Whether or not such protocols are feasible in practice is largely an engineering question. The use of a series of resuspensions may actually be simpler in practice than the creation of continuously variable concentrations of solutes.

For cell suspensions, an important practical limitation to any loading or unloading protocol is solution volume. Continuous dilution of a suspending solution can result in a final volume of many liters. Much of the technology of red-cell freezing has been directed not at cryobiological problems but at control of the volume of the cell suspension during deglycerolization.[60]

5. Cooling Rate

The fact that sufficient cryoprotectant may be present to prevent excessive cell dehydration at any temperature does not mean that cooling rates can be wholly ignored. Since the cell contains freezable water prior to freezing, there is still the possibility of forming intracellular ice if insufficient time is available during freezing for this water to leave the cell. Cooling rates must, therefore, be slow enough to avoid this possibility. Warming rates are generally irrelevant and rapid warming is usually employed purely for the sake of convenience.

6. Choice of Storage Temperature

It is a simple matter, knowing the osmotic tolerance of a cell, to predict the approximate temperature to which it can be frozen as a function of the initial cryoprotectant concentration, assuming that a molal concentration of solute will depress the freezing point by 1.86°C. On this basis, a cell tolerant to four times isotonic osmolality should tolerate freezing to nearly $-10°C$ (5.4 Osm) with a starting concentration of 1.0 m glycerol plus 300 mOsm of salt before the suspending solution will have been concentrated by a factor of four. Glycerol, 2, 4, 6, and 8 m, should protect to, respectively, -17, -32, -47, and $-62°C$. In fact, the 1.86°C/m increment in concentration becomes increasingly misleading at higher concentration and lower temperatures, in part because of deviations from ideal behavior and in part because of increasing solution viscosity so that good storage stability may be seen at a higher temperature than expected. Human red cells, for example, are stable for years at $-60°C$ when frozen with a starting concentration of 5.5 m glycerol.

The duration of safe storage at intermediate temperatures ultimately must be determined empirically. Metabolic reactions may still continue at reduced rates and, since the cells are exposed to a hyperosmotic salt solution, the slow influx of extracellular solute can render animal cells osmotically unstable on thawing. With glycerol, at temperatures below $-60°C$, further concentration of solutes occurs either very slowly or not at all because of the viscosity of the solution which, at $-60°C$, will have been concentrated to about 30 m or roughly 70% v/v. This concentration is also approaching the eutectic of the glycerol-water system. Storage in liquid nitrogen ($-196°C$) or in the vapor phase over liquid nitrogen (about $-120°C$) should confer essentially indefinite storage stability.

V. RAPID (KINETIC) FREEZING

Consideration has so far been limited to freezing at very low cooling rates with the specimen in approximate equilibrium with the extracellular environment such as usually occurs in nature. However, some unprotected cells begin to exhibit some survival as the cooling rate is increased.[61] Ultimately, a cooling velocity may be reached where survival is maximum and, as the cooling rate is further accelerated, survival again falls to zero.

As has been previously discussed, when the cooling rate is increased, freezing may no longer result from a single nucleating event but many small crystals may be formed. If there is insufficient time for water to diffuse to an existing ice crystal, supercooling occurs and ice forms from additional heterogeneous nuclei. At extremely rapid rates of cooling, there may be insufficient time for the finite number of heterogeneous nuclei to grow and most of the water will crystallize spontaneously by homogeneous nucleation as the specimen temperature falls below $-40°C$. The important point, however, is that as the cooling rate increases, insufficient time is available for the completion of events that otherwise would occur were the system in equilibrium. This fact, together with the concept of cell injury through dehydration, leads to a simple and adequate explanation for the survival or lack of survival of cells cooled at higher velocities.

The basic principles of the model are shown in Figure 14. At the top of the figure, the cooling rate is slow. There is adequate time for water to diffuse out of the cell to the ice crystal and the entire system will approach vapor-pressure equilibrium. Intracellular freezing cannot occur (unless the membrane ruptures), since the freezing point of the intracellular solution will be identical to the temperature of the specimen and, presumably, there are no intracellular heterogeneous nuclei. Under these conditions, cell injury must be the result of dehydration and volume reduction.

Turning to the bottom panel in Figure 14, here the cooling velocity is very great and insufficient time is available for much if any of the intracellular water to move out of the cell to contribute to extracellular ice. As a result, the cell interior is progressively supercooled as the specimen temperature falls. As the temperature passes below $-40°C$, homogeneous nucleation takes place and a large number of microcrystals form within the cell. It is probable that most cells do not contain heterogeneous nuclei and that intracellular freezing does not take place above $-40°C$ unless the membrane is damaged and extracellular ice can seed the interior of the cell. There is also some evidence that if the rate of cooling is very rapid so that intracellular nuclei form but do not grow, and if the rate of thawing is sufficiently rapid to prevent significant crystal growth, cells may even survive intracellular freezing. In specimens of more than minute dimensions, these extreme rates of cooling and warming are generally unattainable and, for all practical purposes, intracellular ice should be considered to be generally lethal.

In the center panel of Figure 14 an intermediate situation is illustrated. Here the cooling rate has been sufficiently rapid so that not all of the freezable water has left the cell and the cell may have been spared from dehydration injury. However, there may still have been sufficient dehydration to concentrate cell contents and lower the homogeneous nucleation temperature. As a result, intracellular ice may not develop to a damaging extent even though the cell has not been sufficiently dehydrated to cause dehydration injury.

To achieve this delicate balance between dehydration injury and intracellular freezing, the interplay of several factors is involved. The optimum cooling rate to minimize both dehydration injury and intracellular freezing will depend upon the rate at which water leaves the cell. This, in turn, will depend on the viscosity of the intracellular solution, the surface-to-volume ratio of the cell, the rate of diffusion of water through the cell membrane, the distance between the cell and the nearest ice crystal, and the viscosity of the intervening unfrozen solution. Mazur[63] has analyzed this system and has been able to express mathematically the relationships between most of these parameters. Using his equations, it should be possible to predict the optimum cooling rate provided the values of the various parameters could be determined.

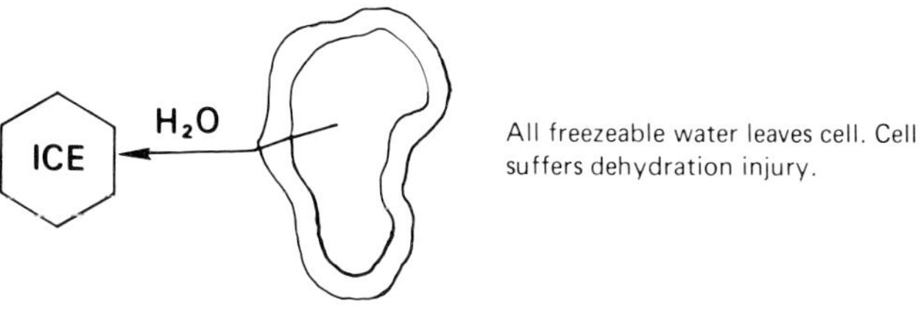

SLOW FREEZING

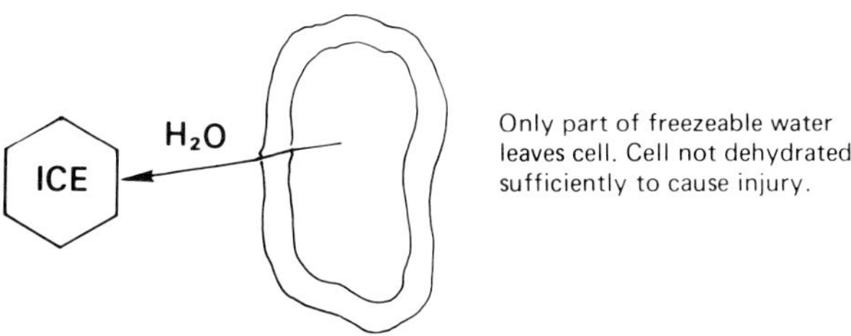

MODERATE RATE FREEZING

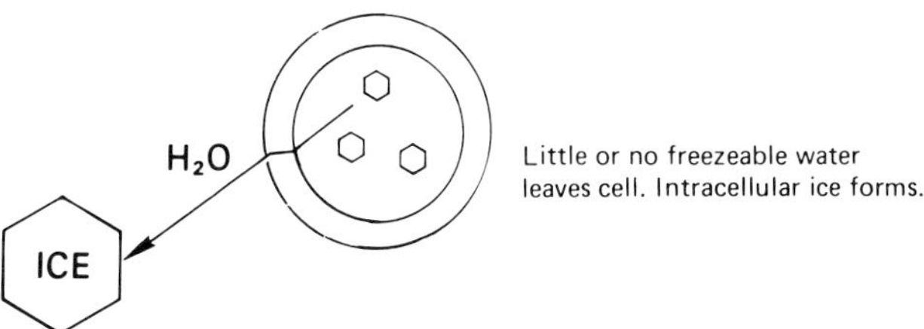

ULTRA-RAPID FREEZING

FIGURE 14. Kinetic freezing injury. During slow freezing (top panel) the cell interior remains in near equilibrium with the external solution. At any temperature, all water that is freezable at that temperature is lost by the cell. Injury is the result of excessive cell dehydration and volume reduction. During very rapid cooling (bottom panel), insufficient time is available for water to leave the cell. At around $-40°C$, homogeneous nucleation occurs and intracellular ice develops, also a generally lethal event. In the center panel, an intermediate situation exists and it is possible that there may be insufficient dehydration to cause dehydration injury, yet sufficient concentration of cell contents to prevent the development of lethal intracellular ice. (From Meryman, H. T. and Williams, R. J., *Crop Genetic Resources: Conservation of Difficult Material*, Withers, L. A. and Williams, J. T., Eds., I.U.B.S., Paris, 1980. With permission.)

Most cells show little or no survival regardless of attempts to optimize cooling rate. The osmotic tolerance of most nonhardy or unprotected cells is such that dehydration injury will occur before there has been sufficient intracellular concentration to prevent intracellular freezing. In such cases, intracellular ice resulting from moderate cooling rates is probably the result of dehydration injury and the seeding of the cell interior through membrane defects rather than from homogeneous nucleation.

Summary — According to the model illustrated in Figure 14, although there are two types of cell injury — dehydration and intracellular freezing — both are really dependent on the single parameter, dehydration. Too much dehydration results in cell injury, possibly as a result of membrane stress. Too little dehydration results in intracellular freezing and an injury which is probably the result either of the mechanical distortion of intracellular structures or of the physical rupture of the cell membrane as a result of the 10% increase in volume as water becomes ice at temperatures where membrane flexibility may be severely restricted.

VI. CRYOPROTECTION DURING KINETIC FREEZING

It should be clear from the preceding discussion that the principal goal of cryoprotection in the kinetic mode is to increase the dimensions of the "window" between excessive and insufficient dehydration. There are several measures that can be taken to increase the probability that a cell will fall into this safe zone.

If the minimum critical volume of a cell is reached when three fourths of its liquid volume is lost, then an unprotected cell will reach critical volume at only four times isotonic. However if a cryoprotectant was present at a concentration of 1 *m*, this critical volume would not develop until an osmolality in excess of 5 Osm had been reached. At slow rates of cooling this would provide protection to only about $-10°C$. By cooling more rapidly, a much lower temperature could be achieved before three fourths of the cell water had been lost. The loss of three fourths of cell water will increase the intracellular cryoprotectant concentration to at least 4 *m*, serving both to lower the homogeneous nucleation temperature and to raise the glass transformation temperature. Even though 4 *m* cryoprotectant may be insufficient to bring these two temperatures to coincidence (see Figure 2), if a viscous cryoprotectant such as glycerol or DMSO is used, the viscosity of the solution at $-50°C$ or below may markedly reduce the rate of crystal growth and permit achieving a very low stabilizing temperature before the development of intracellular ice of damaging dimensions can take place. A cryoprotectant with a larger molar volume, such as DMSO or the glycols, can protect against cell volume loss over and above its pure colligative effect.

A further contribution of cryoprotectants during kinetic freezing is a reduction in the rate of diffusion of water out of the cell because of the increased viscosity produced by the added solute. This reduces the rate of cooling required to obtain just the right amount of dehydration. This, in turn, results in a more uniform cooling rate throughout the specimen and can improve recovery. Some agents appear to be particularly effective in this respect. With blood platelets, for example, the optimum cooling rate in the presence of 0.7 *m* glycerol is 30°C/min,[64] while in the presence of the same concentration of DMSO, the optimum rate is 1°C/min.[65] Either the viscosity of DMSO increases more rapidly than that of glycerol with decreasing temperature, or DMSO is reducing the permeability of water across the membrane.

In summary, to escape injury during kinetic freezing, a cell must not be sufficiently dehydrated to be injured on that basis, but the cell contents must be sufficiently concentrated to prevent the development of ice to an injurious extent. For most cells, this is not attainable in the absence of a penetrating cryoprotectant. It is possible, however, to avoid freezing injury during kinetic freezing with a lower concentration of cryoprotectant than would be required for colligative protection at low cooling rates. Since kinetic freezing produces a situation which is metastable, storage temperatures usually must be low, generally $-120°C$ or below. It is self-evident that maximum warming velocity should be employed during thawing to avoid recrystallization of intracellular ice or additional cell dehydration by extracellular ice.

A. Extracellular Cryoprotectants
Some mention should be made of the extracellular cryoprotectants. These include some

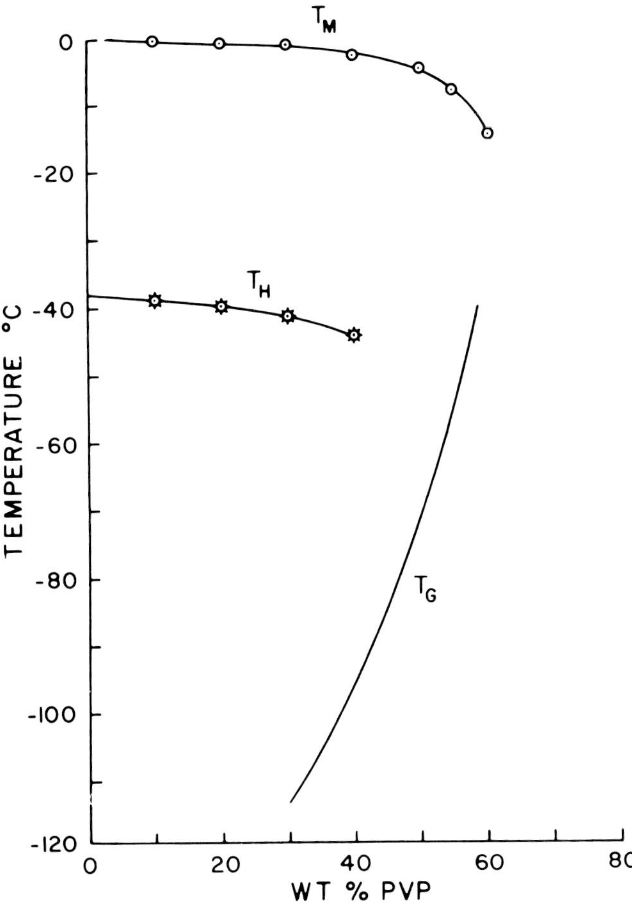

FIGURE 15.   Phase diagram of polyvinylpyrrolidone (PVP). Comparison of this phase diagram with that of glycerol in Figure 2 shows how much more rapidly this polymer decreases melting and homogenous nucleation temperatures and increases the glass transformation temperature as the concentration is increased. If the abscissa were in moles per liter or osmoles the difference would be still more dramatic.

of the sugars and sugar alcohols, but among the most effective are the polymers polyvinylpyrrolidone (PVP), dextran, and hydroxyethyl starch (HES). These compounds, since they presumably do not enter the cell, would not be expected to offer any colligative protection at slow cooling rates and, in fact, this has been shown to be true for human red cells with both sucrose and PVP.[19] Under conditions of kinetic freezing, however, extracellular sugars and polymers can be helpful. This is probably largely the result of their rapidly increasing viscosity with increasing concentration and decreasing temperature, reducing the cooling rate necessary for optimum recovery and thereby permitting more uniformity of cooling rate throughout the sample. With human red cells, for example, the cooling rate for the optimum recovery of cells in plasma containing 7.5% glucose is somewhat in excess of 100°C/sec,[66] whereas the presence of 10% PVP in the plasma reduces the optimum rate to 10°C/sec.[67]

The polymers have an added value in that their osmotic coefficients rise exponentially with concentration so that their ability to prevent water from freezing increases rapidly. Figure 15 reproduces the phase diagram for PVP-water systems and it is apparent that at higher concentrations the compound has a pronounced effect on the several parameters of crystallization (compare with Figure 2 for glycerol). In particular, at concentrations of the order of 50%, the homogeneous nucleation temperature and the glass transformation temperature are approaching each other, putting the attainment of a noncrystallizing solid within easy range. In fact, Skaer et al.[68] have demonstrated that tissue specimens very rapidly frozen in the presence of 50% extracellular PVP display no ice discernible by electron microscopy, either inside or outside the cells. The probable explanation is that the high concentration of extracellular PVP has inhibited the growth of heterogeneous nuclei, thereby preventing the osmotic dehydration of the cells which, since they contain no heterogeneous nuclei, have supercooled to below $-40°C$ where they have undergone homogeneous nucleation. The subsequent cooling has been fast enough to prevent further growth of the submicroscopic intracellular crystals and visible crystallization does not develop. Although immensely valuable as a preparative method for freeze-fracture microscopy, the procedure apparently does not yield viable cells on thawing since, in specimens of useful size, thawing cannot be achieved rapidly enough to prevent recrystallization. This does, however, illustrate the kinds of tricks that may be played utilizing the unusual physical-chemical properties of polymer solutions.

It should be mentioned in passing that the polymeric cryoprotectants also have a special property which can contribute an illusory protection, at least for erythrocytes. Allen et al.[69] have published electron micrographs showing that red cells frozen and thawed in HES had massive membrane defects but still contained their hemoglobin. Williams[70] has shown that HES, PVP, and other cryoprotective polymers lower the surface tension of the suspending medium, establishing an interface with hemoglobin across which that protein cannot pass. In other words, although membrane defects may exist, the interfacial energy difference between the intracellular hemoglobin solution and the extracellular PVP solution maintains a stable interface despite the absence of membrane. In fact, droplets of hemoglobin from lysed, packed red cells dropped into 10% PVP solution maintain their integrity for hours and sink to the bottom of the vessel where their contact angle can be observed and the surface tension differences directly measured. This phenomenon means that red cells rapidly frozen in PVP or other cryoprotective polymers may show an illusory recovery, since cells with damaged membranes may not lose all their hemoglobin. However, on return to PVP-free medium, lysis proceeds to completion.[19] For other cells with more active repair mechanisms, the temporary bandage supplied by these polymers may be of true value.

B. Kinetic Cryoprotection in Practice

According to our model, cryoprotectants used during kinetic freezing serve two purposes. First, they can reduce the rate of diffusion of water from the cell to the ice crystal, thereby reducing the cooling rate required for optimum cell dehydration and enabling the attainment of more uniform cooling rates throughout the specimen. Agents need not penetrate the cell to have this effect. Second, penetrating solutes will provide some colligative support, reducing the amount of cell-volume change at a given intracellular osmolality, reducing the homogeneous nucleation temperature, decreasing the rate of crystal growth, and raising the glass transformation temperature, thus widening the window for survival between excessive cell dehydration and intracellular freezing.

It is for the latter reason that most freezing procedures utilizing the kinetic approach include a penetrating cryoprotectant. Kinetic protocols with glycerol and DMSO generally use a 5 to 10% (0.7 to 1.4 m) concentration. Although good recovery of red cells has been reported with purely extracellular reagents such as PVP and sucrose, many of the cells suffer membrane lesions which do not become apparent until the cells are resuspended in pure saline or plasma.[71]

The requirements for successful freezing by the kinetic mode are

1.   A penetrating cryoprotectant must usually be present prior to freezing.

2. The cells must be partially dehydrated during freezing; not enough to cause dehydration injury but sufficient to concentrate the intracellular solutes so that ice development is avoided.
3. The extracellular solution must have a sufficient viscosity so that the movement of water from the cell to the ice is slow enough to make the protocol manageable in practice.

It should be pointed out that, as Leibo and Mazur[61] showed, some cells have a sufficiently low hydraulic permeability so that the optimal cooling rate is relatively slow, even in the absence of a viscous extracellular solution. In such cases, only the intracellular colligative agent may be necessary to cryoprotect on a kinetic basis. The commonly used intracellular agents, glycerol and DMSO, contribute substantially to both intracellular and extracellular solution viscosity as they are concentrated at low temperature by freezing.

The procedure for designing a program for freezing cells in the kinetic mode can, therefore, be outlined as follows.

1. Choice of Cryoprotectant

Since the cryoprotectant concentration will be low compared to that necessary for purely colligative protection, a wider choice is possible. For many applications, DMSO will be preferable because of its rapid penetration into most cells. Where DMSO toxicity is a problem, glycerol may still be the agent of choice, but information regarding its permeation rate will be essential. Proline has been shown to penetrate plant cells readily and to be an effective cryoprotectant.[54] If the cell in question has a relatively high hydraulic permeability and there is reason to believe that the optimum cooling rate will be high, it may be desirable to add an extracellular solute to increase extracellular viscosity and reduce the optimum cooling rate. Sucrose and glucose are commonly used. Polymers are probably more effective since they affect osmolality less for an equivalent increase in viscosity.

2. Determination of Cooling Regimen

Two approaches to controlled-rate freezing are used. One requires a continuous drop in temperature at a uniform rate to some point, usually below $-40°C$, whence the specimen is transferred to liquid nitrogen. Alternatively, Farrant et al.[72] have reported a simple two-step procedure. The specimen is immersed in a bath at a constant temperature, usually in the vicinity of $-30°C$, held for an empirically determined length of time, then transferred to liquid nitrogen. Where the two methods have been compared, the two-step method produces recoveries equal to the controlled-rate method. In either event, the cooling regimen must be determined empirically by exploring a variety of temperatures and rates in order to optimize recovery.

3. Storage Temperature

Unlike the equilibrium system, a cell suspension frozen by the kinetic mode will be metastable and must be stored at a temperature sufficiently low to prevent the redistribution of water. This means that for long-term storage temperatures must generally be below at least $-120°C$, i.e., in liquid nitrogen or its vapor.

The advantage of kinetic freezing as compared to colligative is solely in the need for a lesser concentration of cryoprotectant. Colligative freezing has the advantage that freezing and cooling rates are less demanding, storage can often be at temperatures of $-80°C$ or higher, accidental freezer failure is not a catastrophe, and the percentage of survival of cells is often much higher. The disadvantage is that very high cryoprotectant concentrations are necessary. This will limit the choice to low-toxicity solutes such as glycerol which, because they generally penetrate cells slowly, present major practical obstacles.

injury from extracellular ice is the result of membrane damage due to osmotically induced cell-volume reduction. We have proposed that in the case of a cell that remains spherical or cylindrospherical there is associated with cell dehydration some degree of resistance to the volume reduction, a resistance implicit in the requirement that membrane area must be reduced in order to accomplish volume reduction.

A membrane resistance to volume reduction requires that there be storage of energy in the membrane which, in turn, implies a decrease in surface tension, some degree of membrane area reduction, an osmotic gradient across the membrane, and a negative intracellular hydrostatic pressure. All of these have been directly measured in selected circumstances. Evidence for this mechanism of freezing injury has been principally developed for plant cells. Evidence that the model may also apply to animal cells is scarce.

Excessive cell dehydration and volume reduction have been shown to be associated with membrane changes — specifically, increase in permeability, loss of membrane material, or irreversible membrane disintegration. In plant cells, the dominant cause of injury is probably either the loss of membrane material or seeding of the cell interior by extracellular ice through membrane defects. Most plants appear to achieve frost tolerance by postponing cell dehydration to a lower temperature, either by increasing intracellular osmolality, rendering some cell water unfreezable, or by being able to maintain intracellular supercooling. At least two plants have evolved a protective mechanism that enables them to lose, store, and recovery membrane lipid. Certain plant species elaborate intracellular compounds that form glasses at relatively high subfreezing temperatures.

Artificial cryopreservation can prevent dehydration injury in three ways. A nontoxic solute such as glycerol, which penetrates the cell, can act as an antifreeze agent to reduce ice formation and solute concentration to less than that which is osmotically damaging. Exploitation of this form of cryoprotection depends on the practicality of introducing and removing high concentrations of cryoprotectant. Alternatively, lesser concentrations of cryoprotectants can be used at a higher cooling rate. The cryoprotectants serve to decrease the rate of diffusion of cell water to the ice crystal as well as providing some antifreeze benefits. An optimum cooling rate will permit low, stabilizing temperatures to be reached before sufficient water has left the cell to cause injury from volume reduction, yet produce sufficient intracellular concentration to restrict the development of intracellular ice. Finally, a combination of high cryoprotectant concentration and hydrostatic pressure can permit vitrification of the entire specimen, thus circumventing all of the effects of ice formation.

REFERENCES

1. **Luyet, B. J. and Gehenio, P. M.,** *Life and Death at Low Temperatures,* Monogr. No. 1, Biodynamica, Normandy, Mo., 1940.
2. **Kylin, H.,** Über die Kälteresistenz der Meeresalgen, *Ber. Deutsch. Bot. Ges.,* 35, 370, 1917.
3. **Lovelock, J. E.,** The mechanism of the protective action of glycerol against haemolysis by freezing and thawing, *Biochim. Biophys. Acta,* 11, 28, 1953a.
4. **Lovelock, J. E.,** The haemolysis of human red blood cells by freezing and thawing, *Biochim. Biophys. Acta,* 10, 414, 1953b.
5. **Mazur, P.,** Causes of injury in frozen and thawed cells, *Fed. Proc.,* 24 (No. 2, Part III), S-175, 1965.
6. **Rasmussen, D. H. and Luyet, B. J.,** Contribution to the establishment of the temperature-concentration curves of homogeneous nucleation in solutions of some cryoprotective agents, *Biodynamica,* 11, 33, 1970.
7. **Meryman, H. T.,** Physical limitations of the rapid freezing method, *Proc. R. Soc. London Ser. B,* 147, 452, 1957.
8. **Mazur, P.,** Physical and chemical basis of injury in single-celled microorganisms subjected to freezing and thawing, in *Cryobiology,* Meryman, H. T., Ed., Academic Press, New York, 1966, 214.

9. **Steponkus, P.,** Destabilization of the plasma membrane of isolated plant protoplasts during a freeze-thaw cycle: the influence of cold acclimation, *Cryobiology,* 19, 671, 1982.
10. **Meryman, H. T.,** Ice crystal formation in frozen tissues, *NMRI Rept. Lect. Rev. Ser.,* No. 53-3, 1953.
11. **Meryman, H. T.,** Freeze-drying, in *Cryobiology,* Meryman, H. T., Ed., Academic Press, New York, 1966, 610.
12. **Meryman, H. T.,** Review of biological freezing, *Cryobiology,* 2, 1, 1966.
13. **Burke, M. J., Gusta, L. V., Quamme, H. A., Weiser, C. J., and Li, P. H.,** Freezing and injury in plants, *Annu. Rev. Plant Physiol.,* 27, 507, 1976.
14. **Williams, R. J.,** Frost desiccation: an osmotic model, in *Analysis and Improvement of Hardiness in Crops,* Olien, C. and Smith, M., Eds., CRC Press, Boca Raton, Fla, 1981, 89.
15. **Takahashi, T.,** Mechanisms of extracellular freezing injury at high subzero temperatures in human polymorphonuclear leukocytes, *Cryobiology,* 18, 622, 1981.
16. **Meryman, H. T,** The exceeding of a tolerable cell volume in hypertonic suspension as a cause of freezing injury, in *The Frozen Cell,* Wolstenholme, G. E. W. and O'Connor, M., Eds., J. & A. Churchill, London, 1970, 51.
17. **Johanssen, N. O. and Krull, E.,** Ice formation, cell contraction and frost killing of wheat plants, *Natl. Swedish Inst. Plant Protect. Contr.,* 12, 345, 1970.
18. **Meryman, H. T., Williams, R. J., and Douglas, M. St. J.,** Freezing injury from "solution effects" and its prevention by natural or artificial cryoprotection, *Cryobiology,* 14, 287, 1977.
19. **Höfler, K.,** Der plasmolytische volumekrische methode und ihre andwendbarkeit, *Ber. Deutsch. Bot. Ges.,* 35, 715, 1917.
20. **Williams, J. M. and Williams, R. J.,** Osmotic factors of dehardening in *Cornus florida* L., *Plant Physiol.,* 58, 243, 1976.
21. **Williams, R. J. and Hope, H. J.,** The relationship between cell injury and osmotic volume reduction. III. Freezing injury and frost resistance in winter wheat, *Cryobiology,* 18, 133, 1981.
22. **Wiest, S. and Steponkus, P.,** Free thaw injury to isolated spinach protoplasts and its simulation at above freezing temperatures, *Plant Physiol.,* 62, 699, 1978.
23. **Siminovitch, D.,** Common and disparate elements in the process of adaptation of herbaceous and woody plants to freezing — a perspective, *Cryobiology,* 18, 166, 1981.
24. **Singh, Y.,** Alterations of membranes in rye cells during lethal freezing and plasmolysis stress: mechanism of freezing injury, *Plant Physiol.,* 69(Suppl.),106, 1982.
25. **Gordon-Kamm, W. J. and Steponkus, P. L.,** Freeze-fracture morphology of the plasma membrane of isolated protoplasts following contraction: influence of cold acclimation, Abstr. Proc. Ann. Meet. Am. Soc. Plant Physiol., Fort Collins, Colo., August, 1983, 96.
26. **Boroske, E. and Elwenspoek, M.,** Osmotic shrinkage of giant egg-lecithin vesicles, *Biophys. J.,* 343, 95, 1981.
27. **Langmuir, I.,** The constitution and fundamental properties of solids and liquids. II. Liquids, *J. Am. Chem. Soc.,* 39, 1848, 1917.
28. **Williams, R. J., Willemot, C., and Hope, H. J.,** The relationship between cell injury and osmotic volume reduction. IV. The behavior of hardy wheat membrane lipids in monolayer, *Cryobiology,* 18, 146, 1981.
29. **Williams, R. J., Hope, H. J., and Willemot, C.,** Membrane collapse as a cause of osmotic injury and its reversibility in a hardy wheat, *Cryobiology,* (Abstr.), 12, 554, 1975.
30. **Gruen, D. W. R. and Wolfe, J.,** Lateral tensions and pressures in membranes and lipid monolayers, *Biochim. Biophys. Acta,* 688, 572, 1981.
31. **Franks, F.,** personal communication.
32. **Adair, G. S.,** Thermodynamic analysis of the observed osmotic pressures of protein salts in solutions of finite concentration, *Proc. R. Soc. London Ser. A,* 126, 16, 1979.
33. **Dick, D. A. T. and Lowenstein, L. M.,** Osmotic equilibrium in human erythrocytes studied by immersion refractometry, *Proc. Roy. Soc. London Ser. B,* 148, 241, 1958.
34. **Baker, H.,** The intracellular pressure of *Nitella* in hypertonic solutions and its relationship to freezing injury, *Cryobiology,* 9, 283, 1972.
35. **Meryman, H. T.,** Freezing injury and its prevention in living cells, *Ann. Rev. Biophys. Bioeng.,* 3, 351, 1974.
36. **Williams, R. J. and Hirsh, A. G.,** Direct demonstration of supercooled intracellular water in frozen plant tissues, *Cryobiology,* 17, 623, 1980.
37. **Nobel, P. S.,** The Boyle-van't Hoff relationship, *J. Theor. Biol.,* 23, 375, 1969.
38. **Zisman, W. A.,** Relation of the equilibrium contact angle to liquid and solid constitution. Contact angle, wetability and adhesion, *Adv. Chem. Ser.,* 43, 1, 1964.
39. **Hamilton, W. C.,** A technique for the characterization of hydrophilic solid surfaces, *J. Coll. Int. Sci.,* 40, 219, 1972.
40. **Langmuir, I. and Waugh, D. F.,** The adsorption of proteins at oil water interfaces and artificial protein lipoid membranes, *J. Gen. Physiol.,* 21, 745, 1938.

41. **Williams, R. J. and Takahashi, T.,** The role of osmotic stress and surface energy in freezing injury of sea urchin eggs, *Comp. Biochem. Physiol.,* 73A, 621, 1982.
42. **Cadenhead, D. A.,** Monomolecular films at the air-water interface, in *Chemistry and Physics of Interfaces,* Ross, S., Ed., Am. Chem. Soc., Washington, D.C., 1971, 27.
43. **Yelenowsky, G.,** Freeze survival of citrus trees in Florida, in *Plant Cold Hardiness and Freezing Stress,* Li, P. H. and Sakai, A., Eds., Academic Press, New York, 1978, 297.
44. **Lindow, S. E., Arny, D. C., Upper, C. D., and Barchet, W. R.,** The role of bacterial ice nuclei in frost injury to sensitive plants, in *Plant Cold Hardiness and Freezing Stress,* Li, P. H. and Sakai, A., Eds., Academic Press, New York, 1978, 249.
45. **Williams, R. J. and Meryman, H. T.,** Freezing injury and resistance in spinach chloroplasts grana, *Plant Physiol.,* 45, 752, 1970.
46. **Hirsh, A. and Williams, R. J.,** unpublished information.
47. **Meryman, H. T.,** A modified model for the mechanism of freezing injury in erythrocytes, *Nature (London),* 218, 333, 1968.
48. **Baust, J. and Miller, L. K.,** Variations in glycerol content and its influence on cold hardiness in the Alaskan carabid beetle, *Pterostichus brevicornis, J. Insect Physiol.,* 16, 970, 1970.
49. **Yancey, P. H., Clark, M. E., Hand, S. C., Bowlus, R. D., and Somers, G. N.,** Living with water stress: evolution of osmolyte systems, *Science,* 217, 1214, 1982.
50. **Morrissey, R. and Baust, J. G.,** The ontogeny of cold tolerance in the gall fly, *Eurosta solidagensis, J. Insect Physiol.,* 22, 431, 1976.
51. **Tanford, C.,** *The Hydrophobic Effect: Formation of Micelles and Biological Membranes,* John Wiley & Sons, New York, 1973, 10.
52. **Carlsen, A. and Wieth, J. O.,** Glycerol transport in human red cells, *Acta Physiol. Scand.,* 97, 501, 1976.
53. **Withers, L. A.,** Storage of plant tissue cultures, in *Crop Genetics Resources — The Conservation of Difficult Material,* Withers, L. A. and Williams, J. T., Eds., University of Reading, Reading, Eng., 1980, 49.
54. **Law, P., Alsop, P., Dooley, D. C., and Meryman, H. T.,** Studies of cell separation: a comparison of the osmotic response of human lymphocytes and granulocyte-monocyte progenitor cells, *Cryobiology,* 20, 644, 1983.
55. **Mironescu, S.,** Hyperosmotic injury in Chinese hamster cells. I. Cell survival in unprotected and DMSO-treated cultures, *Cryobiology,* 14, 451, 1977.
56. **Meryman, H. T., Dooley, D. C., Takahashi, T., Law, P. and Douglas, M. St. J.,** Tolerance of blood cells to osmotic stress, Proc. Int. Cong., ISH-ISBT, Budapest, August, 1982, 419.
57. **Huggins, C. E.,** Frozen blood: principles of practical preservation, *Monogr. Surg. Sci.,* 3, 133, 1966.
58. **Meryman, H. T.,** Cryopreservation of blood and marrow cells: basic biological and biophysical considerations, in *Clinical Practice of Blood Transfusion,* Petz, L. D. and Swisher, S. N., Eds., Churchill Livingstone, London, 1981, 313.
59. **Levin, R. L. and Miller, T. W.,** An optimum method for the introduction and removal of permeable cryoprotectants: isolated cells, *Cryobiology,* 18, 32, 1981.
60. **Meryman, H. T. and Hornblower, M.,** Method for freezing and washing red blood cells using a high glycerol concentration, *Transfusion,* 12, 145, 1972.
61. **Leibo, S. P. and Mazur, P.,** The role of cooling rate in low-temperature preservation, *Cryobiology,* 8, 447, 1971.
62. **Meryman, H. T.,** The relationship between dehydration and freezing injury in the human erythrocyte, *Proc. Int. Conf. Low Temp. Sci. University Hokkaido,* 2, 231, 1967.
63. **Mazur, P.,** Kinetics of water loss from cells at subzero temperatures and the likelihood of intracellular freezing, *J. Gen. Physiol.,* 47, 347, 1963.
64. **Dayian, G. and Rowe, A. W.,** Cryopreservation of human platelets for transfusion. A glycerol-glucose, cooling procedure, *Cryobiology,* 12, 1, 1976.
65. **Djerassi, I., Farber, S., and Roy, A.,** Preparation and *in vivo* circulation of human platelets preserved with combined dimethyl sulfoxide and dextrose, *Transfusion,* 6, 572, 1966.
66. **Meryman, H. T. and Kafig, E.,** Rapid freezing and thawing of whole blood, *Proc. Soc. Exp. Biol. Med.,* 90, 587, 1955.
67. **Doebbler, G. F., Rowe, A. W., and Rinfret, A. P.,** Freezing of mammalian blood and its constituents, in *Cryobiology,* Meryman, H. T., Ed., Academic Press, New York, 1966, 407.
68. **Skaer, H. LeB., Franks, F., and Echlin, P.,** Nonpenetrating polymeric cryofixatives for ultrastructural and analytical studies of biological tissues, *Cryobiology,* 15, 589, 1978.
69. **Allen, E. D., Weatherbee, L., and Permoad, P. A.,** Post-thaw suspension of red cells cryopreserved with hydroxyethyl starch, *Cryobiology,* 15, 375, 1978.
70. **Williams, R. J.,** The surface activity of PVP and other polymers and their antihemolytic activity, *Cryobiology,* 20, 521, 1983.

71. **Meryman, H. T. and Hornblower, M.,** Changes in red cells following rapid freezing with extracellular cryoprotective agents, *Cryobiology,* 9, 262, 1972.
72. **Farrant, J., Walter, C. A., Lee, H., and McGann, L. E.,** Use of two-step cooling procedures to examine factors influencing cell survival following freezing and thawing, *Cryobiology,* 14, 372, 1977.
73. **Fahy, G. M., MacFarlane, D. R., Angell, C. A., and Meryman, H. T.,** Vitrification as an approach to cryopreservation, *Cryobiology,* 21, 407, 1984.

Chapter 3

CRYOBIOLOGY OF ISOLATED PROTOPLASTS: APPLICATIONS TO PLANT CELL CRYOPRESERVATION*

Peter L. Steponkus

TABLE OF CONTENTS

*   This material is, in part, based on work supported by the National Science Foundation Grant No. PCM-8012688 and the U.S. Department of Energy under Contract No. DE-AC02-81ER10917; Department of Agronomy Series Paper No. 1455.

I. INTRODUCTION

In the past decade considerable effort has been directed to the cryopreservation of plant tissue cultures.[39] Prior to 1972, there were only a few reports of the successful cryopreservation of suspension cultures[7,21] and callus cultures.[1] Interestingly, at a plant hardiness workshop held in 1971 at the University of Minnesota, several respected authorities in the field of plant hardiness research expressed doubts that plant tissue cultures could be successfully frozen or even acclimated to low temperatures. Needless to say, such doubts were quickly diminished by numerous reports of the successful cryopreservation of cells, organs, and tissues from a diverse array of plant species[39] and the demonstrated totipotency of cryopreserved cells.[20] Cryopreservation of both cold hardy[23,25] and chilling-sensitive species[5,24] has since been possible.

In most instances, the formulation of successful protocol for the cryopreservation of plant cells, tissues, and organs resulted from empirical approaches. Most often the protocols are variants of the protocol empirically determined to be effective for the cryopreservation of mammalian cell types. In very few instances has the successful protocol evolved from a systematic analysis of freezing injury or the mode of action of various cryoprotective additives. Although the empirical approach is rather inefficient, some successes have resulted. Whereas an analytical understanding of freezing injury and the mode of action of cryoprotectants would provide for a more direct approach, such an analytical understanding has been rather elusive.

For many years, cryoinjury to biological cells has been viewed from the perspective that injury may result from either intracellular or extracellular ice formation. Credit for the origin of this perspective is most appropriately given to Müller-Thurgau[19] who first demonstrated that ice crystals formed within cells when cooling was rapid but confined to the extracellular spaces at slow cooling rates. Subsequently, Siminovitch and Scarth[26] considered differences in the causes of frost injury in nonacclimated and acclimated herbaceous plants on the basis of whether ice formation occurred intracellularly or extracellularly. The same perspective is now commonly used in considering strategies for the cryopreservation of biological cells. This is largely attributable to Mazur's two factor hypothesis of cryoinjury.[15,16] This hypothesis evolved from observations that there is an optimum cooling rate for survival. At suboptimal cooling rates, injury is considered to be due to "solution effects" (i.e., solute concentration, changes in pH, and reduction in cell volume); whereas at supraoptimal rates, injury is ascribed to intracellular ice formation. As a result, strategies that have evolved for cryopreservation of biological cells entail manipulation of the cooling protocol to preclude intracellular ice formation and the addition of various additives to minimize the "solution effects" (see also Chapters 2, 5 to 8, 11, and 12).

The "either/or" perspective has been very helpful in providing experimental approaches and *possible* explanations for cryoinjury. Unfortunately, *the* explanation for cryoinjury remains to be elaborated. It is possible, that the "either/or" perspective is, in part, obscuring the real situation. For example, both intracellular and extracellular ice formation are considered to be *causes* of injury — albeit for different reasons. It is commonly inferred that intracellular ice is injurious but extracellular ice is not a cause of injury. This is in spite of the fact that the manner by which intracellular ice formation effects membrane damage has never been adequately explained. While it is valid to consider that the chemical potential of the intracellular solution equilibrates with that of the partially frozen extracellular medium by *either* intracellular ice formation *or* cell dehydration, it may be erroneous to infer that intracellular ice formation is a *cause* of injury. Although there is little doubt that injury is associated with intracellular ice formation, it is possible that intracellular ice formation is a *result* of damage to the plasma membrane.[31]

The elusiveness of an analytical understanding of cryoinjury may, in part, be the result of another choice of perspectives. Most hypotheses of injury have emerged solely from a perspective of what causes injury with little insight into what constitutes injury. Although cryoinjury is commonly inferred to be the result of injury to the plasma membrane, there are few reports of the specific lesions that are incurred. This is largely because of the difficulty in isolating the plasma membrane for analysis. Isolated protoplasts, however, provide a system in which the plasma membrane can be studied directly *in situ*. Further, isolated protoplasts provide an ideal system for quantitative cryomicroscopic studies.[29] Studies of the cellular and subcellular aspects of freezing injury and cold acclimation of isolated protoplasts[3,4,6,27-37,42,43] are directly applicable to the formulation of cryopreservation protocol for plant cells, tissues, and organs.

Studies of the cryobiology of isolated protoplasts demonstrate that cryoinjury may be manifested as any one of several different forms of injury. These include intracellular ice formation, loss of osmotic responsiveness during cooling, expansion-induced lysis during warming, or altered osmometric behavior. The four forms of injury may be considered as different manifestations of alterations in the semipermeable characteristics of the plasma membrane. Therefore, the objective is to characterize the specific lesions and then elucidate how the interacting thermal, mechanical, and chemical stresses incurred during a freeze-thaw cycle influence the structure and function of the plasma membrane. With such knowledge, the natural process of cold acclimation is better understood. Similarly, such knowledge will allow for the formulation of more effective cryopreservation protocol.

Currently, our understanding of the four lesions varies considerably. Expansion-induced lysis can be discussed at the molecular level, whereas only the phenomenology of the loss of osmotic responsiveness during cooling can be described at this time. Nonetheless, requirements for successful cryopreservation — from the perspective of the plasma membrane — can be considered.

II. INTRACELLULAR ICE FORMATION

At any finite cooling rate, transient supercooling of the intracellular solution will occur. Intracellular ice formation will occur if the supercooled solution is nucleated. Quite often this is observed to occur over the range of -10 to $-20°C$ in a diverse array of cell types.[14] It is generally accepted that extracellular ice rather than intracellular nucleators is responsible for nucleation of the intracellular solution. Apparently the plasma membrane is an effective barrier to external ice crystals — but only above temperatures of -10 to $-20°C$. Thus, Mazur[13,16,17] has reasoned that intracellular ice formation will occur if the cells do not approach osmotic equilibration before reaching the characteristic nucleation temperature.

In 1963, Mazur[13] derived a quantitative model for cellular volumetric behavior during a freeze-thaw cycle given the hydraulic conductivity of the cell to water (L_p), the temperature coefficient of L_p, the internal solute potential, and the ratio of cell surface to volume. The analytical model permits calculation of the extent of supercooling of a cell as a function of cooling rate which can be used to estimate the probability of intracellular ice formation as a function of the extent of supercooling. Generally, a nucleation temperature of -10 to $-15°C$ is assumed and the probability (P) of intracellular ice formation is assigned a value of zero when the degree of supercooling (ΔT) is $< 2°C$ or the fractional osmotic volume (FV_π) < 0.1. P is assigned a value of 1.0 when $\Delta T > 2°C$ and $FV_\pi > 0.1$.[17] Implied is the tacit assumption that at temperatures above the nucleation temperature the probability of intracellular ice formation is zero regardless of the extent of supercooling.

A primary objective of the quantitative model developed by Mazur[13,14,16,17] for the analysis of the cooling rate dependence of intracellular ice formation was to support the contention that diminished cell survival at rapid cooling rates is the result of intracellular ice formation. Qualitatively, the results are consistent with experimental observations in that the predicted incidence of intracellular ice formation at a given cooling rate corresponds with the cell survival observed or, in some instances, the observed incidence of intracellular ice formation.[17] There are, however, instances where the quantitative agreement between the theoretical and experimental results are not consistent. In such analyses, much attention has been given to the validity of the estimates of the parameters that influence volumetric behavior. In contrast, characterization of the nucleation temperature and the probability of intracellular ice formation as a function of supercooling have received relatively little attention. This is unfortunate because Mazur[17] notes that predictions of the cooling rate dependence of the probability of intracellular ice formation are relatively insensitive to the values of ΔT and FV_π selected but are extremely sensitive to the value assumed for the nucleation temperature.

Cryomicroscopy of isolated protoplasts allows for a direct analysis of intracellular ice formation during a freeze-thaw cycle.[3,27,29,31] The cryomicroscope allows for the precise control of temperature and cooling rates. Calculations of cellular volumetric relations during cooling — from which the extent of supercooling and diffusional water permeability may be calculated — are simplified using spherical protoplasts. The use of video image analysis facilitates precise mensuration techniques and, more importantly, the temperature at which intracellular nucleation occurs is observed directly — rather than inferred from survival data. Thus, all of the parameters identified to date as relevant to intracellular ice formation can be determined from direct observations of individual protoplasts during a freeze-thaw cycle.

For protoplasts isolated from nonacclimated leaves of *Secale cereale* L. cv Puma, the incidence of intracellular ice formation is strongly dependent on both the cooling rate and minimum temperature imposed.[3] The critical cooling rate is on the order of 3°C/min. At cooling rates of 5°C/min and greater, the incidence of intracellular ice formation increases greatly as the minimum temperature is decreased over the range of -3 to -20°C. At the slowest cooling rate imposed (3.1°C/min) the majority of the protoplasts approach osmotic equilibrium after cooling to -7.5°C. At faster cooling rates there is a greater extent of osmotic disequilibrium and hence intracellular supercooling which parallels the increase in the incidence of intracellular ice formation.

The critical cooling rate determined for isolated rye protoplasts (3.1°C/min) is the maximum cooling rate at which the incidence of intracellular ice formation will be minimal. When cooled at this rate, the majority of the protoplasts will achieve osmotic equilibrium and hence will not be extensively supercooled before the critical nucleation temperature (-15°C) is attained. For organized tissues, the critical cooling rate will be slower because the extent of supercooling of the intracellular solution is a function of the rate of mass transfer of water relative to the rate of heat transfer. For isolated protoplasts, the plasma membrane is the primary barrier to the diffusion of water to the extracellular ice. This is not the case for organized tissues in which the diffusion of intracellular water to ice loci will be impeded by other factors — most notably cell:cell contact. As heat transfer will not be as greatly affected as mass transfer, a greater extent of intracellular supercooling will occur in cells of organized tissues than in isolated protoplasts when both are cooled at the same rate. Therefore, the optimum cooling rate will be lower. Although Levitt[10] reports that the critical cooling rate of isolated protoplasts is approximately ten times greater than that of organized tissues, he failed to acknowledge this fact in his interpretation[9] of the decreased incidence of intracellular ice formation in acclimated tissues.

As previously discussed, the extent of supercooling, which is largely a function of the cooling rate, is not the sole determinant of intracellular ice formation. Nucleation of the supercooled solution must also occur. In nonacclimated rye protoplasts, nucleation occurs over a relatively narrow temperature range with the median nucleation temperature at -15°C. The nucleation temperature is largely independent of the cooling rate.

As early as 1932, Chambers and Hale[2] deduced that the plasma membrane acts as a barrier to extracellular ice and, as long as the plasma membrane remains intact, the intracellular solution can be supercooled below its freezing point. The reason for the decreased efficacy of the plasma membrane at low temperatures that results in the characteristic nucleation temperature remains to be determined. Mazur[14,17] has suggested that nucleation is largely a consequence of temperature effects on ice crystals. Supposedly, only below $-10°C$ will extracellular ice crystals of a suitable small size be sufficiently stable to pass through aqueous channels in the membrane. Initially Mazur[14] considered alterations in the permeability properties at lower temperatures as a reason for ice nucleation, but discounted this possibility because no marked increase in the permeability to intracellular solutes was observed in yeast cells. More recently, however, the possibility that alterations in the plasma membrane are responsible for nucleation has been reexamined.[22] Prior to this report, our direct observations of isolated protoplasts suggested that the nucleation event is primarily the result of mechanical breakdown of the plasma membrane per se rather than a consequence of temperature dependent characteristics of the external ice.[27-31] Specifically, with high resolution video cryomicroscopy outward flow of the intracellular solution is observed *before* intracellular ice formation and thus effecting contact with extracellular ice matrix.

It is from these observations that the either/or perspective for intracellular ice formation vs. cell dehydration is, in part, questioned. Assuming that the primary cause of nucleation is mechanical breakdown of the plasma membrane, intracellular ice formation is not a cause of injury but is only a manifestation of injury. Thus, if mechanical breakdown occurs when the intracellular solution is supercooled, intracellular ice formation will occur. Alternatively, if the cooling rates are such that supercooling of the intracellular solution is minimal, injury will be manifested as another form of injury. (This form of injury is hypertonic-induced loss of osmotic responsiveness and will be discussed separately.) This suggestion explains the cooling rate dependence of intracellular ice formation, but does not resolve the cooling rate dependence of survival that is often inferred yet not always observed.[17] To reconcile a cooling rate dependence of survival, a cooling rate dependence for mechanical breakdown of the plasma membrane is required. As indicated, the nucleation temperature is largely independent of the cooling rate over the range of -3.1 to $16°C/min$ — but in these studies so was the extent of survival.[31]

Assuming that the characteristic nucleation temperature is a consequence of mechanical failure of the plasma membrane poses an interesting possibility for cryopreservation. Can the characteristic nucleation temperature be lowered? Generally, in most analyses of intracellular ice formation, the nucleation temperature is assumed and considered to be an invariant. Recent observations suggest that this is not appropriate. In studies with isolated rye protoplasts, the incidence of intracellular ice formation is consistently less in protoplasts isolated from acclimated tissues than that observed in protoplasts isolated from nonacclimated tissues.[3] In both nonacclimated and acclimated protoplasts, the extent of supercooling was similar under conditions that resulted in the greatest difference in the incidence of intracellular ice formation — cooling to -15 or $-20°C$ at rates of 10 or $16°C/min$. Further, the hydraulic conductivity determined during freeze-induced dehydration at $-5°C$ was similar for both nonacclimated and acclimated protoplasts. The major distinction between nonacclimated and acclimated protoplasts was the temperature at which nucleation occurred. The median nucleation temperature was $-15°C$ for nonacclimated protoplasts but $-42°C$ for acclimated protoplasts. These results suggest that the decreased incidence of intracellular ice formation following cold acclimation is due to an increased stability of the plasma membrane that decreases the temperature at which nucleation occurs rather than the often supposed increase in water permeability.[9]

Other observations suggest that the nucleation temperature is also influenced by the suspending medium. For example, the nucleation temperature of mouse ova suspended in phosphate buffered saline (PBS) is $-7°C$ whereas those suspended in PBS $+$ 1 M DMSO nucleate below $-40°C$.[8,22] Considering that the stability of the plasma membrane of Friend leukemia cells is increased by the addition of DMSO to the suspending medium,[12] it is possible that DMSO acts, in part, by increasing the stability of the plasma membrane at lower temperatures.

With isolated rye protoplasts, both the tonicity and composition of the suspending medium influence the nucleation temperature. The median nucleation temperature for nonacclimated protoplasts suspended in 0.53 Osm NaCl + $CaCl_2$ solutions is $-15°C$, but is decreased to $-28°C$ when suspended in 1.03 Osm solutions.[3] For acclimated protoplasts suspended in 1.03 Osm solutions of NaCl + $CaCl_2$, the median nucleation temperature is $-42°C$. In 1.03 Osm solutions of KCl + $CaCl_2$, the median nucleation temperature is approximately $-30°C$; in proline it is approximately $-20°C$.

Thus, the formulation of an effective cryopreservation protocol to minimize the incidence of intracellular ice formation must consider not just the cooling rate, but also the composition of the suspending medium and the stage of acclimation of the tissue. The direct observation of the material under investigation by cryomicroscopy provides for the direct determination of the temperature at which nucleation occurs.

III. EXPANSION-INDUCED LYSIS

The predominant form of cryoinjury incurred by protoplasts isolated from nonacclimated tissues occurs following thawing of the suspending medium. Lysis occurs when the protoplasts are osmotically expanding and hence termed "expansion-induced lysis".[34-36] Lysis is a consequence of large area deformations incurred by the plasma membrane as a result of osmotically induced contraction-expansion of the protoplast during a freeze-thaw cycle rather than alterations in the osmotic characteristics. This form of injury is not unique to isolated protoplasts and also is observed in isolated mesophyll cells[32] (see also Chapter 4).

During cell dehydration, the surface area of the plasma membrane is decreased due to subduction of membrane material.[6] Sufficiently large contractions are irreversible because the expansion potential of isolated protoplasts is characterized by a fixed increment rather than a fixed maximum critical surface area. This fixed increment is referred to as the "tolerable surface area increment".[34] In a population of protoplasts the absolute magnitude of the increase in the mean surface area of the population that results in lysis of 50% of the population is denoted as the $TSAI_{50}$ value. The $TSAI_{50}$ varies among the species examined to date and is influenced by the composition of the suspending medium. More importantly, the $TSAI_{50}$ increases during cold acclimation.[31] For example, the $TSAI_{50}$ for protoplasts isolated from nonacclimated leaves of Puma rye is 1000 μm^2 whereas the $TSAI_{50}$ for protoplasts isolated from acclimated leaves is in the range of 2000 to 3000 μm^2 — depending on the composition of the suspending medium.[35]

During a freeze-thaw cycle of protoplast suspensions, the protoplasts contract with the minimum volume attained as a function of the minimum temperature imposed. The volume and corresponding surface area attained at subzero temperatures are consistent with that predicted by Boyle van't Hoff behavior at $0°C$.[3] During warming of the suspension, the protoplasts expand to an extent determined by the osmolality of the suspending medium. For nonacclimated Puma rye protoplasts the mean surface area is 3750 μm^2 in isotonic solutions (0.53 Osm). When a protoplast suspension is frozen to $-2°C$, the mean protoplast surface area is decreased 1000 μm^2. Thus, during subsequent thawing of the suspending medium, lysis occurs in 50% of the population because the TSAI value is exceeded. Based solely on these phenomenological observations, a remedy for mitigation of expansion-induced lysis is apparent. The incidence of this lesion can be reduced simply by limiting the extent of osmotic expansion to a value below the $TSAI_{50}$. This can be accomplished by the addition of solutions of the appropriate osmolality during thawing.

Because exposure to relatively warm subzero temperatures is sufficient to result in expansion-induced lysis to nonacclimated protoplasts, it is reasonable to question the significance of this form of injury to cryopreservation practices that involve substantially lower temperatures. Given that cryoinjury may result from any one of several lethal stresses, a successful cryopreservation protocol must effectively mitigate multiple stresses. The stresses that result in expansion-induced lysis represent the first and last stresses encountered during a freeze-thaw cycle — surface area contraction and expansion. When frozen to lower temperatures, other stresses will prevail and the predominant form of injury will be different — hypertonic induced loss of osmotic responsiveness. Mitigation of this lesion will, however, still result in the cell or protoplast being susceptible to expansion-induced lysis. Thus, successful cryopreservation must mitigate the stresses that result in both lesions. The significance of expansion-induced lysis is also suggested by the fact that post-thaw removal of cryoprotectants frequently results in diminished survival and the immediate transfer to a fresh growing medium free of cryoprotectants is deleterious.[40,41]

Although a remedy for expansion-induced lysis may be prescribed on the basis of phenomenological observations, restricting the extent of osmotic expansion during thawing merely precludes manifestation of injury. The protoplast is predisposed to expansion-induced lysis because of the contraction-induced deletion of membrane material. Therefore more direct approaches to the mitigation of expansion-induced lysis for cryopreservation require an understanding of membrane behavior during osmotically induced contraction and expansion.

The fact that expansion-induced lysis is the result of area expansion of the plasma membrane suggested that membrane disruption is associated with changes in the tension of the plasma membrane.[34] Characterization of the stress-strain relationship (SSR) of the plasma membrane during contraction and expansion.[35,42,43] For small deformations the amount of material in the plane of the membrane is conserved and the membrane behaves as an elastic two-dimensional fluid with an area modulus of elasticity (k_A) of 230 mN $\cdot$ m^{-1}. Because the tension necessary to lyse the plasma membrane is of the order of 4 to 6 mN $\cdot$ m^{-1}, elastic expansion is limited to 2 to 3%. For area deformations greater than this, material is introduced into, or subduced from the plane of the membrane. For area deformations of $\pm$ 15%, the SSR approaches a surface energy law and the resting tension is independent of area and has a value of 100 μN $\cdot$ m^{1}. Therefore, we have proposed the existence of a reservoir into or from which membrane material is transferred during large area contractions or expansions.[42,43]

Existence of the proposed reservoir has been documented by both light and electron microscopic observations.[6,31] Following equilibration in hypertonic (1.20 Osm) solutions, protoplasts appear spherical to the resolution of the light microscope. Scanning electron microscope (SEM) observations demonstrate that the surface texture of the plasma membrane remains smooth and similar to that observed in protoplasts maintained in isotonic (0.53 Osm) solutions. No folding or pleating is observed in transmission electron micrographs (TEM) of thin sections. Instead, numerous vesicles are present in the cytoplasm. The extent of vesiculation is more readily apparent using computer-enhanced image processing of differential interference contrast images.[29] Optical sectioning demonstrates that numerous vesicles occur in the cytoplasm. If the protoplasts are labeled with fluorescein-Concanavalin A prior to contraction, the cytoplasmic vesicles are fluorescent. This is consistent with the inference that the vesicles are a result of the deletion of membrane material from the plasma membrane.

When the morphology of acclimated protoplasts equilibrated in hypertonic solutions is contrasted to that of nonacclimated protoplasts, several differences are apparent.[6a,31] Whereas the plasma membrane of nonacclimated protoplasts is smooth without any folding or pleating, numerous extrusions are observed on the surface of acclimated protoplasts. As viewed with SEM or computer-enhanced DIC microscopy, the extrusions appear as knobs or polyps that are tethered to the surface of the plasma membrane. The knobs range in diameter from 0.2

to 0.6 µm. In thin sections, the extrusions are bounded by the plasma membrane and not merely globules adhering to the surface of the plasma membrane. The interior of the extrusions is densely osmiophilic suggesting a high lipid content.

A similar surface morphology is observed in acclimated protoplasts during a freeze-thaw cycle.[6a,31] The tethered extrusions become apparent during cooling when the protoplasts are contracted and are especially pronounced during warming when the surrounding medium melts. As the protoplasts expand in response to the decreasing osmolality the extrusions are gradually drawn into the plane of the membrane and the surface regains a smooth surface texture. Reabsorbtion of the extrusions and the smooth surface texture of acclimated protoplasts expanded from hypertonic solutions has also been documented with SEM. Thus, the extrusions are observed in both light and electron microscopy and hence not an artifact of the fixation procedures. Further, formation of the extrusions can be elicited by either direct osmotic manipulation or by freeze-induced contraction and is fully reversible in both situations. This indicates that the extrusions are casually related to the increased tolerance of acclimated protoplasts to area deformations.

On the basis of these observations we propose that the behavior of the plasma membrane during contractions is a major factor contributing to the TSAI value and the increased tolerance of acclimated protoplasts to osmotically induced contraction/expansion excursions experienced during a freeze-thaw cycle. Whereas in nonacclimated protoplasts osmotically induced contraction results in an effective deletion of membrane surface area by endocytotic vesiculation, contraction of acclimated protoplasts results in the exocytotic extrusion of the plasma membrane. Material subducted in the form of endocytotic vesicles is not readily reincorporated into the plane of the membrane of acclimated protoplasts results in conservation of membrane area (that which bounds the extrusions) and preferential deletion of lipids (to the interior of the extrusions) which are apparently readily reincorporated.

Measurements of the TSAI value indicate that the composition of the suspending medium can alter the TSAI value.[34] For example, when suspended in equiosmolal solutions of either $NaCl + CaCl_2$ or $KCl + CaCl_2$ and subjected to either osmotically induced expansion or a freeze-thaw cycle, more protoplasts lyse when K^+ rather than Na^+ is the monovalent cation. A more comprehensive study of the influence of monovalent cations demonstrated that the sensitivity of protoplasts to a freeze-thaw cycle followed the series:

$$Li^+ = Na^+ > K^+ = Rb^+ = Cs^+ \text{ and } Cl^- > Br^- > I^-$$

in order of decreasing protoplast survival.

Of more significance to cryopreservation, some solutes can increase the TSAI value.[11] For nonacclimated rye protoplasts, the TSAI value is similar (1000 µm²) for protoplasts osmotically expanded from either isotonic or hypertonic solution whether they are suspended in sorbitol or $NaCl + CaCl_2$. When suspended in proline, the TSAI value for expansion from isotonic solutions is also 1000 µm². If however, the protoplasts are first subjected to hypertonic solutions of proline and then expanded, the TSAI value is nearly doubled to 1900 µm². This value approaches the increased value observed for acclimated protoplasts. In acclimated protoplasts, the increased TSAI value during expansion from hypertonic conditions is largely attributable to the differences in the behavior of the plasma membrane during contraction (endocytotic vesiculation vs. exocytotic extrusion). Several observations suggest that proline is also influencing the contraction event rather that the expansion potential per se. First, proline does not increase the expansion potential from isotonic solutions — only from hypertonic solutions. Second, if protoplasts are contracted in proline but expanded in sorbitol, the beneficial influence of proline is still observed. Also, there is a time dependence for the duration of the exposure to proline in the contracted state.

These observations indicate that the tolerance of the plasma membrane to area contraction/

expansion excursions is a limiting factor to the survival of both isolated protoplasts and isolated cells during a freeze-thaw cycle. This sensitivity of nonacclimated protoplasts is greatly dimished by cold acclimation. Where cold acclimation is not possible, injury resulting from the contraction-induced deletion of membrane material can be precluded by thawing in elevated tonicities to optimize the extent of expansion. Alternatively, the results with proline indicate that the contraction-induced alteration per se is amenable to chemical manipulation.

IV. LOSS OF OSMOTIC RESPONSIVENESS

The third form of cryoinjury observed in isolated protoplasts is the loss of osmotic responsiveness following cooling.[4,30,37] During cooling, the protoplasts exhibit characteristic osmotic behavior and, at a given temperature, attain a volume that is consistent with the Boyle-van't Hoff relation determined at 0°C.[3] During warming, however, the protoplasts are osmotically inactive and remain contracted. Thus, injury occurs during freezing of the suspending medium when the protoplasts are osmotically contracted. Loss of osmotic responsiveness is the predominant form of injury in acclimated protoplasts cooled to the LT_{50} (-25 to -30°C) and in nonacclimated protoplasts cooled significantly below the LT_{50}.

Although it is unknown what alteration in the plasma membrane is responsible for the loss of osmotic responsiveness,* we have reason to believe that this form of injury is related to that which results in intracellular ice formation in nonacclimated protoplasts.[31] Over the range of -3 to -20°C, the incidence of injury manifested as intracellular ice formation and loss of osmotic responsiveness increases with decreasing temperature regardless of the cooling rate. At any given temperature, the combined incidence of injury is constant at all cooling rates over the range of 3.1 to 16°C/min. As the cooling rate to any given temperature is decreased, the proportion of injury manifested as intracellular ice formation is decreased. This decrease, however, is offset by a corresponding increase in the proportion of injury manifested as a loss of osmotic responsiveness. Thus, while the combined incidence of injury is temperature dependent, it is independent of the cooling rate. Only the form of injury is cooling-rate dependent. The constancy of the combined incidence of injury is attributed to a common membrane lesion — mechanical breakdown of the plasma membrane.

The loss of osmotic responsivenes can also be elicited by exposure to hypertonic solutions in the absence of ice or low temperatures.[4] Although such conditions result in the loss of osmotic responsiveness in a manner that parallels the conditions under which injury occurs during a freeze-thaw cycle (in the contracted state), it must be emphasized that this is only a correlation. At this time, our studies of the loss of osmotic responsiveness have been limited to the phenomenology during exposure to hypertonic solutions. Under these conditions, the incidence of injury is determined in the contracted state in order to preclude confounding the estimates of survival due to expansion-induced lysis. This is facilitated by preincubating the protoplasts in fluorescein diacetate and determining the retention of fluorescein in the contracted protoplasts.

From such studies, it is observed that the incidence of injury is a function of the composition and tonicity of the suspending medium, the duration of exposure, and whether the protoplasts are isolated from acclimated or nonacclimated tissues. In both nonacclimated and acclimated protoplasts, the loss of osmotic responsiveness is a function of both solute concentration and duration of exposure. Thus, there is no critical solute concentration threshold. Instead, the time required to attain a given incidence of injury is decreased as the solute concentration

---

* Note added in press: Recently (Gordon-Kamm, W. J. and Steponkus, P. L., *Proc. Natl. Acad. Sci.*, 1984, in press), we have observed that loss of osmotic responsiveness is associated with lamellar to hexagonal$_{II}$ phase transitions in the plasma membrane. These transitions are a consequence of freeze-induced dehydration, rather than subzero temperature per se and their incidence is greatly diminished by the addition of DMSO.

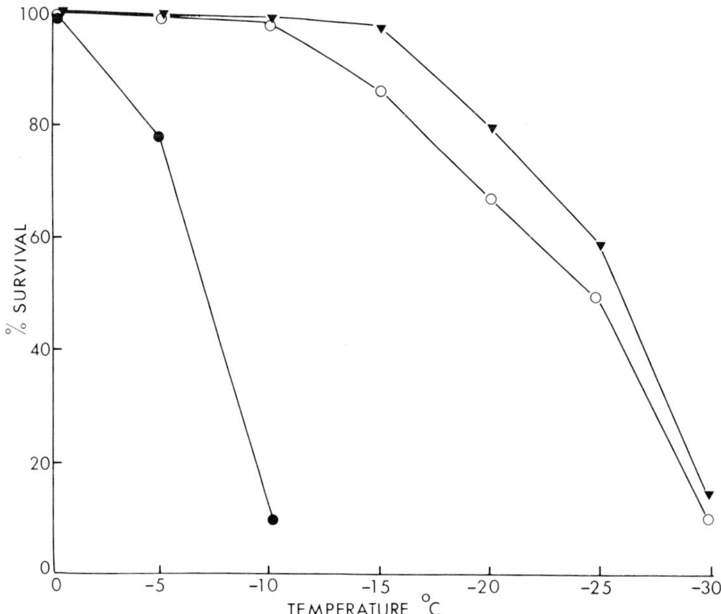

FIGURE 1.   Freezing tolerances of single cells isolated from cold-hardened and non-hardened winter rye leaves. Empty circles: cells isolated from cold-hardened rye. Filled circles: cells isolated from nonhardened rye. Triangles: cold-hardened rye tissues. (From Singh, J. and Miller, R. W., *Plant Physiol.*, 69, 1423, 1982. With permission.)

The use of isolated plant cells for spin-labeling experiments during freezing has distinct advantages over the use of protoplasts for such studies. For example, if the cells are suspended in distilled water, it is possible to remove most of the extracellular water (outside the cell wall) by centrifugation without plasmolysis of the cells. This greatly decreases microwave absorption and heating in the ESR cavity and allows samples containing large numbers of closely packed cells to be inserted in the ESR cavity without loss of the required dielectric properties. Furthermore, freezing of isolated cells under such conditions probably mimics, as closely as possible, extracellular freezing of whole tissues. Singh and Miller[32] successfully incorporated 5NS into single mesophyll cells isolated from hardened and nonhardened winter rye leaves.[33] The extracellular freezing tolerances of these cells were similar to the tissues of their origin (Figure 1) and thus served as a good system to follow spin-probe motional changes during freezing.

The isotropic rapid tumbling nature of the 5NS spin probe in water can be seen in Figure 2A. When the same sample was frozen at $-2°C$, broadening of the spectrum was observed due to spin-exchange interactions as a result of concentration of the hydrophobic probe (Figure 2B). Figure 2C shows that the ESR spectrum of 5NS-labeled rye cells is a composite of two separate signals: a rapidly tumbling component similar to the spectrum of 5NS in water (Figure 2A) and a slower tumbling component representative of 5NS in the cellular membranes which is similar to that observed in rye protoplasts and in egg-yolk lecithin vesicles.[30]

In order to determine whether the rapidly tumbling probe signal arises extracellularly, 0.25 M $NiCl_2$, which does not enter the cell, was added to the cell suspensions. Since paramagnetic nickel ions will broaden the probe signal to extinction by dipole-dipole relaxation processes,[34] the remaining signal represents an average signal arising from intracellular probes intercalated in membranes. Figure 2D indicates that the entire rapid tumbling signal is lost and can therefore be assigned as due to probe molecules external to the cell whereas the membrane signal is intracellular. The external signal is not easily washed away and may

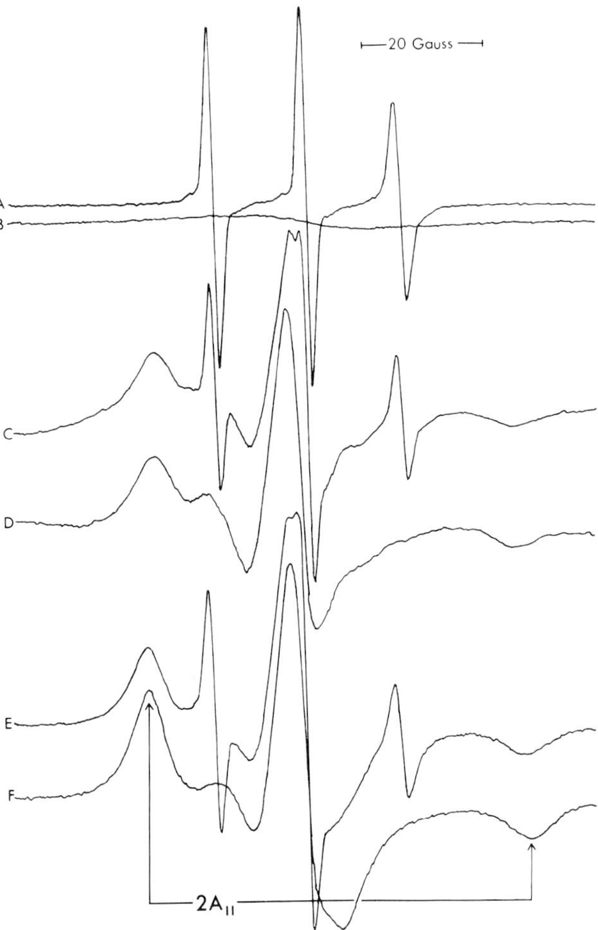

FIGURE 2. Electron spin resonance spectra of 5NS spin probe in isolated rye cells. (A) 5NS in water, (B) same as in A except frozen at
$-2°C$, (C) 5NS in rye cells, (D) same as in C except rye cells were
suspended in 0.25 M NiCL$_2$ after probe was incorporated, (E) same as
C but supercooled at $-2°C$, (F) same as E but freezing initiated at $-2°C$.

originate from 5NS loosely bound within the cell wall. Since the signal of 5NS in water
(Figure 2A) broadens (Figure 2B) by spin exchange due to loss of solubility and micelle
formation of the probe during freezing of water, the external, rapidly tumbling signal was
also useful as an indicator of the initiation of the freezing process.

Essentially, isothermal freezing of rye cells labeled with 5NS could be initiated in the
ESR spectrometer cavity by introduction of a very small grain of dry ice. The cavity
immediately detuned due to a change in the dielectric properties of the sample, and upon
retuning, the rapidly tumbling external signal had disappeared (Figure 2E vs. Figure 2F).
Observations of the changes in membrane order as freezing progressed to lower temperatures
were made by measurements of separation (in gauss) of the high- and low-field resonances
of the spectrum. Plots of the hyperfine splitting (A_{11}) of 5NS in hardened and nonhardened
cells as they undergo extracellular freezing to different temperatures are presented in Figure
3. On initiation of freezing at $-2°C$, an abrupt increase in A_{11} was observed for both
hardened and nonhardened cells. Since extracellular freezing results in both dehydration and
contraction of the isolated cells,[35] the head groups may be affected, causing a change in the

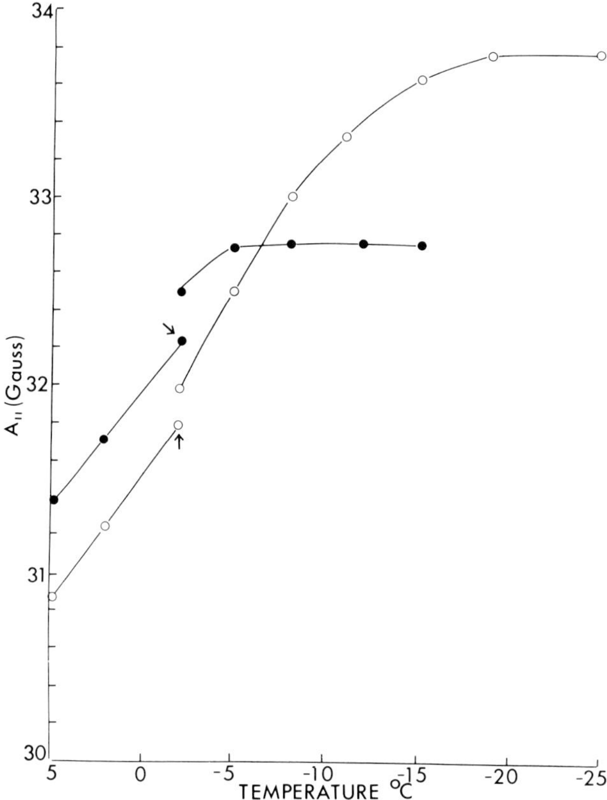

FIGURE 3.   Plots of hyperfine separation (A_{11}) of 5NS in the cells isolated from cold-hardened and nonhardened winter rye during freezing. Empty circles: cold-hardened cells. Filled circles: nonhardened. Arrows indicate the initiation of freezing. (From Singh, J. and Miller, R. W., *Plant Physiol.*, 69, 1423, 1982. With permission.)

total bonding energy of the lipid bilayer. Such a change could lead to an initial closer packing of lipids as reflected by the apparent decrease in fluidity. The results are comparable to increases in molecular order reported for 5NS-labeled liposomes undergoing drying.[31]

Rigid-limit spectra were obtained (powder spectra) as the sample temperatures were lowered further as indicated by an absence of any further increases in A_{11}. Not only was a lower temperature required by hardened cells to produce the rigid limit but the ultimate hyperfine splitting was greater. This result indicated a higher level of order in the hardened rye membranes as compared to nonhardened membranes. This was surprising since it indicated that the membranes of hardened cells can become more ordered at the lethal freezing temperature than nonhardened membranes, in contrast to the observations of greater fluidity in hardened membranes before freezing (lower A_{11} in Figure 3 due to greater fluidity in hardened rye). The most plausible explanation for the results is that at the temperatures where membrane injury occurs in nonhardened cells, the planar phospholipid bilayer lamellar lattice of the cellular membranes is converted irreversibly to other structures which show less average order, whereas at the same subzero temperatures membranes of hardened cells are able to maintain the planar bilayers which undergo closer packing as temperatures are lowered even further. An increased surface curvature resulting from the conversion of planar bilayers to micelles or smaller vesicles would increase the surface area occupied by each head group.[36,37] Since the doxyl group of 5NS, when intercalated between phospholipid molecules, is close to the head groups, an increased surface area per head group would be expected to limit the orderly packing of the methylene carbon atoms in the vicinity of the polar region of the lipid.[38] This, in turn, would be reflected in slower and limited increases in A_{11} in the nonhardened cells. It is also interesting to note here that in both hardened and nonhardened cells good correlation was obtained between the temperature where the rigid limit was reached and the LT_{50} of the cells (compare Figures 1 and 3).

III. ULTRASTRUCTURAL ALTERATIONS OF RYE CELL MEMRANES DURING FREEZING

Microscopic observations of the behavior of plant cells during the imposition of extracellular freezing stress were made extensively as early as 1897 by Molisch[39] and subsequently by Siminovitch and Scarth.[3] However, it was not until the development of a method by Mackenzie et al.,[40] which allowed fixation of tissues for ultrastructural observation in the frozen state, that it was possible to extend observations of cellular changes during freezing to the ultrastructural level. Singh[19] and Singh and Andrews[41] have applied this method to the observation of ultrastructural changes in dark-grown rye seedlings and wheat crowns, respectively, during freezing at lethal temperatures. The results indicated not only that irreversible membrane alterations occur during freezing at lethal freezing temperatures but also that such alterations are very frequently accompanied by loss of the planar bilayered structure of the cellular membranes and appearance of osmiophilic bodies. Furthermore, it was noted that rye cells fixed during lethal plasmolysis showed ultrastructural alterations similar to those of lethally frozen cells.[19] The nature and sequence of the conversion of planar bilayers to an amorphous-like state have not, however, been clearly delineated. The following part of this chapter describes "freeze-fixation" experiments with hardened and nonhardened light-grown rye tissues at the same stage of growth and hardiness as the cells used for the spin-probe studies. The sequence and nature of freeze-induced alterations are then correlated to proposed membrane alterations observed by spin-probe expermients.

The behavior of cold-hardened and nonhardened rye mesophyll cells before freezing and fixed with osmic acid in the frozen state at $-10°C$ is compared under the light microscope (Figure 4). The contracted protoplasm and the collapse of the cell wall is retained after freeze fixation. It is interesting to note that nonhardened cells frozen-fixed at $-10°C$ always appeared more intensely contracted than hardened cells frozen-fixed at the same temperature (Figure 4). However, it was not ascertained if this was the result of some degree of re-expansion of the hardened frozen-fixed cells subsequent to thawing.

In nonhardened rye cells frozen-fixed at $-10°C$ (LT_{50} $-7°C$), different stages or levels of membrane ultrastructural alteration in the frozen state could be observed. The earliest stage is probably fusion and rollup of cellular membranes (Figures 5A and B). Fusion of membranes could be caused by either dehydration or mechanical collapse of the cell in the frozen state or both. In the extracellular frozen state, multibilayered vesicles could be seen to be frequently formed in the cell (Figure 5C). In plasmolysis, injury, similar fusion, and rollup of membranes were observed to occur in the space between the contracted protoplasm and the cell wall.[54] However, in this case, multibilayered vesicles were frequently observed to be *extruded* or detached from the protoplasm. In the case of nonhardened cells frozen-fixed at $-6°C$, or subjected to mild plasmolysis, fusion and subsequent rollup between the tonoplast, plasmalemma, and other membranes were observed to occur.[54] On further extracellular freezing or more intense plasmolysis, even membranes of rolled up, multibilayered vesicles begin to pack so tightly that they lose their lamellar-lattice layers (Figure 5D). The final stage is probably the complete conversion of lamellar bilayered structures into the amorphous state. When nonhardened cells were frozen to $-10°C$ in water and then thawed in the presence of 2% osmic acid, retention of some of the multibilayered vesicles formed during freezing could be observed in lysed cells.[54] It is interesting to note that cells of cold-

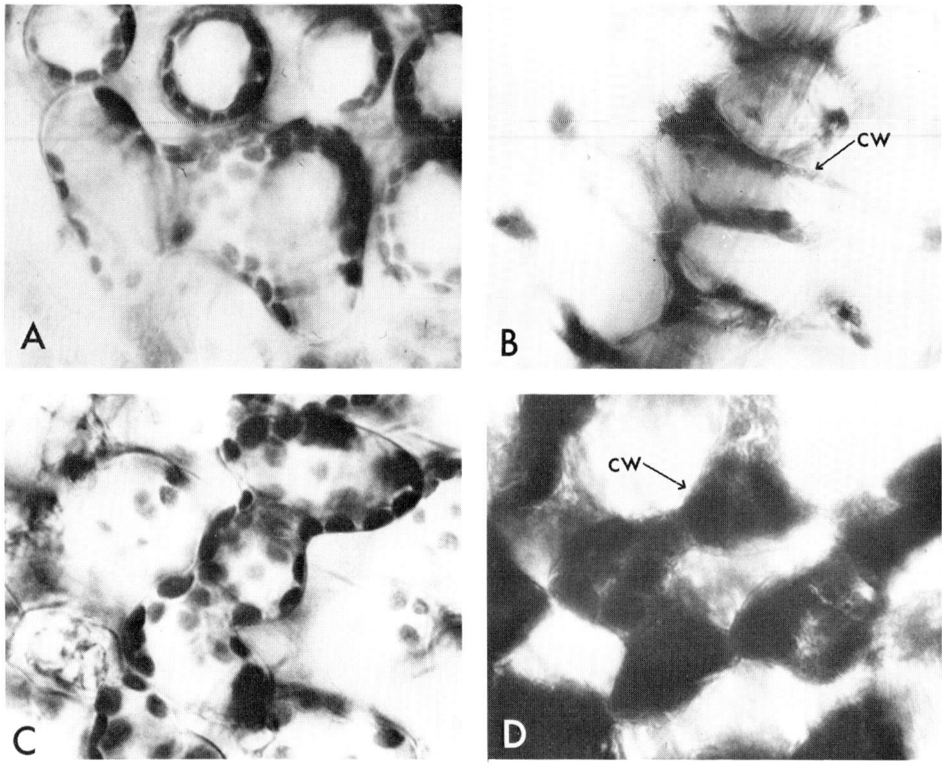

FIGURE 4.   Light microscopic comparisons of cold-hardened and nonhardened rye cells in tissues before freezing and after freeze-fixation in osmic acid at −10°C. (A) Nonhardened rye cells before freezing, (B) nonhardened rye cells frozen fixed at −10°C, (C) cold-hardened rye cells before freezing, and (D) cold-hardened rye cells frozen-fixed at −10°C (cw: cell wall). (Magnifications × 100.)

hardened rye which survived freezing to −10°C showed no such disruption of their membranous architecture when frozen-fixed at −10°C. Some invaginations or infoldings of the plasmalemma were observed (Figure 5E) but no rollup and fusion of cellular membranes to form densely packed multibilayered vesicles or osmiophilic bodies could be discerned. Freeze fixation of hardened tissues at lower freezing temperatures (−15 or −20°C) showed more intense invaginations of the plasmalemma, but only very infrequent occurrences of membrane rollup.[54] We suggest, therefore, that fusion plays a critical, irreversible role in membrane injury, for while the effect of mechanical and dehydrative stresses on the membrane during cell volume and area contraction[16-18] and dehydration can be alleviated by invaginations, fusion and loss of lamellar lattice will inevitably result in a loss of membrane material.

IV. CONCLUSION

The results obtained by ultrastructural analysis of frozen-fixed rye cells support the hypothesis outlined above describing molecular changes in membranes during freezing based on spin-labeling studies. During extracellular freezing to lethal temperatures, membranes of susceptible cells can "roll up" and fuse to form multilayered vesicles. Eventually, the membrane bilayers may lose their phospholipid lamellar lattice altogether. Figure 6 shows the possible routes that such membrane alterations can take, leading to cell damage. There is evidence in model liposome studies to indicate that the water of hydration around phospholipid head groups is not only critical in the maintenance of phospholipid lamellar lattice

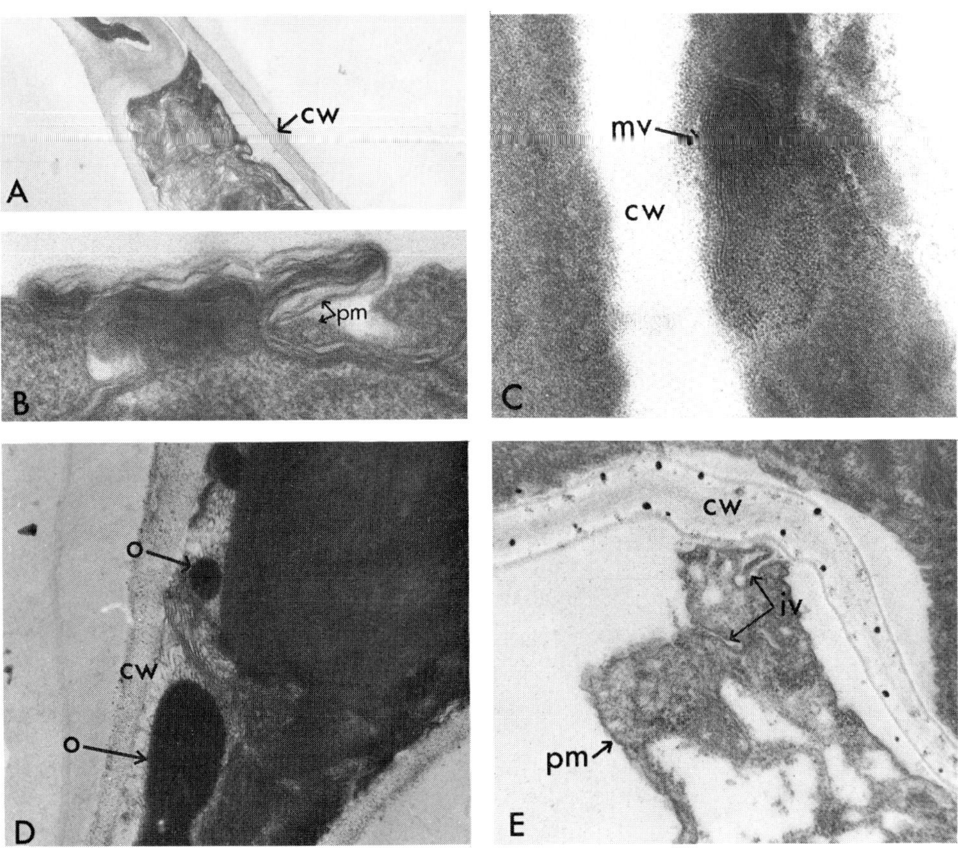

FIGURE 5. Ultrastructure of cold-hardened and nonhardened rye cells froxen-fixed at $-10°C$. (A, B) cw: cell wall; nonhardened cells, rollup of membranes, pm: plasmalemma, magnification $\times$ 6300 and 31,400, respectively; (C) nonhardened cell, formation of multibilayered vesicles, cw: cell wall, mv: multibilayered vesicle, magnification $\times$ 102,000; (D) nonhardened cell, formation of osmiophilic bodies, o: osmiophilic body, magnification $\times$ 24,400; (E) cold-hardened cell frozen-fixed at $-10°C$, iv: invagination, magnification $\times$ 19,400. (B and C from Singh, J. and Miller, R. W., *Plant Physiol.*, 69, 1423, 1982. With permission.)

but is also instrumental in the prevention of fusion of adjacent bilayers.[42] Perturbation of the water status in the vicinity of phospholipid bilayers by either dehydration or increased concentrations of intracellular ions (e.g., Ca^{2+}) such as that encountered during extracellular freezing may cause destabilization of the bilayers leading to fusion.[43-45] The results presented in this chapter indicate that irrespective of whether rollup of membranes or destruction of the phospholipid lamellar lattice occurs during freeze-induced or plasmolysis-induced stress, membrane fusion plays an important role as an initial step in freezing injury in rye cells. Whether perturbation of the membranes is caused by dehydration alone or aided by the mechanical stress of the collapsing cell is at present unknown. Fusion of bilayer membranes of loss of crystalline lamellar lattice at the moment of freezing injury may explain observations of loss of membrane surface area,[4] decompartmentation of cell organelles,[20] and hysteresis in water relaxation times.[21] In ultrastructural observations of osmotically active, single-bilayer, cell-size liposomes during hyperosmotic stress (1 *M* KCl or 2 *M* sucrose) and freeze fixation at $-0°C$, both intrusion and extrusion of the bilayer to form multilamellar vesicles could be discerned.[46] The formation of these vesicles has to be mediated by destabilization and fusion of the bilayer during contraction.

Membranes in cold-hardened cells frozen-fixed at $-10°C$ retained their lamellar lattice.

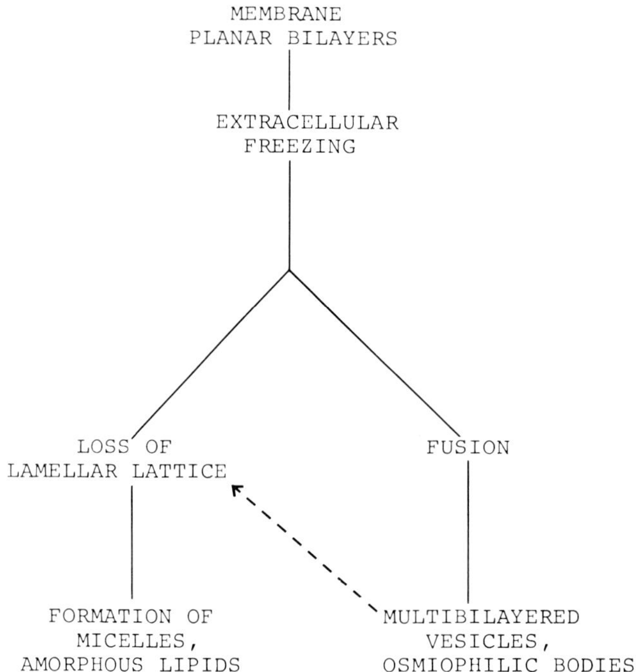

FIGURE 6.   Routes of membrane damage in nonhardened rye cells during extrcellular freezing.

Whereas foldings and invaginations of their plasmalemma were observed, no fusion, or rollup, or formation of osmiophilic bodies by the membranes could be discerned. The ability of membranes of hardened rye cells to avoid fusion, membrane rollup, and irreversible loss of lamellar lattice until lower freezing temperatures were reached may be mitigated by modification of phospholipid composition or by decreased appression of adjacent bilayers of the surface membrane. The type of phospholipid head group, intrinsic membrane proteins, or other factors may play a role. It has been observed that protein incorporation into liposomes stabilizes the latter to freezing[47] and increases in phosphatidyl choline content in wheat can increase hardiness.[48] On the other hand, in black-locust bark cells, which can develop extreme freezing tolerance, no significant changes were observed in the ratio of phosphatidyl choline to phosphatidyl ethanolamine in phospholipids extracted from hardened and nonhardened tissues.[49] However, since these observations were made on total extracts, the effect of phospholipid class changes in local domains cannot be precluded. Studies on model lipid system have indicated that membrane fusion processes can be affected by the ability of different phospholipid classes to coexist in mixtures.[50] The deleterious effect of membrane fusion during freezing may theoretically be alleviated by changes in lipid and/or protein compositions in membranes most affected by the mechanical and dehydrative stresses of extracellular freezing.

Steponkus et al.[51] and Gordon-Kamm and Steponkus[52] observed that internalized vesiculations of the plasmalemma occur during hyperosomotic-induced contractions in nonhardened isolated rye protoplasts. In cold-hardened protoplasts, however, polyp-like exocytotic protrusions of the plasmalemma containing ''tethered'' spheres of osmiophilic substances were observed during hyperosmotic contraction.[52] The authors suggest that internalized vesiculations observed in nonhardened protoplasts represent irreversible membrane loss during the contraction phase, leading to lysis on subsequent re-expansion. In cold-hardened protoplasts, the tethered spheres may represent areas which contain lipid material subducted, or from which lipid could be withdrawn during contraction or re-expansion, respectively (see Chapter 3). However, we point out that unlike a spherical, isolated protoplast, the cell wall in the whole cell collapses with the plasmalemma during freezing. The shape after contraction can be such that large reductions in volume can occur without concomitant large reductions in surface area (concave stellar shape in Figure 4B). Thus, the mechanical stress on the plasmalemma in the whole cell undergoing freeze-induced dehydration may be different from that experienced by an isolated protoplast undergoing hyperosmotic contraction. Our data (described here) suggest that in the cold-hardened cell, large reductions in membrane surface area may not necessarily occur during freezing, and in areas of the cell where it does, invaginations of the membrane can uncouple volume-to-surface-area reductions. Such uncoupling of volume-to-surface area has been shown to be possible even in spherical protoplasts by micropipette aspiration.[53] On the other hand, such invaginations could lead to irreversible membrane rollup and fusion of cellular membranes in nonhardened cells.

REFERENCES

1. **Levitt, J.**, *Responses of Plants to Environmental Stress,* Vol. 1, 2nd ed., Academic Press, New York, 1980.
2. **Steponkus, P. H.**, Cold hardiness and freezing injury of agronomic crops, in *Advances in Agronomy,* Vol. 30, Brady, N. C., Ed., Academic Press, New York, 1978, 51.
3. **Siminovitch, D. and Scarth, G. W.**, A study of the mechanism of frost injury to plants, *Can. J. Res.,* 16, 467, 1938.
4. **Steponkus, P. L. and Wiest, S. C.**, Freeze-thaw induced lesions in the plasma membrane, in *Low Temperature Stress in Crop Plants,* Lyons, J. M., Graham, D., and Raisons, J. K., Eds., Academic Press, New York, 1979, 231.
5. **Dexter, S. T., Tottingham, W. E., and Graber, L. F.**, Investigations of the hardiness of plants by measurements of electrical conductivity, *Plant Physiol.,* 7, 63, 1932.
6. **Palta, J. P., Levitt, J., and Stadelmann, E. J.**, Evaluation of the conductivity method and analysis of ion and sugar efflux from injured cells, *Plant Physiol.,* 60, 393, 1977.
7. **Siminovitch, D., Therrien, H., Wilner, J., and Gfeller, F.**, The release of amino acids and other ninhydrin-reacting substances from plant cells after injury by freezing; a sensitive criterion for the estimation of frost injury in plant tissues, *Can. J. Bot.,* 40, 1267, 1962.
8. **Pomeroy, M. K. and Andrews, C. J.**, Ultrastructural changes in shoot apex cells of winter wheat seedlings during ice encasement, *Can. J. Bot.,* 56, 786, 1978.
9. **Siminovitch, D., Rheaume, B., Pomeroy, K., and Lepage, M.**, Phospholipid, protein and nucleic acid increases in protoplasm and membrane structures associated with development of extreme freezing resistance in black locust cells, *Cryobiology,* 5, 202, 1968.
10. **Gerloff, E. D., Richardson, T., and Stahmann, M. A.**, Changes in fatty acids in alfalfa roots during cold hardening, *Plant Physiol.,* 41, 1280, 1966.
11. **Kuiper, P. J. C.**, Lipids in alfalfa leaves in relation to cold hardiness, *Plant Physiol.,* 45, 684, 1970.
12. **Grenier, G. and Willemot, C.**, Lipid changes in roots of frost hardy and less hardy alfalfa varieties under hardening conditions, *Cryobiology,* 11, 324, 1974.
13. **de la Roche, I. A., Andrews, C. J., and Kates, M.**, Changes in phopholipid composition of a winter wheat cultivar during germination at 2°C and 24°C, *Plant Physiol.,* 51, 468, 1973.
14. **de la Roche, I. A., Pomeroy, M. K., and Andrews, C. J.**, Changes in fatty acid composition in wheat cultivars of contrasting hardiness, *Cryobiology,* 12, 506, 1975.
15. **de la Roche, I. A.**, Increase in linolenic acid is not a prerequisite for development of freezing tolerance in wheat, *Plant Physiol.,* 63, 5, 1979.
16. **Williams, R. J., Willemot, C., and Hope, H. J.**, The relationship between cell injury and osmotic volume reduction. IV. The behavior of hardy wheat membrane lipids in monolayer, *Cryobiology,* 18, 146, 1981.
17. **Williams, R. J. and Hope, H. J.**, The relationship between cell injury and osmotic volume reduction. III. Freezing injury and frost resistance in winter wheat, *Cryobiology,* 18, 133, 1981.
18. **Wolfe, J. and Steponkus, P. L.**, The stress-strain relation of the plasma membrane of isolated plant protoplasts, *Biochim. Biophys. Acta,* 643, 663, 1981.
19. **Singh, J.**, Ultrastructural alterations in cells of hardened and nonhardened winter rye during hyperosmotic and extracellular freezing stresses, *Protoplasma,* 98, 329, 1981.

20. **Stout, D. G., Majak, W., and Reaney, M.,** *In vivo* detection of membrane injury at freezing temperatures, *Plant Physiol.,* 66, 74, 1980.

21. **Rajashekar, C., Gusta, L. V., and Burke, M. J.,** Membrane structural transitions: probable relation to frost damage in hardy herbaceous species, in *Low Temperature Stress in Crop Plants,* Lyons, J. M., Graham, D., and Raisons, J. K., Eds., Academic Press, New York, 1979, 255.

22. **Yoshida, S.,** Freezing injury and phospholipid degradation *in vivo* in wood plant cells. I. Subcellular localization of phospholipase-D in living bark tissues of the black locust tree, *Plant Physiol.,* 64, 241, 1979.

23. **Yoshida, S.,** Freezing injury and phospholipid degradation *in vivo* in woody plant cells. II. Regulatory effects of divalent cations on activity of membrane-bound phospholipase-D, *Plant Physiol.,* 64, 247, 1979.

24. **Miller, R. W., de la Roche, I. A., and Pomeroy, M. K.,** Structural and functional responses of wheat mitochondrial membranes at low temperature, *Plant Physiol.,* 53, 426, 1974.

25. **Vigh, L., Horvath, I., Farkas, T., Horvath, L. I., and Belea, A.,** Adaptation of membrane fluidity of rye and wheat seedlings according to temperature, *Phytochemistry,* 18, 787, 1979.

26. **Pomeroy, M. K. and Raisons, J. K.,** Maintenance of membrane fluidity during development of freezing tolerance of winter wheat seedlings, *Plant Physiol.,* 68, 382, 1981.

27. **Fey, R. L., Workman, M., Marcellos, H., and Burke, M. J.,** Electron spin resonance of 2,2,6,6-tetramethylpiperidine-1-oxyl (TEMPO)-labelled plant leaves, *Plant Physiol.,* 63, 1220, 1979.

28. **Gaffney, B. J.,** Spin-label measurements in membranes, *Meth. Enzymol.,* 32, 161, 1974.

29. **Schreier, S., Polnaszek, C. F., and Smith, I. C. P.,** Spin labels in membranes: problems in practice, *Biochim. Biophys. Acta,* 515, 375, 1978.

30. **Singh, J. and Miller, R. W.,** Spin-label studies of membranes in rye protoplasts during extracellular freezing, *Plant Physiol.,* 66, 349, 1980.

31. **Griffiths, O. H., Dehlinger, P. J., and Van, S. P.,** Shape of the hydrophobic barrier of phospholipid bilayers: evidence of water penetration in biological membranes, *J. Membr. Biol.,* 15, 159, 1974.

32. **Singh, J. and Miller, R. W.,** Spin-probe studies during freezing of cells isolated from cold hardened and nonhardened winter rye: molecular mechanism of membrane freezing injury, *Plant Physiol.,* 69, 1423, 1982.

33. **Singh, J.,** Isolation and freezing tolerances of mesophyll cells from cold-hardened and nonhardened winter rye, *Plant Physiol.,* 67, 906, 1981.

34. **Keith, A. D. and Sipes, W.,** Viscosity of cellular protoplasm, *Science,* 183, 667, 1974.

35. **Steponkus, P. L., Evans, R. Y., and Singh, J.,** Cryomicroscopy of isolated rye mesophyll cells, *Cryo-Lett.,* 3, 101, 1982.

36. **Chrzeszczyk, A., Wishnia, A., and Springer, C. S., Jr.,** The intrinsic structural asymmetry of highly curved phospholipid bilayer membranes, *Biochim. Biophys. Acta,* 470, 161, 1977.

37. **Huang, C. and Mason, J. T.,** Geometric packing constraint in egg phosphatidyl choline vesicles, *Proc. Natl. Acad. Sci. U.S.A.,* 75, 308, 1978.

38. **Marsh, R. P., Philips, A. D., Watts, A., and Knowles, P. F.,** A spin-level study of fractionated egg phosphatidyl choline vesicles, *Biochem. Biophys. Res. Commun.,* 49, 641, 1972.

39. **Molisch, H.,** *Untersuchungen uber das Erfrieren der Pflanzen,* Gustav Fischer, Jena, Yugoslavia, 1897.

40. **Mackenzie, A. P., Kuster, T. A., and Luyet, B. J.,** Freeze-fixation at high subzero temperatures, *Cryobiology,* 12, 417, 1975.

41. **Singh, J. and Andrews, C. J.,** Comparisons of ultrastructural changes during extracellular freezing at −10°C and ice encasement at −1°C in winter wheat crown cells by the method of freeze fixation, *Cryo-Lett.,* 2, 117, 1981.

42. **Parsegian, V. A., Fuller, N., and Rand, R. P.,** Measured work of deformation and repulsion of lecithin bilayers, *Proc. Natl. Acad. Sci. U.S.A.,* 76, 1750, 1979.

43. **Papahadjopoulos, D., Vail, W. J., Jacobson, K., and Poste, G.,** Cochleate lipid cylinders: formation by fusion of unilamellar lipid vesicles, *Biochim. Biophys. Acta,* 394, 483, 1975.

44. **Ohki, S. and Duzgunes, N.,** Divalent cation-induced interaction of phospholipid vesicle and monolayer membranes, *Biochim. Biophys. Acta,* 552, 438, 1979.

45. **Portis, A., Newton, C., Pangborn, W., and Papahadjopoulos, D.,** Studies on the mechanism of membrane fusion: evidence for an intermembrane Ca^{2+}-phospholipid complex, synergism with Mg^{2+}, and inhibition by spectrin, *Biochemistry,* 18, 780, 1979.

46. **Singh, J.,** Alterations in rye cells during lethal freezing and plasmolysis stress: mechanism of freezing injury, *Plant Physiol.,* 69 (Suppl.), 106, 1982.

47. **Strauss, G. and Ingenito, E. P.,** Stabilization of liposome bilayers to freezing and thawing: effects of cryoprotective membrane proteins, *Cryobiology,* 17, 508, 1980.

48. **Horvath, I., Vigh, L., and Farkas, T.,** The manipulation of polar head group composition of phospholipids in the wheat Miranovskaja 808 affects frost tolerance, *Plants,* 151, 103, 1981.

49. **Singh, J., de la Roche, I. A., and Siminovitch, D.,** Membrane augmentation in freezing tolerance of plant cells, *Nature (London),* 257, 669, 1975.

50. **Nicolussi, A., Massari, S., and Colonna, R.,** Effect of lipid mixing on the permeability and fusion of saturated lecithin membranes, *Biochemistry,* 21, 2134, 1982.
51. **Steponkus, P. L., Wolfe, J., and Dowgert, M. F.,** Stresses induced by contraction and expansion during a freeze-thaw cycle: a membrane perspective, in *Effects of Low Temperature on Biological Membranes,* Morris, G. J. and Clarke, A., Eds., Academic Press, New York, 1981, 307.
52. **Gordon-Kamm, W. J. and Steponkus, P. L.,** Morphology of the plasma membrane of isolated protoplasts following osmotic contraction: influence of cold acclimation, *Plant Physiol.,* 69 (Suppl.), 119, 1982.
53. **Wolfe, J. and Steponkus, P. L.,** Mechanical properties of the plasma membrane of isolated plant protoplasts: mechanism of hyperosmotic and extracellular freezing injury, *Plant Physiol.,* 71, 276, 1983.
54. **Singh, J. and Iu, B.,** unpublished data.

I. INTRODUCTION: EMPIRICAL NEED FOR PROTECTIVE SUBSTANCES

" . . . From the pragmatic point of view of preserving viable cells and cultures, . . . when a suspension of cells is stored, at, say $-196°C$, it is highly probable that no loss in viability and no physiological or genetic changes will occur for decades. The problem, then, becomes one of cooling the cells to $-196°C$ and rewarming them without an immediate crippling loss in viability . . . " — P. Mazur.

A. Cold Hardening and Cryoprotection

It is common experience that living plants and plant tissues succumb to freezing at just a few degrees below 0°, the freezing point of pure water. Yet this lethality can be avoided and the plant can withstand very cold frozen temperatures in particular tissues, especially after exposure to the conditions of either "freeze hardening" or treatment with cryoprotective additives. On examination, the effects from these two sets of conditions have aspects in common, including decreased cellular water content and increased content of specific solutes; but in other aspects they are quite different. One can then, by observing the responses to these two sets of conditions, use each set as a model, or basis for comparison, for the other set; the two sets of information could be mutually instructive toward a better physiological and biochemical understanding of the life-preserving processes operating during survival of freezing.

In the freeze-hardening process we observe an adaptive, largely metabolic response to an upper range of cold and freezing temperatures. This chapter, will, however, restrict itself to discussing cryoprotection: the components and conditions imposed by human experimenters as they attempt to achieve, through manipulations (1) long term preservation of plant cells and tissues at ultracold freezing temperatures near or below $-100°C$, and (2) an objective understanding of how this is accomplished.

Clarification may be desirable with respect to usage of three sets of terms describing states of cold toleration by plant tissues: chill resistance, freezing tolerance, and cryogenic stability. Some distinguishing features of a general nature are set out in Table 1.

The three methods (and temperature ranges) of withstanding damage displayed in the table may be quite distinct, yet elements of the protective mechanism of each may be found in the other. A pertinent question here is to what extent do added cryoprotective reagents duplicate or imitate the compounds metabolically mobilized during the natural development of freezing tolerance? The successful use of added cryoprotective compounds, particularly sugars[2] and macromolecular hydrophilic agents[3] appears as a possible imitation of similar cellular constituents that are elaborated in plant tissue during the adaptive circumstances of

Table 1
STATES OF PLANT COLD TOLERANCE

	Chill resistance	Freezing tolerance	Cryogenic stability
Temperature range of primary interest[a] (approx.)	10 to 0°C	0 to −40°C	−130 to −196°C and below
Low temperature defect (unprotected plant)	Metabolic derangement	Structural damage	Structural damage
Necrosis development time in a group of cells in unprotected state	Days	Min—hr	Minutes or less
Protection development time	Days	Days	Minutes
Protective mechanism	Metabolic modification	Metabolic modification	Technically imposed reagents and conditions

[a]   The initiation of protective measures may take place at a higher range of temperatures.

freeze-hardening.[4-6] In point of fact, the use of one or more cryoprotective chemicals has often been found necessary for cell survival. An attempt to describe and bring into perspective the usage of cryoprotective compounds and how they operate is the chief purpose of this chapter.

B. Accomplishing Frozen Stability

Maximow,[7] in 1912, opened the curtain on cryogenic research in discovering that leaf tissue of *Tradescantia discolor*, a tropical plant that is ordinarily killed by exposure to −2°C, could survive colder temperatures by several degrees after addition to the tissue of various simple solutions or the juice from hardened red cabbage leaves. Since that time, the comprehension and imitation of freeze-hardening and the pursuit of theoretical and empirical principles of cell damage prevention have constituted a complex challenge in cryogenic research, one that has only recently begun to be resolved on the basis of clear physical-chemical principles.

Yet, empirically, cells and tissues, when they can be brought in the still-living condition to very low freezing temperatures, can then be stored almost without any further change for very long periods of time (perhaps decades and much longer[8]). Stored in this manner they may later be used for a variety of genetic and experimental purposes described in various chapters in this book. It is noteworthy that the International Board for Plant Genetic Resources has recognized only two methods for achieving reliable, semipermanent "base collections" of plant types, namely, low temperature storage of dependable seeds and cryogenic preservation of plant cultures.[9]

The viable freezing of plant samples to a low temperature is still in a highly empirical state. Given the great stability of tissues at very cold temperatures (e.g., −196°C), the challenge is in manipulating the tissues through the transient freezing and thawing operations with a minimum of ultrastructural and metabolic damage. The usage of appropriate cryoprotective compounds during the freezing process is one key to the viable preservation of plant materials. A continuous search for the types of compounds that would improve the survival of frozen cells and an understanding of their chemical and physical reactions and interactions have been main threads of research in the area of cryobiology.

C. "Viva-Cryopreservation" vs. "Forma-Cryopreservation"

The terms "cryoprotective compounds" and "cryopreservation" used in this chapter are also often used in yet another area of plant tissue preservation and research, but with a distinctly different focus: the preservation of cell morphology, at the gross or ultrastructural level, using plant material that is most often killed in the process.[10] The designated terms will be used here, however, only as they apply to still-living cells or tissues. It is suggested for literature search efficiency, etc., that new distinctive terms, such as ''viva-cryopreservation'' vs. ''forma-cryopreservation'', be adopted to clarify and distinguish these meanings and their usages. In time, superior treatment methods will perhaps be developed that will eliminate the need of making any distinction between the two!

D. Literature Sources

This chapter will attempt to be interpretive, rather than exhaustive, in its coverage of the pertinent research literature. A number of chapters and reviews are suggested as additional information sources for cryoprotective treatments and responses in plants,[11-19] and some other suggested sources contain more general information.[20-28] Still other review sources will be referred to in the pages that follow and be cited in the References.

II. HISTORICAL AND PHYSICAL-CHEMICAL CONCEPTS OF CRYOPROTECTIVE ACTION

''The criteria for a cryoprotective agent are merely that it will penetrate the cell freely and be non-toxic in high concentration'' — H. T. Meryman.[29]

A. Compounds of Cold Hardening, Stress, and Cryoprotection

During the nearly three fourths of a century of studies of experimental cyroprotection since the pioneering investigations of Maximow, the identified compounds found to exert cryoprotection when added to plant tissues have been large in number, complex, and often bewildering in their chemical diversity. Furthermore, the results of analyses of changes in plant constituents during cold hardening[4,5,17,30] have been, by and large, too complex to help in an overall understanding of this problem. Exceptions to this generalization are perhaps seen in an increased concentration of simple osmotic agents,[4] possibly also in the accumulation of specific low molecular weight proteins such as those isolated by Volger and Heber from spinach[31] and the oligosacchardes described by Olien and Smith[6] as well as the compounds proline and the betaines reported to accumulate in plants under stress.[32,33] Maximow,[7] testing extracts from freeze-hardy tissues, detected appreciable protection from glucose, alcohols and polyols, and salts of inorganic and organic acids. Years later, Woodcock et al.[34] reported a large cryoprotective effect at $-190°C$ on sheeps' red blood cells treated with glucose added at 5 to 8% concentrations. Polge et al.,[35] in 1949, in a chance empirical observation, first noticed a cryoprotective effect from the addition of glycerol to human sperm cells, and Lovelock and Bishop,[36] 10 years later, introduced dimethylsulfoxide (DMSO), one of the most effective cryoprotectants yet known for both plant and animal tissues. DMSO was examined because its known property of cell permeation might make it more effective than glycerol in penetrating to the cell's interior. Glycerol was a nonpenetrating compound with respect to the particular cells under study (bovine red blood cells); hence the distinguishing criterion of ''entry into the cell'' was advanced that is often used to categorize substances used for cyroprotection: penetrating vs. nonpenetrating. Interpretations of how cryoprotectants operate are still usually described by this criterion of cell entry, although with reservations.

B. Modes of Action of the Compounds

The center of research interest in the field of cryoprotection — with its great heterogeneity of phenomena and multidisciplined observations — has changed dramatically with time. It has gone from an emphasis on — in fact requirement of — small compounds that penetrate

Table 2
COMPARATIVE PROTECTIVE ACTION OF VARIOUS SOLUTES ON FREEZING INJURY[a]

Slight protection (−7°C)	Medial protection (−10°C)	Medial protection (−15°C)	High protection (−20°C)	Highest protection (−25°C)
Raffinose	D-Glucuronic acid	Ascorbic acid	Sucrose	Glucose
Mannitol	Hydroxyproline	Acetoamide	Sorbitol	β-Methyl-D-
Glucosamine	Glutamic acid	Methyl acetoamide	Lactate	glycoside
Galacturonic acid	Methyl formamide	Dimethyl formamide	Pantothenic acid	Xylose
Dimethyl acetoamide	Formamide	Betaine aldehyde	Methanol	Rhamnose
Acetyl glycine	Dimethyl urea	(Cabbage sap)	Ethylene glycol	Erythritol
α-Aminobutyrate	N-methyl urea	(*Pseudoacacia:* bark	Glyceraldehyde	Xylitol
α-Glycerophosphate	Dimethyl thetine	sap)	Dimethyl sulfoxide	Glycerol
β-Glycerophosphate	Acetyl choline		1,3-Dimethyl urea	Betaine
Water (N.C.): −5°C	N-Methyl pyrrolidone		Methyl glycine	Trimethylamine N-oxide
	Triethylene glycol		Dimethyl glycine	
			Methionine sulfoxide	
			Lactoamide	

[a] Material: cabbage cells, 0.5 M aqueous solutions.

From Sakai, A. and Yoshida, S., *Cryobiology*, 5, 160, 1968. With permission.

the cell, to the opposite pole, exploration of important protective effects derived from even very large nonpenetrating compounds as they affect the properties of primarily, the external solution. In the extensive experiments of Lovelock, the action of glycerol and other small polar compounds, was interpreted as their penetrating and exerting colligative action within the human erythrocytes[37] and sperm cells,[38] with the consequence that, in the proportion that the colligative action of the penetrating compounds maintains water in the liquid state at temperatures below 0°C, an increased volume of cellular solution is maintained, avoiding an excessive concentration of toxic electrolytes in the nonfrozen cellular solution; and a similar influence takes place in the external solution. In this context colligative action is referred to as action by a considered solute, often extraneously added, in lowering the freezing point of the solution in contact with unmelted ice. An increased volume of liquid is available as a solvent for decreasing the concentration (activity) of toxic solutes in solution. When a cryoprotective compound is able to penetrate the cells, the resulting decrease in the equilibrium concentration of salts then applies to the constitutent salts or other potentially toxic compounds within the cell as well as outside the cell. If enough protective compound is present, the salt concentration does not rise to a critically damaging level until the temperature becomes so low that the damaging reactions are slow enough to be tolerated by the cells.

Table 2, compiled by Sakai and Yoshida,[2] lists a number of compounds of a variety of structures and their degrees of effectiveness as cryoprotectants toward cabbage cells, without reference to their penetrability. Low molecular weight neutral compounds such as menthanol, glycerol, acetamide, and DMSO are often considered in the "penetrating" category of protective compounds.[39] Glycerol and other small molecules, however, have, in fact, often been found not to penetrate the cells of plants, depending on the species and conditions, including temperature of application. To cite examples, at 22°C glycerol does not enter the cells of *Haplopappus gracilis,* but it does enter *Acer saccharum,* while sucrose does not penetrate either type of cell.[40] With human red blood cells, the penetration rate by glycerol at 20°C was reported as twice that at 0°C.[41]

Larger molecular weight compounds would not be expected to penetrate the cell. The nonpenetrating compounds may vary in size from glycerol and sucrose, as noted, to large polymeric substances. It has been suggested that nonpenetrating substances act by some other than the colligative mechanism described. The role of larger molecules may be conceived of as to some extent dehydrative, such as by osmotic action.[40] When a large proportion of water is withdrawn from the cell by means of an osmotic differential, less free water is available for intracellular ice crystallization (often identified as a lethal factor),[11,42] or crystallization may be delayed or avoided altogether within the cell. Thus, based on experiments on the freezing of Chinese hamster cells in solutions of DMSO, glycerol, and the large polymer hydroxyethyl starch, at several temperatures, McGann[43] concluded: "Penetrating agents create the environment for a reduction of cell water content at temperatures sufficiently low to reduce the damaging effect of the concentrated solutes on the cells. Nonpenetrating agents osmotically "squeeze" water from the cells primarily during the initial phases of freezing at temperatures between -10 and $-20°C$ when these additives become concentrated in the extracellular regions." In the latter situation the cell wall, impermeable to large compounds, may act as the semipermeable membrane. It is the wall rather than the plasma membrane that then shrinks ("cytorrhysis") as water leaves the cell (see Section V.C). The dehydrative action, supplemented by the further removal of water when ice is formed, can be carried too far and produce an excessive concentration of damaging solutes in the remaining liquid ("solution effect"),[11] or, by another interpretation, the cells collapse under the osmotic pressure difference across the membrane,[44] resulting in death.

However, in point of fact the import of the distinction between penetrating and nonpenetrating mechanisms has been brought into question. Penetrating compounds affect the characteristics of the internal cell solution, particularly in not allowing an excessive exodus of water from the cells too early in the freezing process; and nonpenetrating compounds can exert dehydrative effects on the cell, along with their colligative effect of diluting the solution surrounding the cell. But experimental findings have indicated that neither of these descriptions tells the whole story.

Looking at cells that either glycerol or DMSO can penetrate, when these cells were experimentally exposed to the compounds for very short times during which presumably little penetration would occur, they still exhibited cryoprotection. For example, Lovelock and Bishop[36] in their paper that first introduced the effects of DMSO as a cryoprotectant, exposed bovine red blood cells to 5% DMSO for 30 sec at 0°C as compared with exposure for 2 hr. With the two very different exposure periods, the two sets of cells retained almost the same degree of intactness, near 75%. In a still more explicit demonstration, Mazur and Miller[45] exposed human red blood cells at 0°C to equal osmolal amounts (2.3 Osm) of glycerol, which penetrates, or sucrose, which does not. Both sets gave a high, almost equal survival of freezing to $-196°C$ over the entire range of exposure times from 30 sec to 10 min. The authors also exposed both human and bovine red blood cells to glycerol at 0°C. Even though the bovine cells are 30-fold less permeable to glycerol, both types of cells survived freezing equally well after a 30-sec exposure to 10% glycerol solution. An alteration in membrane properties in interaction with nonpenetrating solutes was suggested.

Solutes such as the polymer hydroxyethyl starch might be considered as totally nonpenetrating water-withdrawing compounds of merely larger molecular weights than nonpenetrating sucrose. The larger molecular weights would, however, render such compounds less osmotically and colligatively effective, considered on a weight basis. Yet, as described in Section III.B, below, in concentrated solutions the compound's colligative action has been shown to be far greater than linear in proportion to the increase in molar concentration.[46] Evidence of cryoprotection (survival) not being closely related to penetration was given above, for small molecules; there is a great deal of additional supporting evidence for this with polymers. Farrant[46] illustrated the nonlinear anomaly as follows: " . . . The sodium chloride mole fraction of 0.014 shown by Lovelock to be the threshold concentration for damage to red blood cells during freezing is reached at above $-3°C$ in the absence of an added substance, at $-14.3°C$ in the presence of glycerol (10 per cent w/w initially), at $-16.1°C$ in sucrose (30 per cent w/w initially) and is not reached by $-20°C$ in the presence of PVP (30 per cent w/w initially)." Rationalization of these observations will be discussed in Section III.B.

The interplay of various factors during freezing may produce (1) a *toxic buildup of solutes* and (2) *intracellular freezing*, the two results that Mazur et al.[47] took into account in proposing the "two factor" explanation of the peak in survival rate observed during a series of experiments where the rate of cooling was varied while freezing several species of cells. During freezing, a disruptive crystallization of ice occurs in the cell at more rapid cooling rates, visualized as the right hand slope of the two-factor curve of Mazur et al. A decrease in liquid water and consequent concentration of electrolytes takes place. At slower rates of cooling, the exposure time, hence toxic effect of concentrated electrolyte solution, is increased, visualized at the left hand slope of the two-factor curve. Cryoprotective substances counteract this toxic effect, hence it follows that the effectiveness of cryoprotective substances is increased — in fact their presence is usually required — at lower rates.[48,49] Mazur and associates[47] clarified this concept, empirically and theoretically, in terms of high concentrations of a cryoprotectant (glycerol) decreasing the toxic impact of solution effects at slow cooling rates, with consequent increased cell survival. This is illustrated in Figure 1A, by the C-D family of curves. The shift in these curves towards the tolerance of lower freezing rates at higher concentrations of cryoprotectant, added to the survival curve of intracellular freezing, results in a shift of the summed maxima of overall survival toward lower cooling rates, as represented in Figure 1B (although there are some exceptions to this concept, as with methanol[50]). A lower rate of cooling is compatible with and fulfills the survival needs of multicellular plant cell clusters and many tissues: for these it is often cooling rates near 1°C/min that make survival possible.[51]

C. Factors Affecting Cryoprotection

A summary follows of factors (some of which were already described, others of which will be described later) that may affect the actions and effectiveness of cryoprotective agents on living cells before and after freezing:

1. The compound(s) used
   a. Chemical nature
   b. Relative lack of toxicity
   c. Molecular size and penetrating ability
   d. Interaction of mixtures of compounds
2. Rate and temperature of addition to cells
3. Shape of freezing rate curve
4. Rate of thawing
5. Rate and temperature of post-thaw dilution

III. EXTENSIONS OF CHEMICAL AND PHYSICAL CONCEPTS OF CRYOPROTECTION

" . . . If . . . polymers do not gain entry to the cell interior and are not bound to the cell membrane, it appears that they must act through a modification of some property of the suspending medium." — W. Connor and M. J. Ashwood-Smith[52]

" . . . If bound water [in the suspending medium] acts as a solvent for dissolved electrolytes, a reduction of damaging salt concentration in the residual liquid can be shown." — M. W. Scheiwe, H. E. Nick, and C. Körber[53]

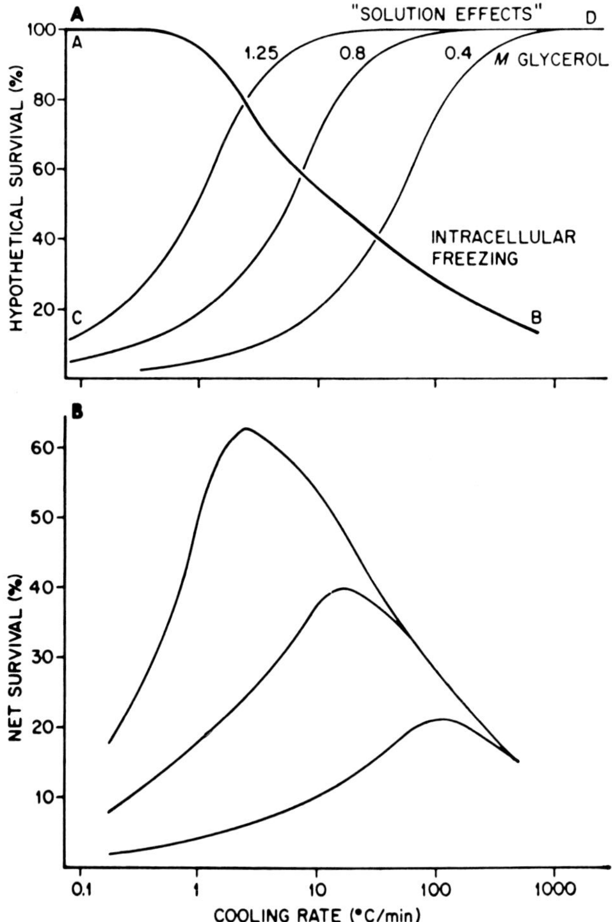

FIGURE 1.   Hypothetical relation between survival, cooling velocity, and additive concentration on the basis of two injurious factors. (A) Separate contributions of intracellular ice formation (curve AB) and solution effects (curves CD) to injury. The three CD curves show the hypothetical effects of three different concentrations of a protective additive. (B) Calculated survival as a result of the contribution of both intracellular freezing and solution effects. (From Mazur, P., Leibo, S. P., Farrant J., Chu, E. H. Y., Hanna, M. G., Jr., and Smith, L. H., *The Frozen Cell*, Wolstenholme, G. E. W. and O'Connor, M. Eds., Churchill, London, 1970, 69. With permission.)

The early concept stressing the penetration by a cryoprotective compound into the protected cell soon turned out to be inadequate, as observed by Lovelock and Bishop[36] even in the first paper on the cryoprotective use of DMSO. Short exposures to DMSO too brief to permeate the cell still showed, as was described, a large, unexplained cryoprotective effect (see Section II.B). A degree of effectiveness by larger, nonpenetrating compounds added to the complexity of interpretation. Even as recently as 1970, a symposium of experts debated the topic ''Must Additives Permeate to Protect?''[24] The results of the discussion were often amorphous and contradictory, with respect to the role of the substances that display cryoprotective action. The subsections that follow will deal with (A) a variety of informative recent findings that require interpretation toward understanding the phenomenon of cryoprotection and (B) some types of information, especially about polymers, that have opened new avenues of interpretation. Section III.C will present supplementary information and concepts pertinent to the actions of cryoprotective substances.

A. Newer Findings

1. Effects of Specific Protein Fractions; Combinations with Sucrose

The 1970 Ciba Conference[24] expressed a dilemma in interpreting the action of cryoprotective molecules, both small and polymeric compounds. A primary conclusion about the action of cryoprotectants was an emphasis on a "salt-buffering" effect from their colligative or colligative-like action: i. e., as the internal and external solutions are concentrated when ice is formed during freezing, the concentrating effect on potentially toxic electrolytes is counteracted by the cryoprotective additives acting colligatively to increase the amount of unfrozen water solvent, so that a lethal concentration of toxic substances is not reached. A cause of concern was how the described effect might operate in the cell's interior when a nonpenetrating cryprotectant was employed.

In this respect, Heber and co-workers' observations of the cryoprotective effect of particular fractions of low molecular weight proteins proved dramatic.[54] They had isolated from hardy spinach leaves some especially active protein fractions (mol 16,000 and 10,000 daltons for two of the fractions, I and II, respectively) that were effective in protecting isolated choroplast membranes (thylakoids) against freezing inactivation at a concentration of less than 0.1% ($<10^{-4}$ M) and were, on a molecular basis, 1000 times as active as sucrose.[31] It seemed reasonable that at such a low concentration, the compounds' action could not be colligative, but rather a molecular effect on a specific configuration or locus of the outer membrane. The interpretation as to molecular loci of freezing injury has been supported by confirming reports.[55-57]

It appears, though, that the proteins alone cannot attain as complete protection against freezing as the combination of a protein with sucrose.[31] The buildup of combinations such as this is suggested as a basis for freeze-hardening as it occurs in the natural state. A great deal of additional investigation will be required to unearth the basis of action of the proteins and their connection to the action of cold hardiness under temperature conditions that activate these systems.

2. Effects of Specific Amino Acids; Loci of Injury; Balance and Mode of Action of Protective Compounds

Heber and co-workers also enunciated a very different principle from previous ones as to the effects of added foreign substances on cells during freezing. Free amino acids were used to study the cryoprotective phenomenon.[55,58] The data showed that, in addition to displaying colligative properties, (1) small added organic molecules can act as agents having, in some manner, quantifiable interactions with cell membranes at specific loci and (2) the ratio of protective compounds to membrane-toxic salts is critical and can be maximized. Studying the effects of amino acids on the thylakoid membranes of spinach chloroplasts, these workers generated a series of hypotheses of the the site of interaction and breakdown of the membranes both from effects of the administered compounds themselves and after freezing in their presence. It was found that in their effects the amino acids fell into several groups.[58] One group of compounds (Group 1, including proline, threonine, γ-aminobutyric acid, and lysine) was protective when freezing thylakoids. Another group (their Group 3, characterized by apolar side chains, such as in phenylalanine) was not protective but, instead, effectively increased the damage from freezing. Like inorganic salts, they became toxic when concentrated, presumably by a disruptive Van der Waals interaction with apolar residues in the membrane. Still another group of amino acids (Group 2, including serine and hydroxyproline) was intermediate in action. The complex results emerging from the use of these compounds, including in synergistic combinations, can be resolved in terms of optimization of the ratio of a *protective* amino acid to a membrane *toxic* compound such as NaCl or one of the described toxic amino acids. Departure from this ratio to either side resulted in inactivation.

Several considerations of the data brought out by Heber and associates indicate that damaging amino acids and inorganic salts need not interact at the same locus of the membrane but may, individually or in groups, have different points of attack, presumably even on both sides of the cell membrane.[59]

They also observed that even though the Group 3 compounds are toxic as described, when they are added at very low concentrations these semipolar compounds (such as 3 mM phenylpyruvate or caprylate) protect against freezing.[59] Heber et al. interpreted this protection as being due to a small, nondisruptive increase in membrane permeability at this low concentration range, thereby allowing entry of otherwise nonpenetrating cryoprotectants to the inside sites of membrane breakdown. A similarly increased permeability in the presence of DMSO has been observed by Zavala and Finkle as an increased rate of entry of glucose and even of the polymer polyethylene glycol 6000.[179] An inferred conclusion from the several damage sites affected by the small compounds described by Heber et al.[55] is that freezing damage is a multiple step occurrence, at several biochemical structural levels. This conclusion was also pointed out by Steponkus et al.[56] (see below) and is implied by the minimization of two different damaging factors during freezing as observed by Mazur et al.,[47] described in Section II. Also, in spinach thylakoids, several enzyme systems critical to the photosynthesis and transport processes of these membranes, but with different inactivation sensitivities, are damaged in progression.[55]

A lucid, detailed rationalization of the synergistic effects of mixtures of particular cryoprotectants was presented in 1971 by Heber et al.[58] It conforms well to more recently presented findings. They hypothesized a cooperative effect on protection from the use of a mixture of cryoprotectants; when one of two or more components of an effective (cryoprotective) mixture is itself mildly toxic, its concentration (damaging effect) in the remaining unfrozen solution is reduced by the colligative effect of the other(s). If more than one of the compounds is toxic but they damage different specific sites, then each may colligatively reduce the mole fraction of the other to the point where, by a sparing action, the respective sensitive sites are not adversely affected (see also References 18 and 59).

Referring back to higher, toxic concentrations of phenylpyruvate-like compounds, Heber et al.[55] determined that at a higher range of concentration the membrane-destabilizing effect of the compounds during freezing was greater than the protective effect of the sucrose or sorbitol added to the system, resulting in cell damage. We see in these experiments an interesting triad of compounds controlling and balancing in a quantitative manner the degree of damage during freezing, namely inorganic salt (e.g., sodium chloride), semipolar compounds (like phenylpyruvate or caproate) and polyols (sucrose or sorbitol). The mechanisms of their interactions are far from understood.

A challenging contradiction of the well-received ideas of Lovelock,[60] based on red blood cells, comes out of findings on plant material from the Heber group. From comparing the profiles of photosynthetic biochemical reactions after freezing thylakoids (in NaCl or phenylpyruvate) with the reaction profiles from normally frozen leaves, Heber et al.,[55] concluded that the damage profile promoted by phenylpyruvate appeared closer to the damage pattern from normal freezing than did the damage from high concentrations of salt. Thus, a new basis for damage in frozen tissue, namely, a *toxic effect of the salts of lipid-soluble acids,* perhaps in addition to the effect of concentrated electrolytes is brought to bear. These complex chemical interactions, as well as the indirect effects from freezing — dehydration and reduction in cell volume (stressed by Meryman[61]) — and the direct effect of ice crystals themselves on cellular structure, will all require careful attention in future studies.

A high degree of complexity in cryoprotective action, with more than one damaging factor operating during freezing, was also suggested by Garber and Steponkus,[62] again from results

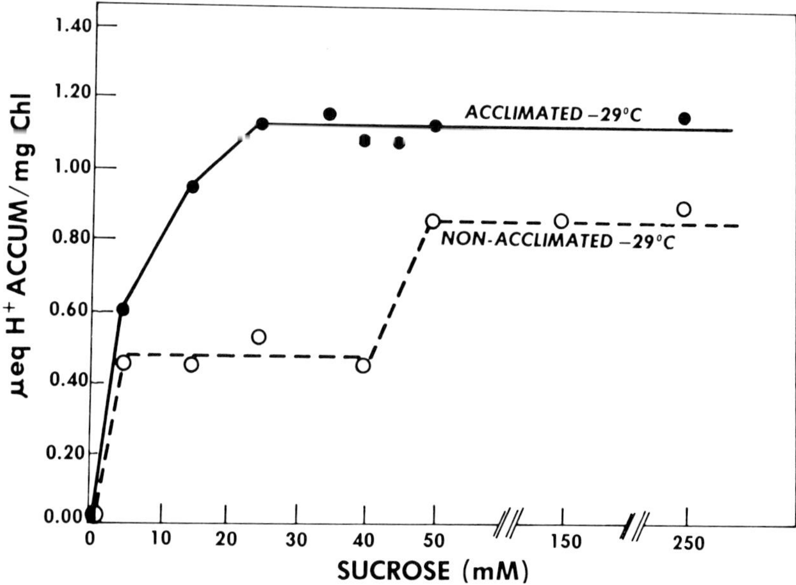

FIGURE 2.   Light induced proton uptake by acclimated and nonacclimated thylakoids frozen to −29°C in varying concentrations of sucrose. (From Steponkus, P. L., Garber, M. P., Myers, S. P., and Lineberger, R. D., *Cryobiology*, 14, 303, 1977. With permission.)

with spinach thylakoids. They were measuring light-dependent proton uptake, related to the Ca^{++} ATPase activity of the membrane. At each of three different freezing temperatures, sucrose showed a plateau of protection of one of the freezing lesions at a finite concentration of sucrose. E.g., at −29°C, a light dependent Δ-pH plateau at 35% of the control was stabilized by sucrose at 10 to 40 mM (Figure 2). For each of the temperatures, protection of a second lesion in proton uptake required a higher sucrose concentration, greater than 50 mM (or 0.9% w/v). They also found that acclimation eliminated one of these levels of physiological sensitivity.[56] Ion uptake by thylakoids from acclimated leaves was maximally activated, as a single plateau, by less than 40 mM sucrose addition. The figure illustrates the two-lesion sucrose-benefited phenomenon and the elimination of one of these in acclimated tissue.

In studies comparing damage from the freezing of chloroplast membranes in the presence of a series of related sugars, Lineberger and Steponkus[63] had determined that in molar cryoprotective ability raffinose > sucrose > glucose, but they later calculated that these sugars were actually equivalent when the molarity terms were corrected by using activity-dependent parameters; the use of this correction is a critical factor when making comparisons of cryoprotective additives.[64]

Wiest and Steponkus[65] concluded that it is neither the colligative reduction of the toxic intracellular or extracellular salt concentration nor the concentration of hydroxyl groups in the nonpenetrating sugars studied that is responsible for cryoprotection but, instead, the total osmolality of the suspending medium; also, it is not the direct effect of a high concentration of salts that is injurious to the cell but rather the cell expansion resulting when the external solution is diluted by washing (see Chapter 3).

3. Other Factors in Cryoprotection and Cell Damage

Other actions of large molecular weight compounds that have been considered as avoiding freezing damage are (1) coating of otherwise sensitive membranes by the polymeric compounds so they are not denatured by the increased concentration of salt solution resulting

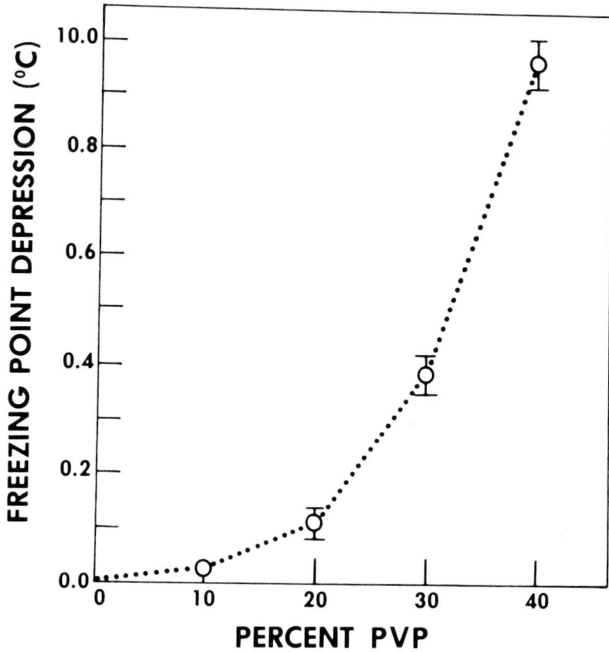

FIGURE 5.   Freezing point depressions of various concentrations of PVP K30 (mol wt 40,000) in distilled water. (From Connor, W. and Ashwood-Smith, M. J., *Cryobiology*, 10, 488, 1973. With permission.)

3. Anomalous Protective Behavior of Polymers

Thus, two mechanisms of polymer action appear to be operating in activating cryoprotection: (1) ability to retain water in the liquid (toxin-diluting) state and (2) imposition of a degree of structure on the water, with modification of both its liquid[88a] and solid state physical-chemical properties. Superimposed on these is, of course, the required near lack of toxicity of the polymer itself.

A high measure of support of the above findings on the formation and importance of polymer-influenced structured water has come from a series of papers from collaborative efforts by German and French workers using model systems. Körber and Scheiwe,[89] using mixtures of DMSO, glycerol, and HES with NaCl-H_2O, showed, by differential thermal analysis, that water is absorbed by HES (0.5 g H_2O/g HES), in a "thermally inert" state (no observed isotherms). This partially immobilized water is nevertheless available to prevent electrolytes from becoming concentrated to a toxic condition, as was also described by Connor and Ashwood-Smith.[52] Even at temperatures in the range -20 to $-40°C$, ordinarily a particularly damaging temperature range for frozen material undergoing either cooling[90,91] (illustrated in Figure 6) or rewarming,[92] 11% HES maintains the salt content of the frozen solution at less than the 4.41% (0.8 *M*) described by Lovelock[60] as lethal. These experiments were confirmed using human red blood cells in the system.[53] In another paper, solutions of greater than 60% concentration HES were shown to remain wholly amorphous (noncrystallized) down to $-153°C$ even at slow cooling rates.[93] However, whether (and how) such large extracellular compounds prevent the crystallization of ice within cells is an unsettled question.

4. Water Stabilization by Small Molecules

Related studies with small molecules in mixtures have been carried out by Boutron and co-workers.[69,70,94] Results of these physical-chemical model studies have indicated advan-

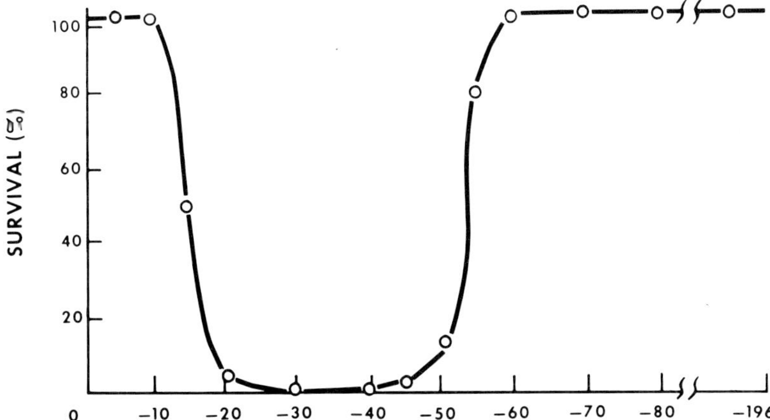

FIGURE 6.   Survival of cells cooled and rewarmed rapidly without prefreezing. An unmounted tissue section held with forceps was rapidly immersed in isopentane baths at various temperatures (°C) from room temperature without prefreezing, and then kept there for 20 sec, before being rewarmed rapidly in water at 30°C. (From Sakai, A., *Plant Physiol.*, 41, 1050, 1966. With permission.)

tages in using mixtures of cryoprotectants (e.g., glycerol-DMSO), including newly tried mixtures. The use of 1,2-propanediol-water (or these plus 1-propanol) gave higher stability of wholly amorphous water and a higher glass-formation tendency than other compounds studied, with no crystallization anticipated within cells even at − 196°C and for quasi-infinite periods of time.[70] From these results, a high degree of tissue stability from the controlled use of such compounds during cryogenic freezing may be anticipated.

C. Special Aspects of Cryoprotective Compounds

1. Toxicity and Purity

A degree of toxicity (nonsurvival) from the addition of cryoprotective compounds to plant tissues has been variously reported[82,95] (see also Chaper 6, Table 4). Such effects have been observed after simply addition and removal of a compound (e.g., DMSO at 10% concentration), with or without freezing,[82] or after a period of growth in medium containing the compound, in the case of DMSO at 1 to 2%.[96,97] or at higher levels.[12,98] Toxicity may be partly due to each of several factors, e.g., osmotic effects, solvent effects, and effects as a specific compound.

Caution must be exercised about whether the effects observed are from the specified compound itself rather than from impurities in it. Matthes and Hackensellner[99] demonstrated the presence of a component in impure DMSO samples that correlates spectrophotometrically with the degree of tissue inactivation by freezing (heart muscle). The absorption curve of impure DMSO is characterized by a distinct peak at 275 nm which may be used for testing the purity of unknown samples. The authors consider that satisfactorily pure DMSO should have a 275 nm absorbance value of <0.03.

2. Genetic Effects of DMSO

A small amount of evidence in higher plants,[100-102] and a large body of evidence in microorganisms and animal cells[103,104] indicates that DMSO in 2 to 10% solution may be involved in generating a variety of genetic and/or epigenetic changes, but also, in contrast, it decreases genetic damage from radiations (see paragraph below). As with so many chemically active and highly solvating molecules added to living organisms, investigation will be required to determine to what extent the reported biological changes may be serious, durable and predictable under the conditions of usage. This may be a particulary pertinent question for DMSO when used as a cryoprotective agent; it is often applied only at ice temperature for a short interval, then chilled to and stored at very low frozen temperatures where both physical and chemical reaction rates are generally so vanishingly slow that all interactions essentially stop.

In addition to the possibly adverse genetic effects of DMSO described above, the compound has a reverse genetic effect in that, by its presence, it protects cells against genetic damage from radiation.[8,102,105,106] Ashwood-Smith and Friedmann[8] reported that in the presence of DMSO (10%), the damaging effects of electromagnetic irradiation on viability and chromosomal changes of hamster cell cultures was reduced by a factor of 3.5 in frozen cells compared to irradiation at 22°C in the absence of DMSO. They calculated a culture storage period, at $-196°C$, of approximately 30,000 years before background radiation would generate lethal and chromosomal damage equivalent to that produced by an acute D_{10} (10% survival) dose of X-rays (600 rad). Protection against irradiation-induced mutation might prove especially significant during the ultracold storage for extended periods of time required for genetic base collections.

IV. INTEGRATION OF EMPIRICAL FINDINGS

'' . . . with the use of a solution diluted enough to be of very low toxicity, and which remains entirely amorphous even with slow cooling and warming rates, all cells would be protected . . . '' — P. Boutran, A. Kaufmann, and Nguyen Van Dang.[107]

This being a section that describes many diverse findings, the interpretations made by the experimenters themselves of their findings will be largely adhered to. However, it may be profitable for the reader to view these descriptions within the framework of the progression of ideas reviewed in Sections II and III.

A. Use of Cryoprotective Compounds

To date over 40 plant species and many types of their tissues have been frozen to $-196°C$ and survived[13,16] (see also Table 4, and Chapters 6 to 12). In most cases cryprotective additives were used in the freezing procedure, most often as single compounds. Lists of such compounds have been compiled for plant and animal tissues.[2,25,26,37,39] Of these compounds, DMSO has dominated the plant literature, with glycerol or sucrose appearing prominently in other experiments. Morris and co-workers[108,109] found, in an extensive sequence of experiments, that methanol (a compound that, like DMSO, penetrates the cell quickly but has been reported as toxic[95] — see Section III.C) is highly protective of green algae during freezing. Dimethylsulfoxamine has been reported as a successful cryoprotectant for *Chlamydomonas.*[110]

The listing of cryoprotective compounds active on plants published by Sakai and Yoshida[2] in 1968 (see Table 2) could currently be amplified primarily by inclusion of a number of highly water soluble oxygenated compounds particularly active in the form of polymers, as was discussed in Section III. After freezing in LN in the presence of protective compounds, higher plants have in several cases given rise to a whole plant after thawing (see Chapters 6 to 8, 11, and 12) in contrast with freezing without any cryoprotectant, a treatment that would usually kill the tissue (see Chapters 7, 9, and 10 for exceptions).

The compounds considered as ''cryoprotectants'' and usually used just prior to freezing (as against growth factors, etc.) have often been found useful in the 5 to 20% range.[13] Yet, as little as 0.3% DMSO has been reported to give maximal protection to conidia of *Neurospora crassa*[111] and, in contrast, the beneficial use of $>30\%$ w/w solutions of high molecular weight polymers has been suggested.[53] Some experimenters have also found it beneficial to add cryoprotective solution a few hours to several days before freezing was

Table 4
PLANT SPECIES THAT HAVE
SURVIVED FREEZING IN
POLYETHYLENE GLYCOL-
GLUCOSE-DMSO (10-8-10%)[a]

Frozen to − 196°C	Frozen to other temperatures (°C)	
Alfalfa (3 lines)[b]	Grape	− 15
Apple	Grapefruit	− 15
Asparagus	Phaseolus	− 15
Carrot[c]	Tomato	− 30
Elm		
Palm (2 lines)[d]		
Rice (8 lines)		
Soybean		
Strawberry[c]		
Sugar cane		
Wheat		

[a]   Method of Ulrich et al.[3]
[b]   Thawed samples grew into flowering plants.[107a]
[c]   Carrot suspension culture; strawberry meristem; all others callus cultures.
[d]   Thawed samples grew into small trees.[107a]

initiated (see Section IV.D.). Evidence from sugar cane suspension cultures treated with a cryoprotective solution (unfrozen) suggested that the major damaging effect from the cryoprotective compounds occurs quickly against susceptible cells in the population, while little additional effect may take place over several hours,[82] confirmed by unpublished observations on alfalfa callus cultures.

B. Cryoprotective Mixtures

A natural outcome of attempts to improve the effect of cryoprotective compounds was to use them in mixtures. Such mixtures have been tried out or adopted by several workers on higher plants,[95,112-118] often, however, without an explanation of the reasons behind using the particular mixture or the proportions decided upon. Presumably some synergistic effect of combinations of different types of cryoprotective compounds,[58,82,94,119] or a sparing action (nonadditivity in their toxicity) was responsible for observed improvements in cryoprotection described in Section III, above.

Systematic studies of the effect of combining cryoprotective compounds have been performed on sugar cane suspensions. Results with mixtures have indicated that: (1) they can have greater effects in maintaining postfreezing viability than single compounds (compared on the basis of the total combined molarity present in the cryoprotective solution); (2) the proportions of the components are important and can be maximized; and (3) a heightened cryoprotective effect is achievable from decreasing the concentrations of introduced compounds possessing detrimental or toxic effects toward cells, at the same time that the overall cryoprotective effect of the combined compounds may be additive (see also Sections III.A. and III.B.4).

Figure 7 shows an effect of varying the proportion of glucose and DMSO during the freezing of sugar cane suspension cultures. After preliminary trials had established a total concentration of about 2 M as near optimum, a molar ratio of 3:1 for DMSO and glucose (approximately 10:8% w/v), respectively, was established as near optimum for survival at

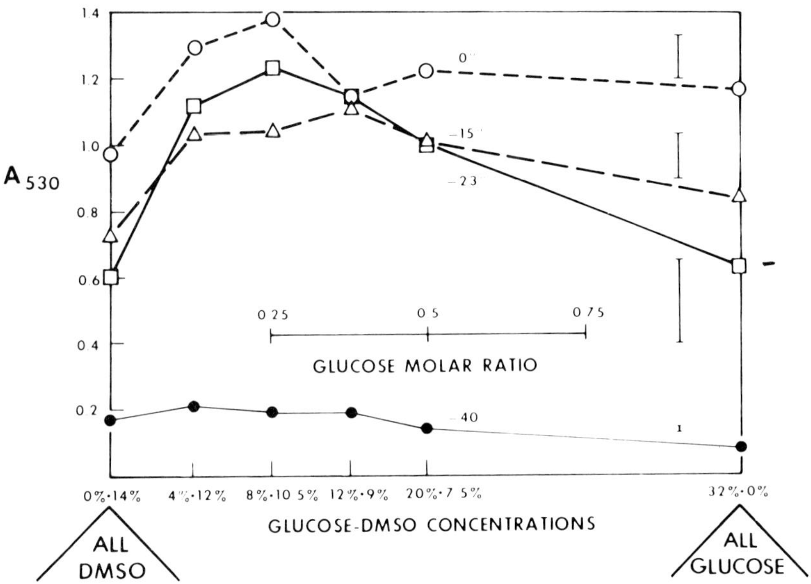

FIGURE 7.   Effects on TTC value brought about by varying glucose-DMSO ratios. Total sample concentration of Glucose + DMSO = 1.9 *M*. Percentage composition of the mixtures (% of each component in the solution, w/v) and molar ratios of the cryoprotectants are both indicated in the abscissa. A_{530} used as in Figure 3. (From Finkle, B. J. and Ulrich, J. M., *Plant. Physiol.*, 63, 598, 1979. With permission.)

the several temperatures tested.[82] Another set of experiments with polyethylene glycol 6000 (mean mol wt 7400 daltons according to the manufacturer, Union Carbide and Carbon Inc.) showed that, when used alone, it is essentially nontoxic at concentrations of up to 25% w/v (see also Section III.A) and, also, noncryoprotective.[3] However, when added to glucose/DMSO mixture, it increases the relative viability index of cryoprotectant-treated unfrozen cells (decreases toxicity), and with frozen cells it increases their viability index by up to 100% (i.e., greatly improves the cells' survival of freezing). These findings have been confirmed by growth experiments with alfalfa callus.[180] A mixture of PEG, glucose and DMSO in the proportions of 10:8:10% w/v has been applied during the freezing of over 20 species of cultured cells, resulting in survival of liquid nitrogen (− 196°C) by over 50% of them (Table 4). The ultrastructural effects of these components, alone and in combinations, have also been studied (Section V). Other research, including with animal tissues and microorganisms, has also indicated cryogenically beneficial effects from the use of mixtures of compounds,[18,20,58] although examples are also published where mixtures were observed not to necessarily have a beneficial effect.[120] Withers and King[117,121] have described the amino acid proline (0.9 to 2.0 *M*) as having a cryoprotective effect when used either alone or to supplement the effects of other cryoprotective agents. As mentioned in Section II.A, proline is one of those compounds reported as being accumulated by plants under stress. Withers[117] also demonstrated how even the dissolving vehicle for cryoprotectants, whether it is water or a complex culture medium that is used, can affect survival.

It has yet to be evaluated how general the positive effects of these several types of cryoprotective mixtures might be for the numerous species and tissues of plants to be cryopreserved. An example of the beneficial effect of a synergistic cryoprotective combination of compounds that may occur in nature is witnessed in the experimentally effective combining of sucrose and a cryogenic small protein fraction present in leaves of spinach and cabbage, as was described in Section III.A.

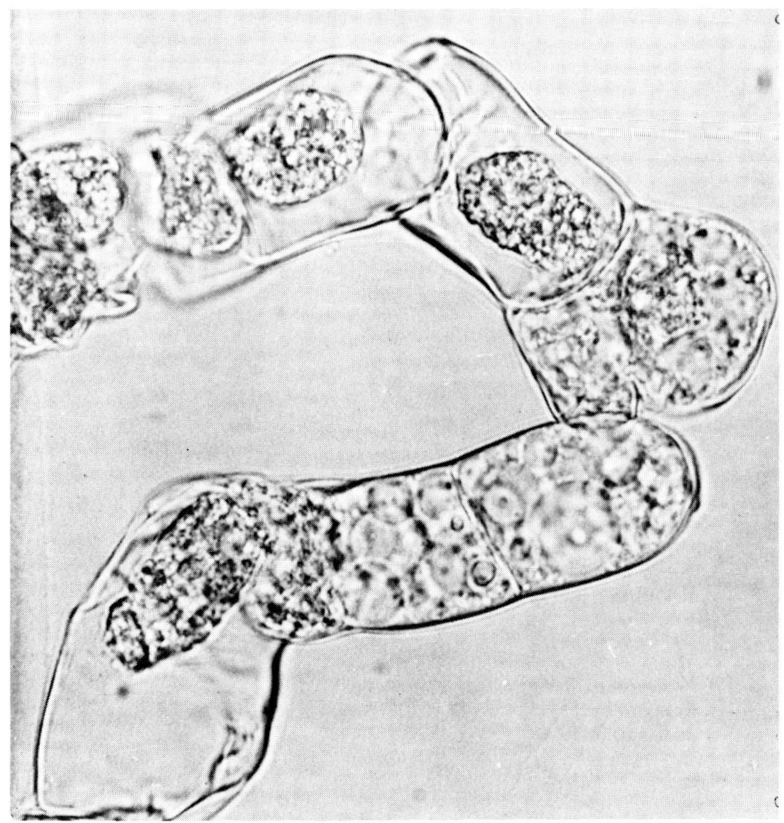

FIGURE 8.   Photomicrograph of unfrozen sugar cane cells exposed to 8:12% glucose-DMSO at near 0°C, then washed. (From Finkle, B. J., and Ulrich, J. M., *Plant Physiol.*, 63, 598, 1979. With permission.)

C. Prior Treatments and Cell Growth Stage

There is a great deal of information that the origin and condition of the tissue to be frozen affects greatly its survival of cryogenic freezing (see Chapter 12). In growing plant tissue the stage of development of the tissue employed is particularly vital. With cultured tissues, important factors are the culture age (time since first isolation of the culture, and since last transfer), growth temperature and degree of aeration (shaking), the stage in the cells' mitotic cycle (learned through synchronous culture[115]), and modifications of the medium. An example of the state of the cultured cells is illustrated in Figure 8. Although almost all the cells appeared turgid at the time of sampling, about one half were physiologically vulnerable in that they became permanently plasmolyzed simply from the addition and washing out of cryoprotective solution. Most of the cells which survived treatment with the solution were then capable of surviving freezing to $-23°C$.[82] Much of the detailed information supporting the importance of culture factors of the type described does not appear in citable literature, rather being within the experience and knowledge of experimental research groups.

The culture medium may be particularly modified by the inclusion of additional hormones[114,122] and osmotic and cryoprotective substances, described below, and by nutrient modification, beneficial or stress-producing.[123,124] Sugars[114] and mannitol[83,125,126] have been used in cultures as osmotic agents that decrease cell size and increase the survival of freezing. DMSO and still other cryoprotective compounds or combinations of compounds have sometimes been added to the growth medium (usually at reduced concentrations) to build up additional cryoprotection[114,118,120,127] (see also Table 3 of Kartha[12] and also Chapter 6).

Cultured cells of soybean[83] and tomato[128] have displayed vigorous growth in media supplemented by cryoprotective soluble polymers, even at high concentrations. Table 3[12] showed that 25% solutions of HES (mol 70,000 to 450,000 daltons), PVP (44,000 daltons), and PEG (4000 daltons) all allowed > 82% cell viability (fluorescence diacetate test) at 24 hr, compared with control cultures;[83] in 50% solutions the viability values were 76, 59, and 5%, respectively.

D. Cryoprotectant Addition and Post-Thaw Treatments

Many of the cryoprotectively effective substances, especially those that penetrate the cell only slowly, or not at all, display osmotic character, that is they withdraw water from and plasmolyze the cells. The accompanying shrinkage of both the cell contents and plasma membrane, particularly the danger of irreversible shrinkage, has often called for a cautious rate of addition of cryoprotective compounds, to avoid an osmotic shocking effect and a damaging excess of pressure across the cell membrane. Taking such precautions was emphasized by Kartha.[12] The amount of attention required to the rate of addition of cryoprotectant is variable in different tissues, and can be checked empirically.

Because of the toxic effects the cryoprotective additives may have on growth[12,129] it often appears desirable to dilute out or remove the cryprotective compounds after the cells are frozen and thawed. In achieving this, the rate of post-thaw dilution (deplasmolysis) has also been assigned a critical role in cell survival[12,40] (see Chapter 12 for some exceptions).

Hence, with cultured cells the sequence of handling events carried out is often as follows: (1) addition (slow) of cryoprotective solution; (2) freezing (most often slow — see Chapter 12) to the final temperature; (3) thawing (rapid); and (4) post-thaw dilution (often gradual).

When steps are taken to remove the cryoprotective solution, it is the overall result of both addition and washing out that must be judged experimentally. To date, little information has appeared on the effects of addition and washing events on plant cells. Finkle and Ulrich[82] and Zavala and Finkle[130] have shown that even with the precaution of slow addition and removal of cryoprotective solution, large, partly irreversible changes take place in the structure of some cells in the population when treated with cryoprotectants (see Figure 8 and Section V). The damage is reflected as a loss in cell survival[82] which, however, varies in degree, as reported even for genetically close mutant lines (rice callus).[131] It seems that much of this lethality from added cryoprotectants might be avoided by a prior screening of cells (e.g., by size, or fluorimetrically, or by selection of the cell cycle stage through synchronous culture[115,132]), perhaps in the presence of osmotic shrinking agents, in order to select the more vigorous or stress-stable population. But with clumps of cells such as one finds in suspension cultures and with callus cultures, cell size selection is difficult or impractical to achieve.

The temperatures of addition and removal of cryoprotective solution can also affect survial and/or cell ultrastructure (see below and also Section V). The optimum temperatures may be empirically evaluated for the particular species and tissue type.

Experimental results indicate that when a post-thaw wash is employed, the temperature of the wash solution is a sensitive and critical aspect of cell survival. With rice callus cultures and with sugar cane suspension cultures a wash step at room temperature contributes greatly to viability, compared with washing at ice temperature.[133] For rice callus the highest survival was achieved from the addition of cryoprotective compounds at ice temperature, followed by their post-thaw removal at room temperature. The results for addition temperature with suspensions of sugar cane were somewhat otherwise[133] (also see Section V.A).

The post-thaw presence of cryoprotective agents may also be handled by alternative methods. One of these is to minimize the amount of these compounds in contact with the test culture during freezing treatment (the ''dry'' (blotted) freezing method of Withers).[134] Alternatively, the cells can be drained after thawing and aseptically placed into direct contact with agar growth medium.[125] The evidence suggests that factors favoring survival or continued growth may be present in the frozen cells and that excessive removal of these factors (as by washing) decreases survival. By placing the thawed cells, without washing, directly on nutrient agar, the agar gel serves to dilute, by diffusion, the residual solution containing toxic cryoprotective substances. Thus a sub-toxic level of these can be achieved for improved growth to take place (see Chapter 12). Such a dilution method has not proved successful with callus of alfalfa; here washing proved necessary for post-thaw growth while blotted (unwashed) pieces of cryoprotectant-treated callus did not survive.[181]

The critical effect of washing temperature that was described for survival of rice and sugar cane cells takes place over a mere 20%C temperature difference. This effect may come to be explained when more detailed information is obtained on such temperature effects. A reasonable rationalization of the effect may lie with the gel-sol transition temperature of the lipoprotein membrane, which for many plant membranes lies specifically within this range.[135] Dealing with a cell membrane in the relatively fluid liquid crystalline state, might (1) reduce the fragility of the membrane under osmotic pressure stress and (2) accommodate a higher rate of equilibration of osmogens through the membrane, alleviating pressure stress[136] and thus counteracting osmotic damage, membrane leakage, and death. Rubenstein has emphasized that in plasmolyzed oat leaf cells a sudden cold rehydration (deplasmolysis) step activates damaging effects.[137] After using a 4°C wash solution as a "cold shock" dilution treatment, he observed a loss of specific proteins from the cell membrane and interference with H^+ transport. Handa et al.[128] found similar metabolite leakages and also loss of viability of tomato suspension cultures grown with high levels of polyethylene glycol, when a sudden reduction in solute content was made in the medium. Deleterious effects from osmotic shock phenomena have received considerable attention for bacterial cells.[138,139]

The role of additives in post-thaw recovery contributes a curious aspect of the mechanisms of cryoprotection. Evidence from Meryman and Hornblower[140] on red blood cells suggests that the cells can be benefitted by cryoprotective additives added after freezing, that the large polymer PVP may not need to enter the cell to be an effective cryoprotectant and, indeed, when added to ice-encased cells (even though it can make contact with them only after thawing), it still protects them. This effect presumably takes place through some extracellular protective mechanism. Data of Pribor[71] on red blood cells supports this, while contrary evidence was obtained in cultured hamster cells by Connor and Ashwood-Smith.[52] Beneficial effects of Ca^{2+} ion in post-thaw treatments have been reported for cabbage sections,[141] tobacco suspension cultures,[142] and onion bulb tissue.[143] Sugawara and Sakai[142] reported that their tobacco cells frozen in a glucose-DMSO mixture survived when 0.125 M $CaCl_2$ was added either before freezing or after thawing. The described beneficial compounds may be involved in cell repair, the ultimate phase of the cryogenic process (also see E, below, and Section V).

E. Effects on Specific Cellular Structures

The subject of the effects that specific cryoprotective compounds exert on plant cell structures could presumably be the heart of the matter of this chapter. Yet it is one of the least substantial as to the experimental information available. What are the specific effects of treatment with particular cryoprotective compounds and of the freezing, thawing, and cryoprotectant removal procedures on important cell structures? The steps of the process may be generally outlined as (1) addition of cryoprotective materials (often leading to concentrating the cell's solutes osmotically), (2) lowering of temperature, resulting in partial crystallization of water outside the cell and perhaps within the cell during freezing, with resultant decrease in volume of extracellular and intracellular liquid phase, concentration of solutes and, under some conditions, migration of cell water to the outside, (3) as the lowering of temperature is continued, further crystallization and recrystallization and further concen-

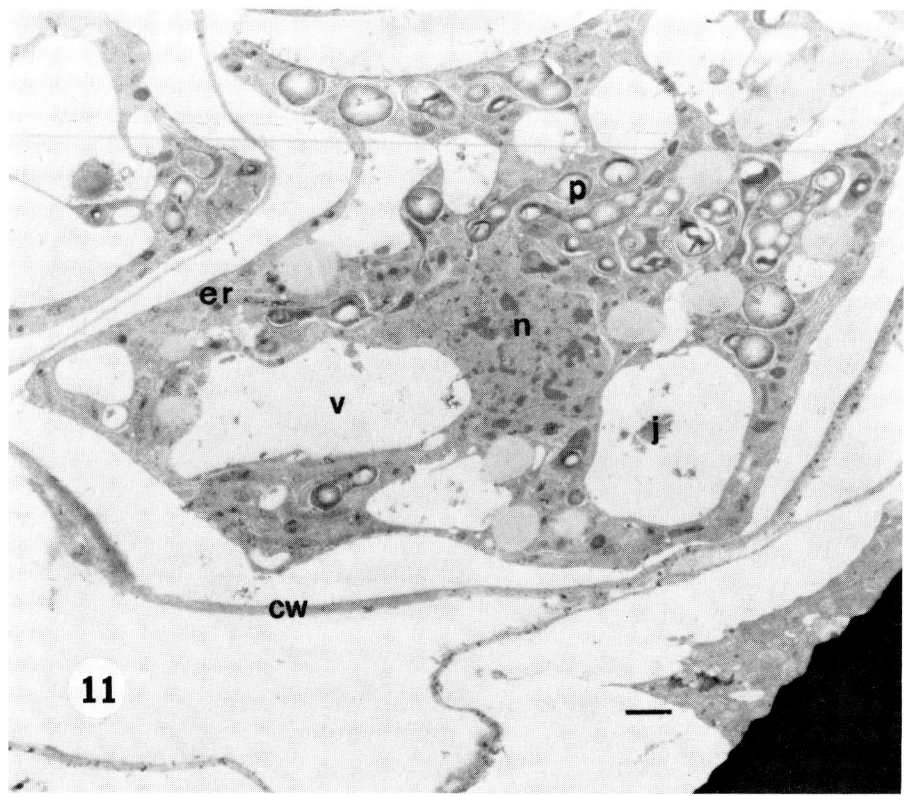

FIGURE 11.   Sugar cane suspension cells exposed to PGD at room temperature; bar = 1.0 μm.

to 0°C for 12 hr displayed vesiculation of endoplasmic reticulum (ER), disassociation of ribosomes from ER, and increases in plastid density. Following 24 hr at this temperature ER was highly vesiculated, free ribosomes were abundant, and mitochondrial cristae were disrupted. Kimball and Salisbury,[164] using seedlings of *Secale cereale* L., *Cynodon dactylon* L Pers., and *Paspalum notatum* Hugge, and Ilker et al.,[165] using the cotyledons of germinating tomato *(Lycopersicon esculentum)*, showed that the chilled tissue systems also developed a pathological morphology. In the cereals studied the plastids appeared to be the most sensitive organelles; other organelles were less altered although mitochondria were swollen, dictyosomes disappeared, and there was an apparent increase in ribosome-associated ER (RER). In contrast, tomato seedlings showed a reduction in protein bodies and cytoplasmic volume, an accumulation of lipoidal bodies, irregular plastids with few thylakoids, and at times, mitochondria with disrupted cristae. These changes may take place in a few hours or after several days. Similar changes may be induced by osmotic treatment and freezing[166] and through chemical fixation when preparing tissue samples for microscopy.[167] Cryoprotectants may cause similarly dramatic ultrastructural alterations; however these changes occur over a period of <1 hr, the approximate time for cryoprotectant addition.[168] The ultrastructural responses observed present some potentially useful information and draw attention to possible key sensitive sites of cryoprotectant and freezing damage. They suggest that: (1) organelles of different species behave differently, (2) organelles within a species are not equally affected, (3) changes may be reversible under some conditions, and (4) cell response may be rapid.

Because ultrastructural changes may be induced by different factors, some or all of which may be present during cryogenic treatment, it is difficult to distinguish among induced ultrastructural changes (that is, those caused by temperature or cryoprotectant stress). Com-

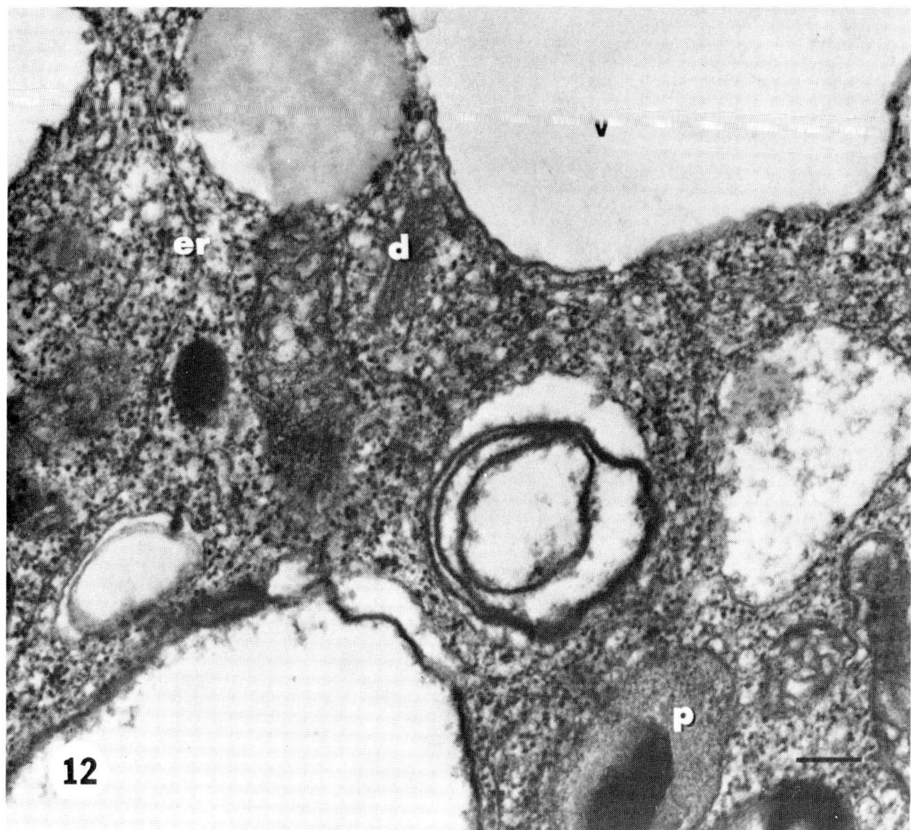

FIGURE 12.   Sugar cane suspension cells exposed to PGD. Note that many membranes are still intact; bar = 0.25 μm.

bining of stresses may exacerbate the observed individual effects, thus it becomes difficult to draw reasonable, accurate conclusions about how cryoprotectants interact with cells and protect cellular components.

C. Cryoprotectant Interactions with Cells

It is apparent from the figures that the presence of cryoprotectants results in apparent cellular disorganization. Yet, cellular disorganization does not always result in cell death, and cryoprotectants are, indeed, often necessary when frozen to −196°C, thawed, and recovered, even though there are exceptions.[49,169] The effects of cryoprotectants on the structure of plant cells when treated for preservation of viability have not been extensively studied.[130,146,161,168,170] Two basic approaches have been used in ultrastructural investigations, (1) to fix cells in the presence of cryoprotectants, or (2) to freeze cells according to the normal cryogenic routine and to freeze substitute them with the cryoprotectant *in situ*. Such studies can provide clues as to how cryoprotectants function to promote viable freezing.

Withers[161] and Zavala and Finkle[168] have observed cells that were fixed with the cryoprotectants *in situ*. In unfrozen cryoprotected cells of sugar cane, cellular alterations have been reported following the addition of polyethylene glycol, glucose, and dimethylsulfoxide, alone or in combinations.[130] A most interesting observation made on these treated cells was that plasmolysis and cytorrhysis were greatest in PGD treated cells. This maximally cryoprotective combination[3] also offered the best preservation of cellular membranes (Figure 12). Why this is so is not clear but preliminary experiments indicate that P in the presence of D helps to maintain membrane structure.[168] It would be of interest to determine the physiological status of different membrane fractions following exposure to different cryoprotectants.

In excised pea meristems, Haskins and Kartha[171] found no change in the fine structure after incubation in 5% DMSO for as long as 7 days. However, the gross morphology of the apical dome was altered. Withers and Davey[156] studied DMSO treated, frozen, and freeze-substituted cells of sycamore. The cells showed a highly disturbed ultrastructure. Plasmolysis was minimal, cytoplasm was electron dense, and the cytoplasmic and organelle membranes were observed in negative contrast. Cells frozen without cryoprotectants often have no recognizable structural features. Effective cryoprotectants preserve some structural integrity even though apparent cellular organization is greatly deranged.

D. Recovery from Cryoprotectant Addition

Gazeau[172] observed that cryoprotectant-induced changes are readily reversible. In wheat coleoptiles measurable growth occurred within 24 hr after exposure to glycerol, DMSO, and sucrose. In rapidly growing cultured sugar cane suspensions, we have found that cells recovered their pretreatment morphological appearance 8 to 12 hr after addition and removal of PGD.[180] While stepwise recovery from cryoprotective treatment has not been studied in detail it is not difficult to consider that the sequence of cellular recovery involves processes similar to those that have been described for chilled tissues.[163]

E. Freezing and Thawing after Treatment with Cryoprotectants; and Recovery

Electron micrographs of cryoprotected, frozen and thawed plant cells show many cells with nearly normal morphology immediately after treatments. Even in those instances where cryoprotectants are not necessary to preserve viability, normal cellular morphology returns to many cells immediately after thawing.[169] However in other cells, some ultrastructural alterations attributable to cooling may be exacerbated by cryoprotectants: that is, ER becomes highly vesiculated, mitochondria generally disrupted and hypertrophied, plastids fragmented or polymorphic, and nuclei more invaginated.[116,130,161,168] Usually a lag phase in growth is observed following deep cryogenic freezing.[82,125,171-173] The lag phase may be attributed to the time required to repair cellular damage[172,174] or for the remaining live cells to undergo a number of divisions before measurable growth resumes, or both. Post-thawing repair may require cellular reorganization or even a period of redifferentiation before growth can continue.

A general observation in studies dealing with recovery from cryogenically induced stresses is that membrane recovery, both morphological and physiological, is a critical step in reestablishing osmotic and ionic balance and subcellular compartments.[57,143,174]

The results of Cella et al.[174] support this. They found that the degree of success in obtaining protoplasts from frozen rice suspension cells was variable: when the cells were processed immediately after thawing there was a better yield of protoplasts than when processed 12 hr later. After that (postrepair?), the cells again began to give higher yields of protoplasts. Post-thaw experiments with animal cells also suggest that repair is membrane associated,[175] that during the repair process cells stop progressing through the cell cycle,[144] and that repair is not dependent on protein synthesis.[175]

Learning to promote the specific processes of repair through cryoprotective prefreezing and post-thawing manipulations (as illustrated in Section IV) must be a major goal in enhancing the plant cell's recovery from the traumas inflicted by freezing and thawing. Much remains here to be explored.

Perhaps the most remarkable observation of all is that cells do undergo tremendous ultrastructural changes during cryogenic treatments, yet they are able to recover.

VI. NEW AVENUES TOWARD IMPROVED CRYOPROTECTION

"To find protective additives that are more effective than any now known . . . will require far better understanding of the mechanisms by which solution effects kill, and the mechanisms by which additives protect." — P. Mazur[11]

In summing up, it seems that there remain two major questions that still need definite answers, even following the excellent research that has already been done on them, namely: how do the several types of cryoprotectants prevent damage and how does a cell or tissue recover from the damage that does occur? These will be dealt with briefly, as well as a few other questions and goals of cryoprotection that in the authors' view require emphasis

A. Cryoprotective Avoidance of Damage

As has been evident in this chapter, we still do not have an exact sense of how cryoprotective compounds operate, or even the location of the critical site(s) of protection. A goal of cryoprotection would be to achieve no damage, or only minor (repairable) damage. We have seen evidence in Section III.A for the presence of more than one site of damage in even a single type of membrane and it may be that several sites need to be protected by several different compounds operating by different mechanisms.

A neatly packaged answer to freezing damage would be to totally avoid, or minimize, even at very cold temperatures, the crystallization of ice with its disruptive consequences. Means to accomplish this have been suggested. One would be to add cryoprotectant at just above the freezing temperature and continue with additions incrementally during cooling as the freezing point is successively lowered to very cold temperatures, thereby holding reagent toxicity (reactivity) to a minimum[176] (and proceed in reverse during warming[177]). Another method, recently emphasized, is the use of high concentrations of large soluble polymers to maintain a noncrystallizing, electrolyte-dissolving solution in liquid form to very low temperatures so that a concentrating effect on toxic compounds would be avoided even at the coldest temperatures (Section III.B). Thus the original emphasis on small, penetrating solutes, or even nonpenetrating osmotic agents, to minimize the damage from crystallization has given rise to new directions of using nonpenetrating polymeric agents that may prevent ice crystallization altogether. A great deal of additional exploration towards making the best use of such large polymeric compounds lies before us. In fact, to date, the effects of high concentrations of high molecular weight compounds in avoiding freezing damage have been tested with animal tissues and suspensions but almost not at all with plant material. It is too early to know to what extent such compounds will make the viva-cryoprotection of plant cells more successful and more predictable.

A powerful concept was expressed by Mazur[11] and by Connor and Ashwood-Smith[52] that internal membranes (including, presumably, organelle membranes) may be protected by *in situ* naturally occurring protective polymers that are concentrated to increased effectiveness as free water decreases in the freezing cell. The concept suggests that, possibly, only the outer membrane need be protected by external chemical manipulation. However, not all cells (of all species, or even of all tissues within a single plant) may have the appropriate constituents, making such cells poorer candidates for cryogenic freezing by exterior-only treatment methods. Such cells may require, in addition, a penetrating reagent to colligatively resist the internal crystallization of ice. This requirement would, in addition to the sparing action effect of reagent combinations described by Heber et al.[58] (see Section III.A), help explain a need for mixtures of cryoprotectants to achieve maximum effectiveness. There may come a stage where, with greater knowledge and understanding of the roles of cryoprotective compounds, specific components will be added for specific purposes (as with automobile motor oil additives) toward achieving more universal solutions to the problem of preserving cell vitality at exceedingly low storage temperatures.

Two other potentially damaging phases of the freezing process — thawing and post-thawing treatments — are more often taken for granted than the cooling phase (or, at least, are little discussed) in terms of cryoprotection, and they may present very different elements of cell damage (see Section IV.D). These will require a great deal of increased attention to understanding the respective processes and taking appropriate precautions to deal with them, to minimize damage. For instance the post-thaw removal of cryoprotective compounds (*whether* they are removed, at what rate, at what temperature, and whether compounds and conditions that alleviate damage (or counteract loss of key growth factors) are used in the wash solution) presents a complex set of variables affecting osmotic trauma to membranes, availability of cell growth substances and, possibly, enzyme-complex molecular damage.

The effectiveness of such studies can be enhanced and their impact increased by detailed studies of the effects on cell ultrastructure. We have seen that even closely related lines of cells have different degrees of sensitivity to damage (Section IV.D) and that the different cryoprotective compounds have profoundly different effects on cells. These relationships need a great deal more study and interpretation of their complex interactions in protection (and disruption) of cell structure and the primary molecular sites of damage. For the latter, enzyme local localization techniques could be of great value.

B. Cryoprotective Recovery from Damage

The concept of maximizing the freeze-thaw conditions to optimize post-thaw repair of damaged sites has, thus far, hardly been considered. There are indications that recovery from damage does take place. (Sections V.D and E). Presumably the damaging effects observed, in addition to being prevented by particular prethaw and post-thaw treatments, could also be counteracted, and healing be encouraged, by conplementary reagents and conditions presented to the tissue during thawing and post-thawing manipulations. Studying how the degraded molecules and structures of stressed and damaged tissues are reorganized and reassembled and how the repair process could be enhanced, is on open field of future research.

C. Storage by Cryopreservative Procedures

As a greater capability and better understanding of cryoprotective procedures become evident, one begins to visualize their use on a routine, reliable basis in frozen gene banks. The methods used by Withers and King[178] and Ulrich et al.[3] (see Section IV.B) show promise of this. To make such a practice feasible, a few set procedures would be needed for most of the plant material to be preserved, perhaps a maximum of 2 to 3 standardized cryoprotective mixtures and freezing protocols for, as an estimate, 90% of the tissue to be routinely handled. If too many special procedures are required, routine cryopreservation of large numbers of clones would not be feasible. With better understanding and solutions to the basic problems of damage to and healing of cryogenically frozen tissues, such a goal might be accomplished as an economical, convenient, utilitarian[9] operation for long-term storage of genetic and somatic tissues and for experimental needs.[119]

D. Assessment of Particular Cryoprotective Agents

A primary criterion of cryoprotective compounds is that they should themselves be relatively nondamaging. An example — a particularly useful and intriguing one — is DMSO and the roles of both the positive and negative genetic effects attributed to it (Section III.C). The possible defects and uncertainties ascribed to DMSO must be balanced against the advantages to be gained in its usefulness for long-term cryogenic storage of plant tissues. The defects attributed to this useful compound may prove to be will-o-the-wisps, or a serious drawback, or something in between to be taken into account and dealt with (like the effects of cosmic rays). New work should be directed toward establishing on a solid basis the degree of genetic reliability to be secured during the cryogenic treatments and storage conditions that will normally be used and assessing the need for — and possible alternatives available to — using DMSO in cryoprotective procedures.

ACKNOWLEDGMENTS

The authors gratefully acknowledge the efforts and suggestions of A. P. MacKenzie, J. L. Bomben, D. S. Reid, R. Scherrer, and D. Cummings in reviewing this chapter.

REFERENCES

1. **Mazur, P.,** Physical and chemical basis of injury in single-celled microorganisms and subjected to freezing and thawing, in *Cryobiology,* Meryman, H. T., Ed., Academic Press, New York, 1966, chap. 6.
2. **Sakai, A. and Yoshida, S.,** The role of sugar and related compounds in variations of freezing resistance, *Cryobiology,* 5, 160, 1968.
3. **Ulrich, J. M., Finkle, B. J., Moore, P. H., and Ginoza, H.,** Effect of a mixture of cryoprotectants in attaining liquid nitrogen survival of callus cultures of a tropical plant, *Cryobiology,* 16, 550, 1979.
4. **Sakai, A.,** Seasonal variations in the amounts of polyhydric alcohol and sugar in fruit trees, *J. Hortic. Sci.,* 41, 207, 1966.
5. **Siminovitch, D., Rheaume, B., Pomeroy, K., and Lepage, M.,** Phospholipid, protein, and nucleic acid increases in protoplasm and membrane structures associated with development of extreme freezing resistance in black locust tree cells, *Cryobiology,* 5, 202, 1968.
6. **Olien, C. R. and Smith, M. N.,** Protective systems that have evolved in plants, in *Analysis and Improvement of Plant Cold Hardiness,* Olien, C. R. and Smith, M. N., Eds., CRC Press, Boca Raton, Fla. 1981, chap. 4.
7. **Maximow, N. A.,** Chemische Schutzmittel der Pflanzen gegen Erfrieren, *Ber. Dtsch. Bot. Ges.,* 52, 293, 504, 1912.
8. **Ashwood-Smith, M. J., and Friedmann, G. B.,** Lethal and chromosomal effects of freezing, thawing, storage time, and X-irradiation on mammalian cells preserved at $-196°$ in dimethyl sulfoxide, *Cryobiology,* 16, 132, 1979.
9. International Board Plant Genetic Resources, *Report of the Working Group on In Vitro Conservation,* Williams, J. T., Ed., IBPGR, Rome, 1983.
10. **Skaer, H. L. B., Franks, F., and Echlin, P.,** Non-penetrating polymeric cryofixatives for ultrastructural and analytical studies of biological tissues, *Cryobiology,* 15, 589, 1978.
11. **Mazur, P.,** Cryobiology: the freezing of biological systems, *Science,* 168, 939, 1970.
12. **Kartha, K. K.,** Cryopreservation of germplasm using meristem and tissue culture, in *Application of Plant Cell and Tissue Culture to Agriculture & Industry,* Tomes, D. T., Ellis, B. E., Harney, P. M., Kasha, K. J., and Peterson, R. L., Eds., University of Guelph, Guelph, Ontario, Canada, 1982, 139.
13. **Withers, L. A.,** *Tissue Culture Storage for Genetic Conservation,* Technical Report of the International Board for Plant Genetic Resources, Rome, 1980.
14. **Withers, L. A.,** Low temperature storage of plant tissue cultures, in *Plant Cell Cultures II, Advances in Biochemical Engineering,* Vol. 18, Feichter, A., Ed., Springer-Verlag, Basel, 1980, 101.
15. **Withers, L. A.,** Storage of plant tissue cultures, in *Crop Genetic Resources, the Conservation of Difficult Material,* Withers, L. A. and Williams, J. T., Eds., International Board for Plant Genetic Resources, Rome, 1980, 49.
16. **Morris, G. J.,** Plant cells, in *Low Temperature Preservation in Medicine and Biology,* Ashwood-Smith, M. J., and Farrant, J., Eds., University Park Press, Baltimore, 1980, 253.
17. **Levitt, J.,** Chilling, freezing, and high temperature stress, in *Responses of Plants to Environmental Stresses,* Vol. 1, 2nd ed., Academic Press, New York, 1980.
18. **Finkle, B. J. and Ulrich, J. M.,** Effects of combinations of cryoprotectants on the freezing survival of sugarcane cultured cells, in *Plant Cold Hardiness and Freezing Stress, Mechanisms and Crop Implications,* Li, P. H. and Sakai, A., Eds., Academic Press, New York, 1978, 373.
19. **Finkle, B. J.,** Freezing preservation, in *Biochemistry of Fruits and Their Products,* Vol. 2, Hulme, A. C., Ed., 1971, 653.
20. **Ashwood, M. J. and Farrant, J., Eds.,** *Low Temperature Preservation in Medicine and Biology,* University Park Press, Baltimore, 1980.
21. **Meryman, H. T. and Williams, R. J.,** Mechanisms of freezing injury and natural tolerance, and the principles of artificial cryoprotection, in *Crop Genetic Resources, the Conservation of Difficult Material,* Withers, L. A. and Williams, J. T., Eds., International Board for Plant Genetic Resources, 1980, 5.
22. **Meryman, H. T.,** Cryoprotective agents, *Cryobiology,* 8, 173, 1971.
23. **Luyet, B. and Rapatz, G.,** A review of basic researches on the cryopreservation of red blood cells, *Cryobiology,* 6, 425, 1970.
24. **Wolstenholme, G. E. W. and O'Connor, M., Eds.,** *The Frozen Cell,* Ciba Foundation Symposium, Churchill, London, 1970.
25. **Karow, A. M. Jr.,** Cryoprotectants — a new class of drugs, *J. Pharm. Pharmacol.,* 21, 209, 1969.
26. **Doebbler, G. F.,** Cryoprotective compounds, review and discussion of structure and function, *Cryobiology,* 3, 2, 1966.
27. **Rowe, A. W.,** Biochemical aspects of cryoprotective agents in freezing and thawing, *Cryobiology,* 3, 12, 1966.
28. **Meryman, H. T., Ed.,** *Cryobiology,* Academic Press, New York, 1966.

29. **Meryman, H. T.,** Modified model for the mechanism of freezing injury in erythrocytes, *Nature (London),* 218, 333, 1968.
30. **Simonovitch, D. and Briggs, D. R.,** The chemistry of the living bark of the black locust tree in relation to frost hardiness. I. Seasonal variations in protein content, in *Arch. Biochem.,* 23, 8, 1949.
31. **Volger, H. G. and Heber, V.,** Cryoprotective leaf proteins, *Biochim. Biophys. Acta,* 412, 335, 1975.
32. **Yelenosky, G.,** Accumulation of free proline in citrus leaves during cold hardening of young trees in controlled temperature regimes, *Plant Physiol.,* 64, 425, 1979.
33. **Goas, G., Goas, M., and Larher, F.,** Accumulation of free proline and glycine betaine in *Aster tripolium* subjected to a saline shock: a kinetic study related to light period, *Physiol. Plant.,* 55, 383, 1982.
34. **Woodcock, A. H., Thistle, M. W., and Cook, W. H., and Gibbons, N. E.,** The ability of sheep's erythrocytes to survive freezing, *Can. J. Res.,* D, 29, 206, 1941.
35. **Polge, C., Smith, A. V., and Parkes, A. S.,** Revival of spermatozoa after vitrification and dehydration at low temperatures, *Nature (London),* 164, 666, 1949.
36. **Lovelock, J. E. and Bishop, M. W. H.,** Prevention of freezing damage to living cells by dimethyl sulfoxide, *Nature (London),* 183, 1394, 1959.
37. **Lovelock, J. E.,** The protective action of neutral solutes against haemolysis by freezing and thawing, *Biochem. J.,* 56, 265, 1954.
38. **Lovelock, J. E. and Polge, C.,** The immobilization of spermatozoa by freezing and thawing and the protective action of glycerol, *Biochem. J.,* 58, 618, 1954.
39. **Nash, T.,** Chemical constitution and physical properties of compounds able to protect living cells against damage due to freezing and thawing, in *Cryobiology,* Meryman, H. T., Ed., Academic Press, New York, 1966, 179.
40. **Towill, L. E. and Mazur, P.,** Osmotic shrinkage as a factor in freezing injury in plant tissue cultures, *Plant Physiol.,* 57, 290, 1976.
41. **Mazur, P. and Miller, R. H.,** Permeability of the human erythrocyte to 1 and 2 M solutions at 0 or 20°C, *Cryobiology,* 13, 507, 1976.
42. **Molisch, H.,** Investigations into the freezing of plants (1897) (transl.), *Cryo-Lett.,* 2, 98, 1981.
43. **McGann, L. E.,** Differing actions of penetrating and cryoprotectiye agents, *Cryobiology,* 15, 382, 1978.
44. **Meryman, H. T., Williams, R. J., and Douglas, M. St. J.,** Freezing injury from "solution effects" and its prevention by natural or artificial *Cryobiology,* 14, 287, 1977.
45. **Mazur, P. and Miller, R. H.,** Survival of frozen-thawed human red cells as a function of the permeation of glycerol and sucrose, *Cryobiology,* 13, 523, 1976.
46. **Farrant, J.,** Is there a common mechanism of protection of living cells by polyvinylpyrrolidone and glycerol during freezing?, *Nature (London),* 222, 1175, 1969.
47. **Mazur, P., Leibo, S. P., Farrant, J., Chu, E. H. Y., Hanna, M. G., Jr., and Smith, L. H.,** Interactions of cooling rate, warming rate and protective additive on the survival of frozen mammalian cells, in *The Frozen Cell,* Wolstenholme, G. E. W. and O'Connor, M., Eds., Churchill, London, 1970, 69.
48. **Farrant, J., Walter, C. A., Lee, H., Morris, G. J., and Clarke, K. J.,** Structural and functional aspects of biological freezing techniques, *J. Microsc.* 111, 17, 1977.
49. **Nei, T.,** Structure and function of frozen cells: freezing patterns and post-thaw survivals, *J. Microsc.,* 112, 197, 1978.
50. **Morris, G. J., Coulson, G., and Clarke, A.,** The cryopreservation of *Chlamydomonas, Cryobiology,* 16, 401, 1979.
51. **Bajaj, Y. P. S. and Reinert, J.,** Cryobiology of plant cell cultures and establishment of gene-banks, in *Plant, Cell, Tissue, and Organ Culture,* Reinert, J. and Bajaj, Y. P. S., Eds., Springer-Verlag, Basel, 1977, 757.
52. **Connor, W. and Ashwood-Smith, M. J.,** Cryoprotection of mammalian cells in tissue culture with poly mechanisms, *Cryobiology,* 10, 488, 1973.
53. **Scheiwe, M. W., Nick, H. E., and Körber, C.,** An experimental study on the freezing of red blood cells with and without hydroxyethyl starch, *Cryobiology,* 19, 461, 1982.
54. **Heber, U.,** Freezing injury in relation to loss of enzyme activities and protection against freezing, *Cryobiology,* 5, 188, 1968.
55. **Heber, U., Tyankova, L., and Santarius, K. A.,** Effects of freezing on biological membranes *in vivo* and *in vitro, Biochim. Biophys. Acta,* 291, 23, 1973.
56. **Steponkus, P. L., Garber, M. P., Myers, S. P., and Lineberger, R. D.,** Effects of cold acclimation and freezing on structure and function of chloroplast thylakoids, *Cryobiology,* 14, 303, 1977.
57. **Palta, J. P. and Li, P. H.,** Alterations in membrane transport properties by freezing injury in herbaceous plants: evidence against rupture theory, *Physiol. Plant.,* 50, 169, 1980.
58. **Heber, U., Tyankova, L., and Santarius, K. A.,** Stabilization and inactivation of biological membranes during freezing in the presence of amino acids, *Biochim. Biophys. Acta,* 241, 578, 1971.
59. **Heber, U., Volger, H., Overbeck, V., and Santarius, K. A.,** Membrane damage and protection during freezing, *J. Am. Chem. Soc.,* 1979, p. 159.

60. **Lovelock, J. E.,** The haemolysis of human red blood cells by freezing and thawing, *Biochim. Biophys. Acta,* 10, 414, 1953.

61. **Meryman, H. T.,** The exceeding of a minimum tolerable volume in hypertonic suspension as a cause of freezing injury, in *The Frozen Cell,* Wolstenholme, G. E. W. and O'Conner, M. Eds., Churchill, London, 1970, 51.

62. **Garber, M. P. and Steponkus, P. L.,** Alterations in chloroplast thylakoids during an *in vitro* freeze-thaw cycle, *Plant Physiol.,* 57, 673, 1976.

63. **Lineberger, R. D. and Steponkus, P. L.,** Influence of sugars on freezing injury in chloroplast thylakoids, *Cryobiology,* 12(Abstr.), 565, 1975.

64. **Lineberger, R. D. and Steponkus, P. L.,** Cryoprotection by glucose, sucrose, and raffinose to chloroplast thylakoids, *Plant Physiol.,* 65, 298, 1980.

65. **Wiest, S. C. and Steponkus, P. L.,** Freeze-thaw injury to isolated spinach protoplasts and its simulation at above freezing temperatures, *Plant Physiol.,* 62, 699, 1978.

66. **Meryman, H. T.,** Review of biological freezing, in *Cryobiology,* Meryman, H. T., Ed., Academic Press, New York, 1966, 1.

67. **Williams, R. J. and Meryman, H. T.,** Freezing injury and resistance in spinach chloroplast grana, *Plant Physiol.,* 45, 752, 1970.

68. **Skaer, H. Le B., Franks, F., Asquith, M. H., and Echlin, P.,** Polymeric cryoprotectants in the preservation of biological ultrastructure, *J. Microsc.,* 110, 257, 1977.

68a. **MacKenzie, A. P.,** Non-equilibrium freezing behaviour of aqueous systems, *Phil. Trans. R. Soc. London Ser. B,* 278, 167, 1977.

69. **Boutron, P. and Kaufmann, A.,** Stability of the amorphous state in the system water-1,2-propanediol, *Cryobiology,* 16, 557, 1979.

70. **Boutron, P., Delage, D., Rousitt, B., and Körber, C.,** Ternary systems with 1,2-propanediol — a new gain in the stability of the amorphous state in the system water-1,2-propanediol-1-propanol, *Cryobiology,* 19, 550, 1982.

71. **Pribor, D. B.,** PVP contrasted with dextran and a multifactor theory of cryoprotection, *Cryobiology,* 11, 60, 1974.

72. **Pribor, D. B.,** Biological interactions between cell membranes and glycerol or DMSO, *Cryobiology,* 12, 309, 1975.

73. **Ruwart, M. J., Holland, J. F., and Haug, A.,** Fluorimetric evidence of interactions involving cryoprotectants and biomolecules, *Cryobiology,* 12, 26, 1975.

74. **Klein, E., Lyman, R. B., Jr., Peterson, L., Berger, R. I., and Smith, G. K.,** Effect of dimethyl sulfoxide on lipoproteins stored at low temperatures, *Cryobiology,* 3, 328, 1967.

75. **Pushkar, N. S., Sheenberg, M. G., and Oboznaya, E. I.,** On the mechanism of cryoprotection by polyethylene oxide, *Cryobiology,* 13, 142, 1976.

76. **Moiseyev, V. A., Nardid, O. A., and Belous, A. M.,** On a possible mechanism of the protective action of cryoprotectants, *Cryo-Lett.,* 3, 17, 1982.

77. **Ashwood-Smith, M. J. and Warby, C.,** Protective effect of low and high molecular weight compounds on the stability of catalase subjected to freezing and thawing, *Cryobiology,* 9, 137, 1972.

78. **Miller, J. S. and Cornwell, D. G.,** The role of cryoprotective agents as hydroxyl radical scavengers, *Cryobiology,* 15, 585, 1978.

79. **Farrant, J. and Woolgar, A. E.,** Human red cells under hypertonic conditions; a model system for investigating freezing damage, 2 sucrose, *Cryobiology,* 9, 16, 1972.

80. **Fahy, G. M.,** Analysis of ''solution effects'' injury: rabbit renal cortex frozen in the presence of dimethyl sulfoxide, *Cryobiology,* 17, 371, 1980.

81. **Rowe, A. W. and Lenny, L. L.,** Cryopreservation of granulocytes for transfusion: studies on human granulocyte isolation, the effect of glycerol on lysosomes, kinetics of glycerol uptake and cryopreservation with dimethyl sulfoxide and glycerol, *Cryobiology,* 17, 198, 1980.

82. **Finkle, B. J. and Ulrich, J. M.,** Effects of cryoprotectants in combination on the survival of frozen sugarcane cells, *Plant Physiol.,* 63, 598, 1979.

83. **Franks, F., Fowler, D., and Echlin, P.,** Effects of water soluble synthetic polymers on plant cell growth and division, *Cryo-Lett.,* 1, 153, 1980.

84. **Fink, A. L. and Cartwright, S. J.,** Cryoenzymology, in *Crit. Rev. Biochem.,* 11, 145, 1981.

85. **Farrant, J. and Woolgar, A. E.,** Possible relationships between the physical properties of solutions and cell damage during freezing, in *The Frozen Cell,* Wolstenholme, G. E. W. and O'Conner, M., Eds., Churchill, London, 1970, 97.

86. **MacKenzie, A. P.,** Macromolecular cryoprotection, *Cryobiology,* 17, 612, 1980.

87. **McClendon, J. H.,** The osmotic pressure of concentrated solutions of polyethylene glycol 6000, and its variation with temperature, *J. Exp. Bot.,* 32, 861, 1981.

88. **Levitt, J.,** Winter hardiness in plants, in *Cryobiology,* Meryman, H. T., Ed., Academic Press, New York, 1966, 495.

88a. **MacKenzie, A. P. and Rasmussen, D. H.,** Interactions in the water-polyvinylpyrrolidone system at low temperatures, in *Water Structure at the Water-Polymer Interface,* Jellinek, H. H. G., Ed., Plenum Press, New York, 1972, 146.

89. **Körber, C. and Scheiwe, M. W.,** The cryoprotective properties of hydroxyethyl starch investigated by means of differential thermal analysis, *Cryobiology,* 17, 54, 1980.

90. **Lovelock, J. E.,** The denaturation of lipid-protein complexes as a cause of damage by freezing, *Proc. R. Soc. London Ser.* B, 146, 427, 1957.

91. **Sakai, A.,** Survival of plant tissue at super-low temperatures. IV. Cell survival with rapid cooling and rewarming, *Plant Physiol.,* 41, 1050, 1966.

92. **Sakai, A., Otsuka, K., and Yoshida, S.,** Mechanism of survival in plant cells at super-low temperatures by rapid cooling and rewarming, *Cryobiology,* 4, 165, 1968.

93. **Körber, C., Scheiwe, M. W., Boutron, P., and Rau, G.,** The influence of hydroxyethyl starch on ice formation in aqueous solutions, *Cryobiology,* 19, 478, 1982.

94. **Boutron, P. and Kaufmann, A.,** Stability of the amorphous state in the system water-glycerol-dimethylsulfoxide, *Cryobiology,* 15, 93, 1978.

95. **Samygin, G. A. and Matveeva, N. M.,** Protective action of glycerol and of other substances readily penetrating the protoplasts during freezing of plant cells, *Soc. Plant Physiol.,* 14, 879, 1967.

96. **Bajaj, Y. P. S.,** Effect of dimethyl sulfoxide on the growth, RNA, and protein metabolism in bean callus cultures, *In Vitro,* 6, 242, 1970.

97. **Davis, D. G., Wergin, W. P., and Dusbabek, K. E.,** Effects of organic solvents on growth and ultrastructure of plant cell suspension, *Pest. Biochem. Physiol.,* 8, 84, 1978.

98. **Srivastava, K. C. and Smith D. G.,** Effect of cryo-protective agents on the growth and ultrastructure of *Saccharomyces cerevisiae, Cytobios,* 29, 29, 1980.

99. **Matthes, G. and Hackensellner, H. A.,** Correlations between purity of dimethyl sulfoxide and survival after freezing and thawing, *Cryo-Lett.,* 2, 389, 1981.

100. **Hervás, J. P. and Giménez-Martin, G.,** Dimethyl sulphoxide effect on division cells, *Experientia,* 29, 1540, 1973.

101. **Ihrke, C. A. and Kronstad, W. E.,** Genetic recombination in maize as affected by ethylenediaminetetraacetic acid and dimethyl sulfoxide, *Crop Sci.,* 15, 429, 1975.

102. **Bose, S. and Basu, S. K.,** Preliminary studies on the mitotic irregularities in onion (*Allium cepa* L.) after single and combined treatments with dimethyl sulfoxide and x-rays, *Indian Biol.,* 9, 44, 1977.

103. **Yarus, M.,** Why do organic solvents enhance mistakes in aminoacyl tRNA synthesis?, *Arch. Biochem. Biophys.,* 174, 350, 1976.

104. **Friend, C. and Freedman, H. A.,** Effects and possible mechanism of action of dimethylsulfoxide on friend cell differentiation, *Biochem. Pharmacol.,* 27, 1309, 1978.

105. **Kaul, B. L.,** Studies on radioprotective role of dimethyl sulfoxide in plants, *Rad. Bot.,* 10, 69, 1970.

106. **Barnett, B. M.,** Radioprotective effects of dimethyl sulfoxide in *Drosophila melanogaster, Rad. Res.,* 51, 134, 1972.

107. **Boutron, P., Kaufmann, A., and Van Dang, N.,** Maximum in the stability of the amorphous state in the system water-glycerol-ethanol, *Cryobiology,* 16, 372, 1979.

107a. **Finkle, B. J., Ulrich, J. M., and Tisserat, B.,** Responses of several lines of rice and date palm callus to freezing at − 196°C, in *Plant Cold Hardiness and Freezing Stress, Mechanisms and Crop Implications,* Vol. 2, Li, P. H. and Sakai, A., Eds., Academic Press, New York, 1982, 643.

108. **Morris, G. J. and Canning, C. E.,** The cryopreservation of *Euglena gracilis, J. Gen. Microbiol.,* 108, 27, 1978.

109. **Morris, G. J. and McGrath, J. J.,** A cryomicroscope study of the morphology of *Chlamydomonas reinhardii* during freezng and thawing, *Cryobiology,* 17(Abstr.), 624, 1980.

110. **Gresshoff, P. M.,** *Chlamydomonas reinhardi* — a model plant system, II. Cryopreservation, *Plant sci. Lett.,* 9, 23, 1977.

111. **Barnhart, E. R. and Terry, C. E.,** Cryobiology of *Neurospora crassa* II. Alteration in freeze response of *Neurospora crassa* conidia by additives, *Cryobiology,* 8, 328, 1971.

112. **Latta, R.,** Preservation of suspension cultures of plant cells by freezing, *Can. J. Bot.,* 49, 1253, 1971.

113. **Sugawara, Y. and Sakai, A.,** Survival of suspension-cultured sycamore cells cooled to the temperature of liquid nitrogen, *Plant Physiol.,* 54, 722, 1974.

114. **Finkle, B. J., Sugawara, Y., and Sakai, A.,** Freezing of carrot and tobacco suspension cultures, *Plant Physiol.,* 56(Suppl.), 80, 1975.

115. **Withers, L. A.,** The freeze-preservation of synchronously dividing cultured cells of *Acer pseudoplatanus* L., *Cryobiology,* 15, 87, 1978.

116. **Gazeau, C. M.,** Accroissement de la résistance au froid (jusqu á − 30°C) de plantules de Blé (*Triticum aestivum* L.) avec des substances protectrices. Effects de ces traitements sur les ultrastructures des cellules méristématiques racines, *Ann. Sci. Nat. Bot. (Paris),* 1, 97, 1979.

117. **Withers, L. A.,** The cryopreservation of higher plant tissue and cell cultures — an overview with some current observations and future thoughts, *Cryo-Lett.,* 1, 239, 1980.

118. **Bajaj, Y. P. S.,** Survival of anther-, and ovule-derived cotton callus frozen in liquid nitrogen, *Curr. Sci.,* 51, 139, 1982.

119. **Farkas, D. L. and Malkin, S.,** Cold storage of isolated class C chloroplasts. Optimal conditions for stabilization of photosynthetic activities, *Plant Physiol.,* 64, 942, 1979.

120. **Nag, K. K. and Street, H. E.,** Freeze preservation of cultured plant cells, I. The pretreatment phase, *Physiol. Plant.,* 34, 254, 1975.

121. **Withers, L. A. and King, P. J.,** Proline: a novel cryoprotectant for the freeze preservation of cultured cells of *Zea mays* L., *Plant Physiol.,* 64, 675, 1979.

122. **Grout, B. W. W., Westcott, R. J., and Henshaw, G. G.,** Survival of shoot meristems of tomato seedlings frozen in liquid nitrogen, *Cryobiology,* 15, 478, 1978.

123. **Ben-Amotz, A. and Gilboa, A.,** Cryopreservation of marine unicellular algae, II. Induction of freezing tolerance, *Marine Ecol. Prog. Ser.,* 2, 221, 1980.

124. **Morris, G. J. and Clarke, A.,** The cryopreservation of *Chlorella.* Accumulation of lipid as a protective factor, *Arch. Microbiol.,* 119, 153, 1978.

125. **Withers, L. A. and Street, H. E.,** Freeze preservation of cultured plant cells, III. The pregrowth phase, *Plant Physiol.,* 39, 171, 1977.

126. **Pritchard, H. W., Grout, B. W. W., Reid, D. S., and Short, K. C.,** The effects of growth under water stress on the structure, metabolism and cryopreservation of cultured sycamore cells, in *Biophysics of Water,* Franks, F., Ed., John Wiley & Sons, New York, in press.

127. **Gazeau, C. M.,** Effets de divers traitements cryoprotecteurs sur le développement de plantules de Blé (*Triticum aestivum* L.) et sur les ultrastructures des cellules méristématiques des racines, *Ann. Sci. Nat. Bot. (Paris),* 1, 3, 1979.

128. **Handa, A. K., Bressan, R. A., Handa, S., and Hasegawa, P. M.,** Characteristics of cultured tomato cells after prolonged exposure to medium containing polyethylene glycol, *Plant Physiol.,* 69, 514, 1982.

129. **Dougall, D. K. and Wetherell, D. F.,** Storage of wild carrot cultures in the frozen state, *Cryobiology,* 11, 410, 1974.

130. **Zavala, M. E. and Finkle, B. J.,** Effects of cryoadditives on cultured sugarcane cells, *J. Cell Biol.,* 87, 328a, 1980.

131. **Finkle, B. J., Ulrich, J. M., Schaeffer, G. W., and Sharpe, F., Jr.,** Cryopreservation of cells of rice and other plant species, in *Potentials of Cell and Tissue Culture Techniques in the Improvement of Cereal Plants,* Hu Han and Raday, N. C., Eds., Int. Rice Res. Institute, Manila, Philippines, in press.

132. **McGann, L. E., Kruuv, J., and Frey, H. E.,** Effect of hypertonicity and freezing on survival of unprotected synchronized mammalian cells, *Cryobiology,* 9, 107, 1972.

133. **Finkle, B. J. and Ulrich, J. M.,** Cryoprotectant removal temperature as a factor in the survival of frozen rice and sugarcane cells, *Cryobiology,* 19, 329, 1982.

134. **Withers, L. A.,** Freeze preservation of somatic embryos and clonal plantlets of carrot, *Plant Physiol.,* 63, 460, 1979.

135. **Lyons, J. M., Raison, J. K., and Steponkus, P. L.,** The plant membrane in response to low temperature: an overview, in *Low Temperature Stress in Crop Plants, The Role of the Membrane,* Lyons, J. M., Graham, D., and Raison, J. K., Eds., Academic Press, New York, 1979, 1.

136. **Van Zoelen, E. J. J., Van Der Neut-Kok, E. C. M., De Gier, J., and Van Deenen, L. L. M.,** Osmotic behaviour of *Acholeplasma laidlawii* B cells with membrane lipids in liquid-crystalline and gel state, *Biochim. Biophys. Acta,* 394, 463, 1975.

137. **Rubenstein, B.,** Regulation of H$^+$ excretion, *Plant Physiol.,* 69, 939, 945, 1982.

138. **Leder, I. G.,** Interrelated effects of cold shock and osmotic pressure on the permeability of the *Escherichia coli* membrane to permease accumulated substrates, *J. Bacteriol.,* 111, 211, 1972.

139. **Green, J. A.,** The effects of nutrient limitation and growth rate in the chemostat on the sensitivity of *Vibrio cholerae* to cold shock, *Fed. Eur. Microbiol. Soc. Microbiol. Lett.,* 4, 217, 1978.

140. **Meryman, H. T. and Hornblower, M.,** Changes in red cells following rapid freezing with extracellular cryoprotective agents, *Cryobiology,* 9, 262, 1972.

141. **Levitt, J.,** The moment of frost injury, *Protoplasma,* 48, 289, 1957.

142. **Sugawara, Y. and Sakai, A.,** Effect of Ca$+$$+$ on the survival of cultured plant cells after freezing and thawing, *Low Temp. Sci. Ser. B,* 33, 28, 1975.

143. **Palta, J. P., Levitt, J., and Stadelmann, E. J.,** Freezing injury in onion bulb cells, II. Post-thawing injury or recovery, *Plant Physiol.,* 60, 398, 1977.

144. **Law, P., Lepock, J. R., and Kruuv, J.,** Effect of protective agents on amount and repair of sublethal freeze-thaw damage in mammalian cells, *Cryobiology,* 16, 430, 1979.

145. **Morris, G. J., Clarke, K. J., and Clarke, A.,** The cryopreservation of *Chlorella.* III. Effect of heterotrophic nutrition on freezing tolerance, *Arch. Microbiol.,* 114, 249, 1977.

I. INTRODUCTION

Maintenance and preservation of germplasm collections representing a wide array of genetic diversity of cultivated plants and their wild relatives are considered to be essential for plant breeding programs. In other words, it is imperative to preserve all available germplasm resources for future needs. However, the demand placed on the world for increased food production tends to result in the production of high-yielding varieties of crop plants with a narrow genetic base. This approach, although attractive on a short-term basis, might have serious repercussions in the event of a major disease outbreak, or pest infestation, or environmental perturbation. For crop species whose germplasms are maintained in the form of seeds, the seed viability can be retained for long periods by storing them in sealed containers at $-18°C$ or less and at a moisture content of between 5 and 7%.[1] Seeds which can be stored in this manner and in which longevity is increased in a predictable manner by a decrease in temperature and moisture content have been described as showing "orthodox" viability characteristics.[2] In contrast, there are a number of species such as coffee, cacao, oil palm, coconut, rubber, etc. which produce seeds that cannot be dried without immediate injury and viability loss. Such seeds are classified as "recalcitrant" (see Table 1).[2]

Although genetic diversity in collections is extremely important, the trueness-to-type of each accession preserved is to be given due consideration. In species where the segregation results in the expression of undesirable traits in part of the population, maintenance of germplasm in the form of seeds might cast some doubt as to its validity unless the sample is rigorously tested to ensure the maintenance of homogeneity. Furthermore, since nature deals with a complexity of microorganisms in various forms of association with cultivated plants, some of the seed collections could carry internally seed-borne pathogens such as fungi, bacteria, and viruses which, in due course, would result in the deterioration or loss of viability of the seed collections. At the extreme end of the scale, there are certain crop species which do not produce viable seeds, thus making the preservation of those species very difficult by conventional means.

With regard to the crops that are vegetatively propagated preservation of germplasm presents special problems. Among the vegetatively propagated species, some annuals exhibit a high degree of heterozygosity and do not produce seeds. Such plants are conventionally maintained in the form of tubers and cuttings which possess only a very limited life span. Maintaining the vegetatively propagated crops in the field is the practice of choice often followed with disastrous consequences. In addition to being labor-intensive and expensive, this approach often results in loss of valuable genetic material due to diseases, pests, and other unforeseen problems. In view of the various problems associated with the preservation aspect of seed as well as vegetatively propagated plant material, alternative strategies are to be sought, and it is in this context that the application of in vitro techniques have to be examined.

In the domain of plant tissue culture (see Chapter 1) meristem culture techniques have occupied a unique position in terms of practical application. The technique has found its place not only in the clonal propagation, but also in the production of disease-free plants, especially those infected with viruses.[3-4] Furthermore, since the constituent cells of shoot apical meristems are less differentiated and genetically more stable, the progeny regenerated by in vitro culture of meristems have, so far, resulted in the maintenance of genetic stability in greater proportions when compared to other methods of in vitro plant regeneration. A possible exception would be the stability attained through somatic embryogenesis. Here, however, evidence for true-to-type preservation is still inadequate for arrival at any meaningful conclusion. These unique attributes make plant meristems ideal candidates for the preservation of germplasm. On the basis of information available, the area of germplasm

Table 1
ECONOMICALLY IMPORTANT PLANTS WITH RECALCITRANT SEEDS[30]

Family	Scientific name	Common name
Anacardiaceae	*Mangifera indica*	Mango
Bombacaceae	*Durio zibethinus*	Durian
Dipterocarpaceae	*Dryobalanops aromatica*	Borneo camphor
Erythroxylaceae	*Erythroxylum coca*	Cocaine
Euphorbiaceae	*Hevea brasiliensis*	Rubber
Fagaceae	*Castanea* spp.	Chestnut
Fagaceae	*Quercus* spp.	Oak
Gramineae	*Zizania aquatica*	Wild rice
Guttiferae	*Garcinia mangostana*	Mangosteen
Juglandaceae	*Carva illinoensis*	Pecan
Juglandaceae	*Juglans* spp.	Walnut
Lauraceae	*Cinnamomum* sp.	Cinnamon
Lauraceae	*Persea americana*	Avocado
Meliaceae	*Swietenia* sp.	Mahogany
Moraceae	*Artocarpus heterophyllus*	Jackfruit
Palmaceae	*Cocos nucifera*	Coconut
Rubiaceae	*Coffea arabica*	Coffee
Rubiaceae	*Coffea canephora*	Coffee
Rutaceae	*Citrus* spp.	Orange, lemon, grapefruit, etc.
Sterculiaceae	*Theobroma cacao*	Cacao
Theaceae	*Thea sinensis*	Tea
Zingiberaceae	*Elettaria cardamomum*	Cardamom

preservation discussed here falls distinctly into two categories: short- to medium-term preservation and long-term preservation by cryogenic methods.

II. SHORT- TO MEDIUM-TERM STORAGE OF GERMPLASM

Based on the regeneration of genetically identical plants in a disease-free condition from cultured meristems or shoot-tips, it becomes possible to maintain plantlets in vitro in a viable condition for varying periods of time. However, one of the basic requirements to be considered before this kind of in vitro storage is attempted would be to test high-frequency regeneration of plantlets from the in vitro cultured meristems or shoot-tips. Although plantlet formation has been accomplished with a number of agronomically important species, there are a large number of species, especially those classified as recalcitrant (Table 1) which have received little attention in this regard. Introducing these species to in vitro conditions and maintaining them in a state of growth alone will be highly desirable from the point of view of germplasm conservation. The advantages of storing meristem cultures in vitro over seed banks are particularly salient in the case of woody materials; shorter recycling times are needed to maintain the collections. With seeds even of annual plants, a complete growing season of several months is needed and care must be taken to prevent inbreeding depression, genetic drift, and contamination with pollen from other accessions in order to maintain the genetic integrity of the samples.[5] With shrubs and trees, which have regeneration cycles of many years, the contrast becomes still more impressive, and the need for in vitro storage strategies becomes more pressing.

Several approaches may be followed for maintaining germplasm collections in vitro (Table 2). Plant species which have an inherent slow rate of growth and proliferation could simply be maintained in that manner in vitro. An example of such a storage system may be found in the case of coffee. Coffee plantlets regenerated from in vitro cultured meristems on a

Table 2
IN VITRO STORAGE STRATEGIES FOR SHORT- TO MEDIUM-TERM PRESERVATION OF PLANT GENETIC MATERIAL

Species	In vitro storage methods	Storage duration	Ref.
Apple (*Malus domestica*)	Shoot-tips of apples cultured on basal medium + 5 mg/ℓ benzylaminopurine and stored at 1°C or 4°C	1 year	29
Cassava (*Manihot esculenta*)	Nodal cuttings from meristem-derived plantlets maintained in large vessels on nutrient medium with low osmotic concentration and activated charcoal	2 years	9
	Meristem-derived plants on filter paper bridges in tubes containing nutrient medium, at 20°C, 16 hr photoperiod, 100—200 lx intensity	1 year	30
Coffee (*Coffea arabica*)	Meristem-derived plants on 1/2 MS medium lacking in sucrose, plantlets kept in tubes at 26°C, 16 hr photoperiods at 4000 lx intensity	2½ years	6
Forage grasses *Dactylis* spp. *Festuca* sp. *Phleum* sp.	Plantlets maintained at 2—4°C	ca. 1 year	31
Lolium multiflorum	Shoot-tip plantlets stored at 2—4°C, 300 lx, 3 hr photoperiod	1 year	32
Forage legumes *Medicago* sp.	Plantlets in vitro stored at 2—6°C	15—18 months	33
Trifolium sp.	Plantlets in vitro stored at 2—6°C	15—18 months	33
Grape (*Vitis rupestris*)	Maintenance of shoot-tip cultures at 9°C	1 year	34
Potato (*Solanum tuberosum*)	Storage of meristem-derived plants at reduced temperature in the range of 6—10°C with a mean survival of 61% at 6°C; survival further increased to 83% when stored at 12°C day/6°C night	1 year	35
Strawberry (*Fragaria* spp.)	Meristem-derived plantlets stored at 4°C in dark with addition of 1 or 2 drops of culture media every 3 months	6 years	36
Sugarbeet (*Beta vulgaris*)	Cultures stored at 12°C	18 weeks	37
Sweet potato (*Ipomoea batatas*)	Meristem-derived plantlets on MS medium with 1.0 μM each of BA and NAA + 5% each of sucrose and mannitol, at 26°C, 16 hr photoperiods and 4000 lx intensity	Drastic growth reduction within 3 months	30

nutrient medium could be maintained for several months and storage duration could then be increased to 2 or 2-1/2 years by growing the plantlets on a medium with reduced nutrients and lacking in sucrose.[6] On the other hand, for plant species which proliferate excessively in vitro, their uninterrupted growth would impose on cultures limiting factors such as nutrients, aeration, space, and elaboration of various chemicals into the culture medium. Therefore, most attempts in the in vitro storage of plant material have forcused on imposing some sort of stress on the cultures to achieve reduced growth or minimal conditions of growth. Growth limitation in culture has been achieved in several ways: (1) by inducing osmotic stress through the application of osmotica such as sucrose or mannitol; (2) by supplying the carbon source at sub- or supra-optimal levels; (3) by maintaining the cultures at reduced temperature and light; (4) by incorporating growth retardants into the culture medium, and (5) by introducing compounds which are thought to confer some degree of chilling resistance to the cultures. Combinations of these methods have also been used with varying degrees of success (Table 2).

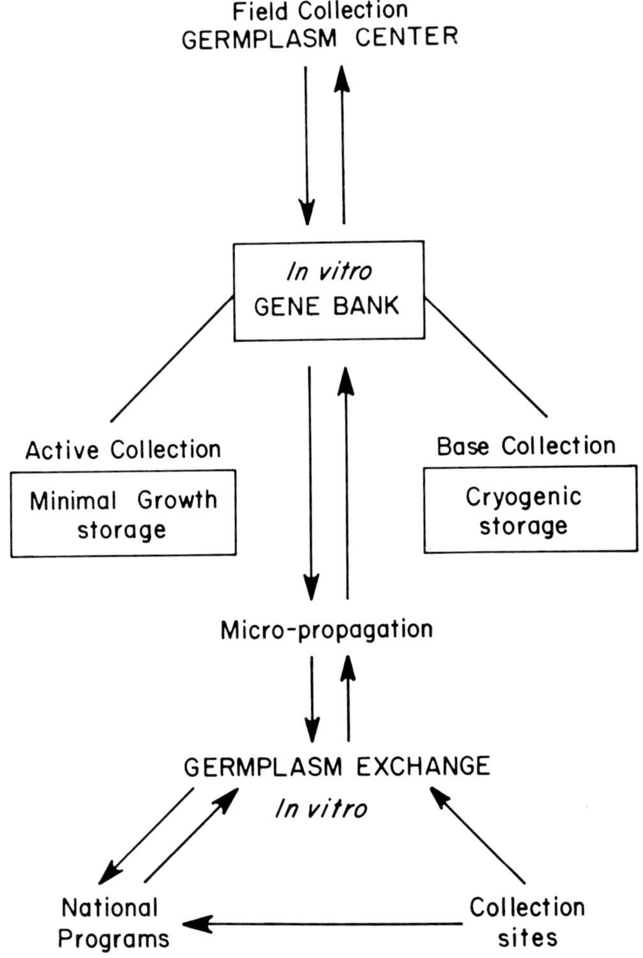

FIGURE 1.   Schematic representation for the utilization of in vitro cultures and cryopreservation in the conservation and international exchange of cassava clones.[9]

The advantages of using in vitro storage techniques are enormous, especially for crops which are vegetatively propagated and for species whose seeds cannot be stored by conventional means. Since the germplasm is conserved in a viable but slow-growing state, it could be examined periodically for any deterioration in vigor and quality and could also be grown into plants in case of immediate demand. As shown for potato and cassava, plant material grown in vitro is also useful for international exchange of genetic material.[7-8] So far, over 1300 cassava accessions of CIAT's germplasm collection have been transformed into in vitro cultures for storage, and it has been anticipated that nearly 6000 clonal accessions can be stored in a 5 × 6 × 3 m room, using 5 vessels of 50 × 140 mm per accession.[9] Roca envisages a dual conservation system for cassava comprising a base collection under cryogenic storage and an active or working collection under minimal growth conditions in vitro (see Figure 1 for storage strategy). In spite of the obvious advantages, these in vitro storage methods can only satisfy short- to medium-term needs. Problems arising out of loss of vigor of plantlets, microbial contamination of cultures, depletion of nutrients from a medium, and possibly, progressive changes of the genome, although surmountable, cannot be ignored. Also, the size of the task of maintaining all valuable in vitro collections of vegetatively propated species is indeed considerable, in particular if they do not have the propensies for mass proliferation upon reculture.

III. LONG-TERM PRESERVATION OF GERMPLASM (CRYOPRESERVATION)

As mentioned in the preceding section, it is highly desirable to have storage techniques available which permit long-term preservation of germplasm, and so, in recent years, attention has been focused on the utilization of cryogenic techniques. Cryopreservation is based on the reduction and subsequent arrest of metabolic functions of biological material by the reduction in temperature approaching that of liquid nitrogen while maintaining viability. At the temperature of liquid nitrogen ($-196°C$) almost all the metabolic activities of cells are at a standstill and they can be preserved in such a state for extended periods. Only very few biological materials, however, can be frozen to subzero temperatures without adversely affecting cell viability. Detailed information concerning cryopreservation of several types of plant material can be found in various chapters in this book. However, it should be mentioned that the discovery of chemicals with cryoprotective properties (see Chapter 2) paved the way for the development of effective cryopreservation techniques. A major breakthrough was the detection of glycerol as a cryoprotectant for freezing avian spermatozoa in the now-classical work of Polge et al.[10] Their observations generated widespread enthusiasm among persons interested in low-temperature preservation as it applies to biology and medicine. Since then, a number of low-molecular-weight compounds have been identified as potential cryoprotectants, the most commonly used ones being dimethylsulfoxide (DMSO) and glycerol. Dimethylsulfoxide was originally used to prevent freezing damage to human and bovine red blood cells and bull spermatozoa.[11] Cryopreservation of plant cells and organs is still in its infancy and only limited information is available concerning the various aspects of temperature reduction in relation to viability loss. Although there is a general awareness of the potential application of this technique to the long-term preservation of germplasm, in reality the actual number of species to which this technique is successfully applied is rather small (Table 3). The role of cryoprotectants in the viable freezing of plant cells is discussed in Chapter 5.

A. Cryopreservation of Plant Meristems
1. General Considerations

One of the basic requirements prior to cryopreservation of plant meristems, is the availability or development of an efficient and reliable protocol for faithful regeneration of plants from in-vitro-cultured meristems or shoot-tips. This consideration would obviously narrow the choice of plant material currently available for cryopreservation unless research efforts are extended to include many other plant species, so far untested. The expected regeneration process should preferably originate from the meristem tissue and not from the component cells via a callus-mediated organogenetic pathway since the latter has been proved to induce genetic abnormalities in the population of the regenerants.[12] Any form of genetic aberration affecting the trueness-to-type of the germplasm should be viewed with extreme caution. It would be desirable to use shoot-tips or meristems from plantlets maintained in vitro since replicates of such cultures could be employed for in vitro storage. In addition to this, the high degree of variation in viability obtained with cryopreserved meristems[13-14] could be minimized by uniform application of cultural conditions to the donor plants. The first published report of the cryopreservation of meristems of in-vitro-maintained plantlets concerned the strawberry.[15] Recently, similar attempts with two cultivars of potato have also shown promising results.[16] The main obstacle underlying wide application of this approach lies in the lack of availability of in vitro systems which permit mass proliferation of shoots in culture. If only single-shoot cultures were used for the isolation and cryopreservation of meristems, large numbers of plants would need to be maintained.

A number of simple neutral solutes such as glycerol, other hydroxylic compounds, DMSO, and pyridine-*N*-oxide[17] protect living cells against damage during freezing and thawing. Such solutes lower the temperature at which freezing first occurs and alter the crystal habit of ice when it separates. The colligative properties of the cryoprotectants minimize the deleterious action of electrolyte concentration resulting from conversion of water into ice.[17] The essential characteristics of cryoprotectants and the manner in which they interact with the cell during freezing are dealt with in Chapters 2 and 5.

The point that is of immediate relevance in the cryopreservation of plant meristems is the extent of cytotoxic effects these chemicals exert on the material. Most of the cryoprotectants exhibit varying degrees of cytotoxicity at higher concentrations. Cytotoxicity is manifested in various degrees, ranging from complete killing to modification in the morphogenetic response of cells in culture. In previous experiments, preculturing the meristems prior to cryopreservation on plant regeneration medium, supplemented with nontoxic levels of cryoprotectants (DMSO), improved the postfreezing viability considerably.[15,18] The extent of sensitivity of plant meristems to a few cryoprotectants is summarized in Table 4.[4] In all cases, higher concentrations of these compounds exerted cytotoxicity, the extent of which depended upon the plant species in question and the duration of exposure. Cytotoxicity can occur also when meristems are exposed to low concentrations over long periods. The results presented in Table 4 simply serve to illustrate the range of effects of these chemicals. In practice, the meristems are not incubated at such elevated levels for such long durations. Some of the chemicals, DMSO and methanol, permeate more rapidly into the cells than a compound such as glycerol. Accordingly, the concentration of the cryoprotectants and the duration of the cells exposure to them need to be controlled in such a manner as not to cause sudden plasmolysis. The methods of application and removal of cryoprotectants are discussed in detail in Chapter 2. Reducing the water content in cells prior to freezing can considerably improve their survival subsequent to freezing and thawing. For example, gradual osmotic dehydration of cassava meristems with sorbitol, increasing from an initial osmolality of 0.4 to 1.0 M over a 24-hr period, followed by treatment with nontoxic levels of DMSO (5%), conditions the cells to go through the freezing and thawing process more successfully than without such treatments.[19] However, the effect of osmotic dehydration induced in itself could adversely affect the viability of cells of certain species since they fail to deplasmolyse upon return to isotonic solutions. Therefore, the tolerance levels of cells to induced dehydration need to be determined before this approach is considered. Chapters 2 and 3 deal extensively with the effect of cell-volume reduction as it applies to the maintenance of cell viability.

Another important factor which is very critical in cryopreservation is the rate of cooling or freezing methods. The following section deals with this aspect as it relates to the cryopreservation work done so far with meristems of a few plant species.

2. Freezing Methods
a. Slow Freezing
Methods of freezing plant meristems slowly are based on the physicochemical events occurring during freezing. Mazur[20] has identified the factors to which a cell is subjected during freezing and thawing. With progressive temperature reduction, the cell and its external medium initially supercool, followed by ice formation in the external milieu. In the case of plant cells, the cell wall and the plasma membrane act as barriers and prevent the ice from seeding the cell interior at temperatures above ca. $-10°C$, and thus the cell remains unfrozen but supercooled. As the temperature is further lowered, an increasing fraction of extracellular solution is converted into ice, resulting in the concentration of extracellular solutes. Since the cell remains supercooled and its aqueous vapor pressure exceeds that of the frozen exterior, the cell equilibrates by loss of water to external ice (dehydration). Slowly cooled cells reach equilibrium with the external ice by efflux of water and will remain shrunken,

Table 3
A BRIEF SUMMARY OF CRYOPRESERVATION PROTOCOL FOR MERISTEMS OR SHOOT-TIPS

Species	Specimen preparation and cryoprotectants	Freezing methods	Viability	Ref.
Apple (*Malus domestica*)	Hardened for 20 days at −3°C	Slow freezing at −30 to −40°C, LN storage	100% regrowth and 77% budding success	38
Other 15 *Malus* spp.	Buds from hardy trees	Slow freezing to −40°C without cryoprotectants, LN storage	100% regrowth	39
	Suspended in water containing 10% DMSO	Prefrozen slowly to −10 and −70°C, LN storage	100% regrowth	39
Carnation (*Dianthus caryophyllus*)	5% DMSO	Rapid freezing in LN	15—33% callus,shoots	24
	Meristems from cold-hardened donor plants, 5% DMSO	ca. 50°C/min	100% plant regeneration	26
	10% DMSO + 5% glucose	Slow freezing to −15°C or below, LN storage	100% plant regeneration	40
Cassava (*Manihot esculenta*)	Glycerol 10% + sucrose 5%	Rapid freezing in LN	13% plantlets, 8% callus + roots	41
	DMSO, glycerol, sucrose, sorbitol, mannitol, etc., in various concentrations and combinations	Rapid freezing in LN	Unreliable, only callus formation in a few isolated instances	4, 13
	15% DMSO + 3% sucrose	Each meristem frozen in 2—3 μℓ of cryoprotectant solution on aluminum foil at 0.5°C/min to −25 or −30°C, LN storage	16—86% regrowth, plantlets, callus, and callus + leaves	13
Chickpea (*Cicer arietinum*)	Precultured for 24 hr on medium with 4% DMSO and frozen in 4% DMSO	Slow freezing at 0.6°C/min to −40°C, LN storage	40% plant regeneration	42
	Precultured 3—5 days on nutrient medium, frozen in 5% each of sucrose, glycerol, and DMSO	Rapid freezing in LN	27—36% regrowth	43
Pea (*Pisum sativum*)	2-day preculture on medium with 5% DMSO, frozen in 5% DMSO	Slow freezing at 0.6°C/min to −40°C, LN storage	Over 60% plant regeneration	18

Species	Treatment	Freezing method	Results	Ref.
	Precultured 2 days on medium with 5% DMSO or cold-hardened at 0°C for 10 days followed by 2 days preculture on 5% DMSO medium	Slow freezing at 0.5°C/min to −40°C, LN storage	100% survival	44
Peanut (*Arachis hypogaea*)	3—5 day preculture on medium, 5% each of sucrose, glycerol, and DMSO	Rapid freezing in LN	23—31% regrowth shoots, roots, or both)	43
Potato (*Solanum goniocarlyx*)	72 hr preculture on nutrient medium, frozen in 10% DMSO	Rapid freezing in LN	20% survival (callus + shoots)	45
(*Solanum tuberosum*)	5% glycerol	Rapid freezing in LN	18% survival	46
	5% each of glycerol, sucrose, and DMSO	Rapid freezing in LN	26% plant regeneration	47
	Precultured on nutrient medium 73 hr; frozen in 10% DMSO	Rapid freezing in LN	60% survival and plant regeneration	48
	Precultured on nutrient medium for 2 days, frozen in 10% DMSO	Slow freezing at 0.3—0.4°C/min to −30° or −40°C, LN storage	12—80% shoot regeneration depending upon cultivars	49
(*Solanum etuberosum*)	Precultured for 2 days on nutrient medium, frozen in 10% DMSO	0.3 to 0.4°C/min to −40°C, LN storage	60—100% shoot regeneration	14
Strawberry (*Fragaria* x *ananassa*)	Runner-tip meristems frozen in 3% sucrose + 10—16% DMSO	Prefrozen at −20 to −30°C, immersed in LN	60—80% shoot regeneration	50
(*Fragaria* x *ananassa*)	Meristems from in-vitro propagated shoots precultured 2 days in 5% DMSO, frozen in 5% DMSO	Slow freezing at 0.84°C/min to −40°C, LN storage	95% plant regeneration, profuse multiple shoot regeneration	15
Tomato (*Lycopersicon esculentum*)	10—15% DMSO	Liquid nitrogen vapor freezing (20—50°C/min)	30% callus and plants	51

Table 5
VIABILITY AND GROWTH RESPONSES OF CASSAVA MERISTEMS CRYOPRESERVED BY DROPLET-FREEZING METHOD

Prefreezing temperature	Cryoprotectants	Viability	Growth responses
−25°C	15% DMSO + 3% sucrose	16—80%	Plantlets, callus, and callus + leaves
−30°C	15% DMSO + 3% sucrose	14—86%	Plantlets, callus, and callus + leaves
−40°C	15% DMSO + 3% sucrose	16—25%	Mostly callus or callus + leaves, plantlets in a limited number of cases
Nil (rapid freezing)	5% DMSO	0	Callus formation in a few isolated instances, no plant regeneration

Note: Meristems were frozen at a cooling rate of 0.5°C/min to the indicated temperatures and stored in liquid nitorgen for 1 hr. No viability in controls frozen without cryoprotectants.

be prefrozen prior to liquid nitrogen storage. Although maximal survival and plant regeneration were noted for meristems cooled only to −20°C, this temperature was not acceptable for cryopreservation since the freezing was not complete due to the high concentration of cryoprotectants used. At −30°C, the meristems were completely frozen and yet maintained a high rate of viability (>90%), while at −40°C the viability dropped to a mere 3%. The results obtained from meristems stored in liquid nitrogen after prefreezing to −25, −30, and −40°C are presented in Table 5. The viability of meristems prefrozen to −25° or −30°C, followed by liquid nitrogen storage, were almost identical. The meristems frozen to −40°C prior to liquid nitrogen storage exhibited the lowest rate of survival. A high degree of variation in viability was observed between experiments, the reason for which is unclear and the explanation hard to give, except to attribute it to the heterozygosity of the plant. Also, cassava is a tropical plant and very sensitive to chilling. It is also pointed out here that similar variations in viability results have been reported with cryopreserved potato-shoot apices.[14] Another problem yet to be resolved with cryopreserved cassava meristems is that of increasing the number of whole plantlets regenerated in vitro from meristems exhibiting signs of viability and regrowth in the form of callus and leaf differentiation. It is emphasized here that in order for the technique to be of immediate application for the cryogenic storage of cassava collections, the plant recovery rate needs considerable improvement. Research efforts are directed, in this laboratory, toward finding alternate means of reducing the toxicity level of DMSO (15%) which in itself is attributable to the modification of morphogenetic responses.

B. Storage, Viability, and Variability

At the temperature of liquid nitrogen, all the metabolic activities of meristem cells are at a standstill and therefore the meristems can be preserved in such a state for extended periods of time. In the examples shown in Table 3, meristems of only a few plant species have been stored in liquid nitrogen for any significant period of time, the reason being that if the meristems can withstand liquid nitrogen temperature for the initial storage duration of even a few minutes, their prolonged storage should be of no serious consequence with regard to the retention of viability. We have stored pea and strawberry meristems in liquid nitrogen for over 2 years and were able to regrow plants from them at a very high success rate (Figures 6 and 7). In the case of cryopreserved strawberry meristems, a fluctuation in viability has been noticed after various periods of liquid nitrogen storage.[15] This fluctuation has been attributed to the undefined variation in the physiological status of the meristems used in

A   B

FIGURE 6. Plant regeneration from pea meristems stored in liquid nitrogen for over 2 years. The meristems, precultured on 5% DMSO-supplemented nutrient medium for 48 hr, were frozen at a cooling rate of 0.6°C/min to −40°C and stored in liquid nitrogen. After thawing at 37°C, (A) the meristems were induced to differentiate into plantlets[18] and (B) grown to maturity.

different experiments and not to the loss of viability subsequent to liquid nitrogen storage as had been thought initially.

A large number of strawberry plants raised from cryopreserved meristems were field-tested for a period of over three generations of vegetative cycle. All the plants survived the extreme winter conditions prevalent in Saskatchewan, where a temperature drop to as low as −40°C or more is not unusual. No evidence of "winter kill" was apparent in any of the plants, since they all produced, on average, eight to ten runners per plant every growing season. Moreover, the plants exhibited vigorous growth habits and produced fruits of excellent quality. No abnormalities were noticed in any of the plants examined. Based on this encouraging observation, it could be stated with reasonable certainty that the cryopreservation process per se does not lead to any preferential selection within an existing cell population in the strawberry. Similarly, Towill[16] examined 1000 regenerants from cryopreserved meristems of two cultivars of potato (*Solanum tuberosum*), Red Pontiac, and Norland, in order to detect any variants that might have arisen as a result of the cryopreservation process. He also concluded that the cryopreservation process does not lead to the production of abnormal plants since all plants were normal except one. The production of abnormal plants in low frequency is not uncommon to any progeny derived from in-vitro-manipulated cells, tissues, or organs and therefore should cause no serious concern.

IV. CONCLUSIONS

Preservation of valuable germplasm of cultivated crops and their wild relatives is imperative to the maintenance of a broad genetic base in order to meet future demands. It has been demonstrated conclusively that meristem culture is a reliable tool in producing virus-free plants in a genetically unaltered condition. Therefore, meristem culture could be used in the short- to medium-term preservation of germplasm of recalcitrant and vegetatively propagated crops. Plants species which possess an inherent slow rate of growth and proliferation can be maintained in that manner in vitro. Cultures which grow or proliferate rapidly

FIGURE 7.   Morphogenetic response of strawberry meristems stored in liquid nitrogen for over 2 years. The meristems, precultured for 48 hr on 5% DMSO-supplemented nutrient medium, were frozen at a cooling rate of 0.84°C/min to −40°C and stored in liquid nitrogen. (A) The thawed meristems were cultured in vitro to induce shoot formation. (B and C) Each regenerated shoot, on a mass proliferation medium, forms hundreds of new shoots. These individual shoots (C) could be rooted to form plantlets or used in further mass propagation.[15] (D) The plantlets were grown in the field for over three generations of vegetative cycle.

may be stored in vitro for a short to medium term by imposing on the cultures various forms of stress to bring about growth limitation.

In recent years, there has been increased interest in combining meristem culture and cryobiological techniques for long-term preservation of crop species. Meristems or shoot-tips of a few plant species — namely, apple, potato, pea, chickpea, strawberry, cassava, etc. — have been successfully cryopreserved. Meristems of pea and strawberry were maintained in liquid nitrogen for over 2 years with no decline in viability. The results of the field performance of strawberry and potato plants raised from cryopreserved meristems are very encouraging. The work done so far in the cryopreservation of meristems offers some room for optimism with regard to the application of the technique in the long-term preservation of germplasm. At the same time, caution should be expressed concerning the general applicability of the technique to include all the plant species.

REFERENCES

1. International Board of Plant Genetic Resources, *Report of Working Group on Engineering, Design and Cost Aspects of Long-term Seed Storage Facilities,* IBPGR, Rome, 1976.
2. **Roberts, E. H.,** Predicting the storage life of seeds, *Seed Sci. Technol.,* 1, 499, 1973.
3. **Kartha, K. K.,** Meristem culture and cryopreservation — methods and applications, in *Plant Tissue Culture,* Thorpe, T. A., Ed., Academic Press, New York, 1981,181
4. **Kartha, K. K.,** Cryopreservation of germplasm using meristem and tissue culture, in *Application of Plant Cell and Tissue Culture to Agriculture and Industry,* Tomes, D. T., Ellis, B. E., Harney, P. M. J., Kasha, K. J., and Peterson, R. L., Eds., University of Guelph, Guelph, Ontario, Canada, 1982, 139.
5. **Hawkes, J. G.,** Genetic conservation of "recalcitrant species" — an overview, in *Crop Genetic Resources, The Conservation of Difficult Material,* Withers, L. A. and Williams, J. T., Eds., Int. Union Biol. Sci. Series B42, IUBS, Paris, 1980, 93.
6. **Kartha, K. K., Mroginski, L. A., Pahl, K., and Leung, N. L.,** Germplasm preservation of coffee (*Coffea arabica* L.) by *in vitro* culture of shoot apical meristems, *Plant Sci. Lett.,* 22, 301, 1981.
7. **Roca, W. M., Bryan, J. E., and Roca, M. R.,** Tissue culture for the international transfer of potato genetic resources, *Am. Potato J.,* 56, 1, 1979.
8. **Roca, W. M.** Genetic resources unit: tissue culture section, *Annual Report,* CIAT, Columbia, 1979, 1.
9. **Roca, W. M., Reyes, R., and Beltrań, J.,** Effect of various factors on minimal growth in tissue culture storage of cassava germplasm, in Proc. 6th Symp. Int. Soc. Trop. Root Crops, Lima, Peru, February 21, 1983
10. **Polge, C., Smith, A. U., and Parkes, A. S.,** Revival of spermatozoa after vitrification and dehydration at low temperatures, *Nature (London),* 164, 666, 1949.
11. **Lovelock, J. E. and Bishop, M. W. H.,** Prevention of freezing damage to living cells by dimethylsullfoxide, *Nature (London),* 183, 1394, 1959.
12. **D'Amato, F.,** The problem of genetic stability in plant tissue cell cultures, in *Crop Genetic Resources for Today and Tomorrow,* Frankel, O. and Hawkes, J. G., Eds., Cambridge University Press, New York, 1975, 333.
13. **Kartha, K. K., Leung, N. L., and Mroginski, L. A.,** *In vitro* growth responses and plant regeneration from cryopreserved meristems of cassava (*Manihot esculenta* Crantz), *Z. Pflanzenphysiol.,* 107, 133, 1982.
14. **Towill, L. E.,** *Solanum etuberosum:* a model system for studying the cryobiology of shoot-tips in the tuber-bearing *Solanum* species, *Plant Sci. Lett.,* 20, 315, 1981.
15. **Kartha, K. K., Leung, N. L., and Pahl, K.,** Cryopreservation of strawberry meristems and mass propagation of plantlets, *J. Am. Soc. Hortic. Sci.,* 105, 481, 1980.
16. **Towill, L. E.,** Improved survival after cryogenic exposure of shoot-tips derived from *in vitro* plantlet cultures of potato, *Cryobiology,* 20, 567, 1983.
17. **Nash, T.,** Chemical constitution and physical properties of compounds able to protect living cells against damage due to freezing and thawing, in *Cryobiology,* Meryman, H. T., Ed., Academic Press, New York, 1960, 197.
18. **Kartha, K. K., Leung, N. L., and Gamborg, O. L.,** Freeze-preservation of pea meristems in liquid nitrogen and subsequent plant regeneration, *Plant Sci. Lett.,* 15, 7, 1979.
19. **Kartha, K. K.** Unpublished data, 1983
20. **Mazur, P.,** Freezing injury in plants, *Ann. Rev. Plant Physiol.,* 20, 419, 1969.
21. **Farrant, J.,** General observations on cell preservation, in *Low Temperature Preservation in Medicine and Biology,* Ashwood-Smith, M. J. and Farrant, J., Eds., Unversity Park Press, Baltimore, 1980, 1.
22. **Gamborg, O. L., Miller, R. A., and Ojima, K.,** Nutrient requirements of suspension cultures of soybean root cells, *Exp. Cell Res.,* 50, 151, 1968.
23. **Murashige, T. and Skoog, F.,** A revised medium for rapid growth and bioassays with tobacco tissue cultures, *Physiol. Plant.,* 15, 473, 1962.
24. **Seibert, M.,** Shoot initiation from carnation shoot apices frozen to $-196°C$, *Science,* 191, 1178, 1976.
25. **Luyet, B. J.,** The vitrification of organic colloids and of protoplasms, *Biodynamica,* 1, 1, 1937.

26. **Seibert, M. and Wetherbee, P. J.,** Increased survival and differentiation of frozen herbaceous plant organ cultures through cold treatment, *Plant Physiol.,* 59, 1043, 1977.

27. **Withers, L. A.,** Freeze-preservation of somatic embryos and clonal plantlets of carrot (*Daucus carota*), *Plant Physiol.,* 63, 460, 1979.

28. **Kartha, K. K. and Gamborg, O. L.,** Elimination of cassava mosaic disease by meristem culture, *Phytopathology,* 65, 826, 1975.

29. **Lundergran, C. and Janick, J.,** Low temperature storage of *in vitro* apple shoots, *HortScience,* 14, 514, 1979.

30. **Kartha, K. K.,** Genepool conservation through tissue culture, in Proc. *COSTED Symp. Tissue Culture Economically Important Plants,* Rao, A. N., Ed., University of Singapore, Singapore, 1981, 213.

31. **Dale, P. J.,** Meristem tip culture in herbage grasses, *Fourth Int. Cong. Plant Tissue Cell Culture,* (abstr.), Int. Assoc. Plant Tissue Cult., University of Calgary Press, Calgary, Canada, 1978, 106.

32. **Dale, P. J.,** A method for *in vitro* storage of *Lolium multiflorum* Lam, *Ann. Bot. (London),* 45, 497, 1980.

33. **Cheyne, V. A. and Dale, P. J.,** Shoot-tip culture in forage legumes, *Plant Sci. Lett.,* 19, 303, 1980.

34. **Galzy, R.,** Recherches sur la croissance de *Vitis rupestris* Scheele Sain et Court Nve cultivé *in vitro* à différentes témperatures, *Ann. Phytopathol.,* 1, 149, 1969.

35. **Henshaw, G. G., Stamp, J. A., and Westcott, R. J.,** Tissue culture and germplasm storage, in *Plant Cell Culture: Results and Perspectives,* Sala, F., Parisi, B., Cella, R., and Ciferri, O., Eds., Elsevier, Amsterdam, 1980, 277.

36. **Mullin, R. H. and Schlegel, D. E.,** Cold storage maintenance of strawberry meristem plantlets, *HortScience,* 11, 100, 1976.

37. **Hussey, G. and Hepher, A.,** Clonal propagation of sugar beet and the formation of polyploids by tissue culture, *Ann. Bot. (London),* 42, 477, 1978.

38. **Sakai, A. and Nishiyama, Y.,** Cryopreservation of winter vegetative buds of hardy fruit trees in liquid nitrogen, *HortScience,* 13, 225, 1978.

39. **Katano, M., Ishihara, A., and Sakai, A.,** Cryopreservation of shoot apices of apple winter buds, unpublished data, 1982.

40. **Uemura, M. and Sakai, A.,** Survival of carnation (*Dianthus caryophyllus* L.) shoot apices frozen to the temperature of liquid nitrogen, *Plant Cell Physiol.,* 21, 85, 1980.

41. **Bajaj, Y. P. S.,** Clonal multiplication and cryopreservation of cassava through tissue culture, *Crop Impr.,* 4, 198, 1977.

42. **Kartha, K. K. and Gamborg, O. L.,** Meristem culture techniques in the production of disease-free plants and freeze-preservation of germplasm of tropical tuber crops and grain legumes, in *Diseases of Tropical Food Crops,* Maraite, H. and Meyer, J. A., Eds., Université Catholique, Louvain, Belgium, 1978, 267.

43. **Bajaj, Y. P. S.,** Freeze-preservation of *Arachis hypogaea* and *Cicer arietinum, Indian J. Exp. Biol.,* 17, 1405, 1979.

44. **Kumu, Y., et al.,** unpublished data, 1983.

45. **Grout, B. W. W. and Henshaw, G. G.,** Freeze-preservation of potato shoot-tip cultures, *Ann. Bot. (London),* 42, 1227, 1978.

46. **Bajaj, Y. P. S.,** Initiation of shoots and callus from potato-tuber sprouts and axillary buds frozen at −196°C, *Crop Impr.,* 4, 48, 1977.

47. **Bajaj, Y. P. S.,** Tuberization in potato plants regenerated from freeze-preserved meristems, *Crop Impr.,* 5, 137, 1978.

48. **Stamp, J. A.,** Freeze-preservation of Shoot-Tips of Potato Varieties, M.Sc. thesis, University of Birmingham, Birmingham, England.

49. **Towill, L. E.,** Survival at low temperatures of shoot-tips from cultivars of *Solanum tuberosum* group Tuberosum, *Cryo. Lett.,* 2, 373, 1981.

50. **Sakai, A., Yamakawa, M., Sakata, D., Harada, T., and Yakuwa, T.,** Development of a whole plant from an excised strawberry runner apex frozen to −196°C, *Low Temp. Sci. Ser. Bull.,* 36, 31, 1978.

51. **Grout, B. W. W., Westcott, R. J., and Henshaw, G. G.,** Survival of shoot meristems of tomato seedlings frozen in liquid nitrogen, *Cryobiology,* 15, 478, 1978.

Chapter 7

CRYOPRESERVATION OF SHOOT-TIPS OF FRUIT TREES AND HERBACEOUS PLANTS

Akira Sakai

TABLE OF CONTENTS

I. MECHANISM OF SURVIVAL OF VERY HARDY CELLS COOLED TO THE TEMPERATURE OF LIQUID NITROGEN

Various organisms, when they are sufficiently desiccated, can then be cooled to the temperature of liquid nitrogen (LN) without injury. Extracellular freezing, too, is considered an effective method of dehydrating living cells. Very hardy woody plants, sufficiently dehydrated by extracellular freezing, survived immersion in LN[1-3] or in liquid helium[4] when subsequently rewarmed slowly in air at 0°C.

Still another method for maintaining viability at the temperature of LN, using ultrarapid cooling and rapid rewarming, was presented by Luyet.[5] This method consisted of preventing the growth of the intracellular ice nuclei which are usually formed during the cooling process by rapidly passing through the temperature zone of intracellular ice-crystal growth. Many attempts have been made to utilize this method. The expected results were only obtained, however, with a few plant materials which were especially desiccated,[6] or with very hardy cortical cells.[1,7] In extremely hardy cells which survived slow freezing below −70°C, this method was found to be effective in maintaining viability at the temperature of LN, especially when combined with the prefreezing method.[1,2]

The mechanism of survival of plant cells at superlow temperatures will be illustrated, using results obtained on the cortical cells from winter mulberry twigs. A unicellular layer of cortical tissues mounted between cover glasses with 0.05 mℓ water survived slow freezing at any temperature above −115°C for 16 hr. These cells frozen slowly to −90°C survived subsequent rapid rewarming by direct immersion in water at 30°C, although cells from less hardy plants are well known to be sensitive to rapid thawing. However, these tissue sections mounted with water between cover glasses were all killed by intracellular freezing when directly immersed into a bath kept at −20°C or below. Thus, it is reasonable to consider that cell death was caused only by intracellular freezing. In such a sense, it is favorable to consider the mechanism of injury of such material on the basis of survival value which then simplifies the interpretation of the results observed.

An unmounted section of cortical tissue (20 to 30 μm thick, 1 to 2 mm wide and 2 to 3 mm long), held with a forceps, was immersed directly into LN (cooling rate: 1500°C/sec) and then rewarmed rapidly by immersion in water at 30°C or slowly in air at 0°C. These ultrarapidly cooled cells still remained alive when rewarmed rapidly, while the cells were all killed when rewarmed slowly. By contrast, when even extremely hardy cortical cells were immersed in LN when mounted between cover glasses with 0.03 mℓ of water (cooling rate: 150°C/sec), all cells were killed when rapidly rewarmed. No survival was obtained even when suspending water was frozen at −2°C, which is nearly the same as the freezing point of the tissue sap, by lethal intracellular freezing occurring at comparatively high temperatures following freezing of the suspending water.[8] However, in cells (cooling rate: 400°C/sec) immersed in LN after prefreezing at −5°C or below for 10 min, survival increased abruptly when the cells were rewarmed rapidly. The survival of yeast cells cooled ultrarapidly was as low as 5 and 10%, even when rewarmed ultrarapidly (40,000°C/min).[9] The very low survival of yeast cells both cooled and rewarmed very rapidly is most likely explained by the fact that they were suspended in water, since these cells are apt to freeze intracellularly as pointed out above.

The effects of prefreezing and rewarming rates upon the survival of cells immersed in LN are summarized in Figure 1. Almost all of the cells prefrozen below −20°C survived immersion in LN, regardless of rewarming methods used. However, if the cells were only partially dehydrated by prefreezing at temperatures above −15°C, the rewarming rate seriously influenced their survival. Especially in cells prefrozen at −5°C, all were killed by slow rewarming in air at 0°C (rewarming rate: 1°C/sec), but all survived rapid rewarming in water at 30°C (rewarming rate: 400°C/sec). These facts suggest that in the sufficiently

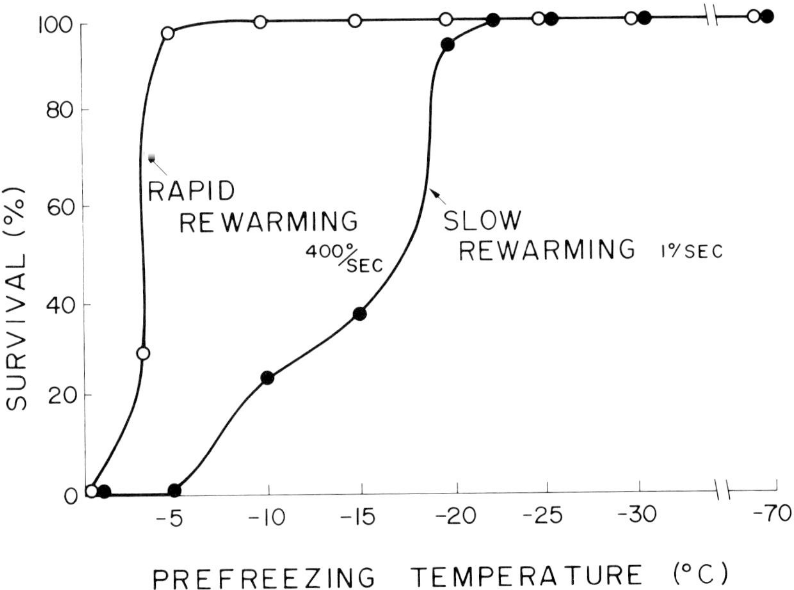

FIGURE 1.   Effect of rewarming rate upon the survival of cells immersed in LN following prefreezing at various temperatures. The tissue sections were mounted between cover glasses with 0.05 mℓ water. Tissue sections were rewarmed rapidly in water at 30°C (400°C/sec) or slowly in air at 0°C (1°C/sec). (From Sakai, A., *Plant Physiol.*, 41, 1050, 1966. With permission.)

prefrozen cells, below −20°C, almost all of the freezable water in the cells is withdrawn by extracellular freezing. Thus, fully freeze-dehydrated cells are not injured by rapid immersion in LN and even the subsequent slow rewarming at 0°C. On the other hand, in partially dehydrated cells, freezable water still remains, to some degree, in the cells after prefreezing, and the intracellular ice nuclei formed in the course of rapid cooling in LN probably grow and cause injury during the subsequent slow rewarming in air, while during rapid rewarming there might be no time for the ice nuclei to grow to a damaging size. The temperature range in which cells are injured during slow rewarming in air at 0°C (2.6°C/sec) following removal from LN was determined. The tissue sections in the process of rewarming were rapidly rewarmed by immersion in water at 30°C immediately after reaching a definite temperature. As shown in Figure 2, almost all of the cells were killed within 6 sec at the temperature range between −40 and −30°C. Also, the temperature range at which injury occurs during rewarming was determined by keeping them for the length of time in baths kept at different temperatures ranging from −5 to −70°C following removal from LN. The results are summarized in Figure 3. The main characteristic of survival is that in a limited temperature range survival abruptly decreases with the rising temperature from −60 to −50°C, from −55 to −45°C, and from −45 to −40°C in the cells kept for 60, 10, and 3 min, respectively. In the cells kept at various temperatures for 5 sec, only a slight decrease in survival was noted in the temperature range from −40 to −30°C. In this experiment, it took about 3 sec to rewarm from −60 to −30°C. When rewarmed immediately after reaching −30°C, all were alive, while the survival rates of the cells kept at −35°C decreased with length of time, and most of the cells were killed in 30 sec (Figure 3). To clarify further the effect of rewarming rate on survival, the tissue sections immersed in LN after prefreezing at −10°C were transferred rapidly to a bath at −60°C. These cells were then rewarmed at various rates through the temperature range from −55 to −2.5°C. When these cells were passed rapidly through the temperature range within about 2.2 sec, they

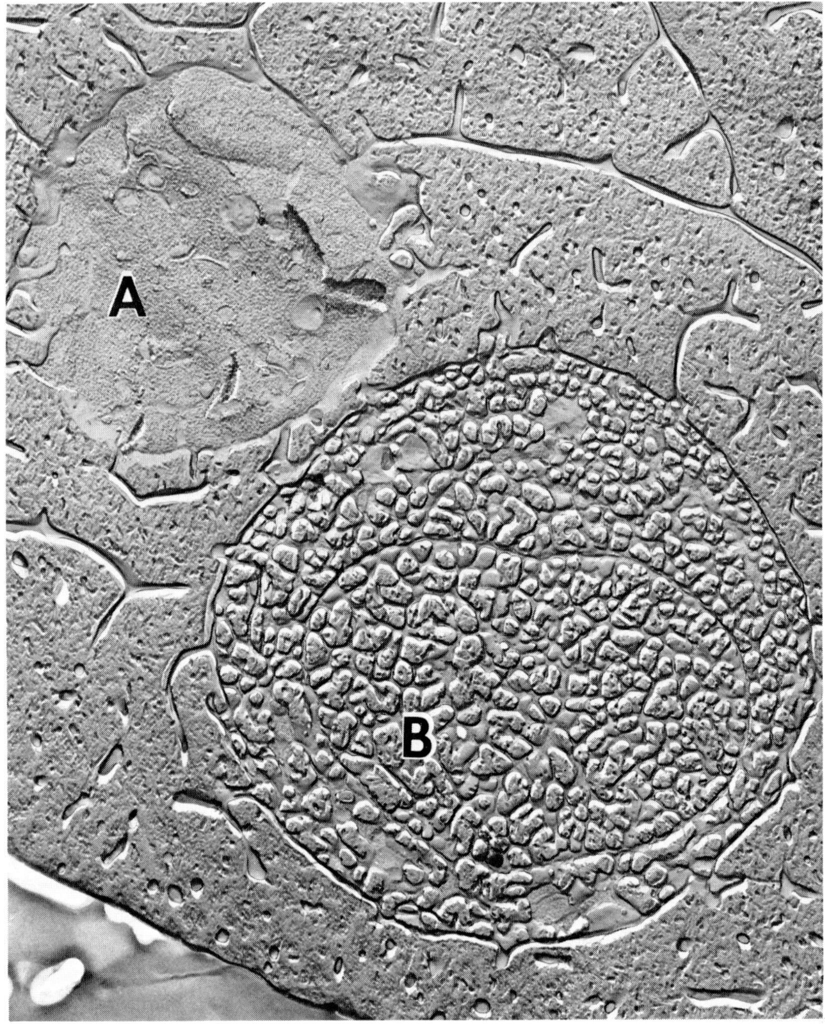

FIGURE 6. Electron micrograph of a freeze-etched tumor cell primarily frozen at $-27°$ by spray freezing. (Magnification $\times$ 7500.) (From Shimada, K. and Asahina, E., *Contrib. Inst. Low Temp. Sci. Hokkaido Univ. Ser. B*, 30, 65, 1972. With permission.)

rate of intracellular crystals and by preventing lethal intracellular freezing. The cooling rate necessary for maintaining viability with rapid cooling and rewarming varies considerably depending upon the size of specimens, the quality and quantity of the suspending medium, the degree of prefreezing or dehydration from the use of hypertonic solutions, and the degree of freezing tolerance of the cells concerned. It was also demonstrated that prefreezing temperature for maintaining viability after immersion in LN is related to relative freezing resistance, hardier samples require less prefreezing.[3]

II. CRYOPRESERVATION OF APICAL MERISTEMS

Shoot apex culture has become an indispensable tool in clonal propagation and in the elimination of viral pathogens from many crop plants.[17] Since the consituent cells of the shoot apex are less differentiated and have more uniform ploidy than those of mature tissues, plants regenerated by the in vitro culture of the shoot apex without the callus-mediated

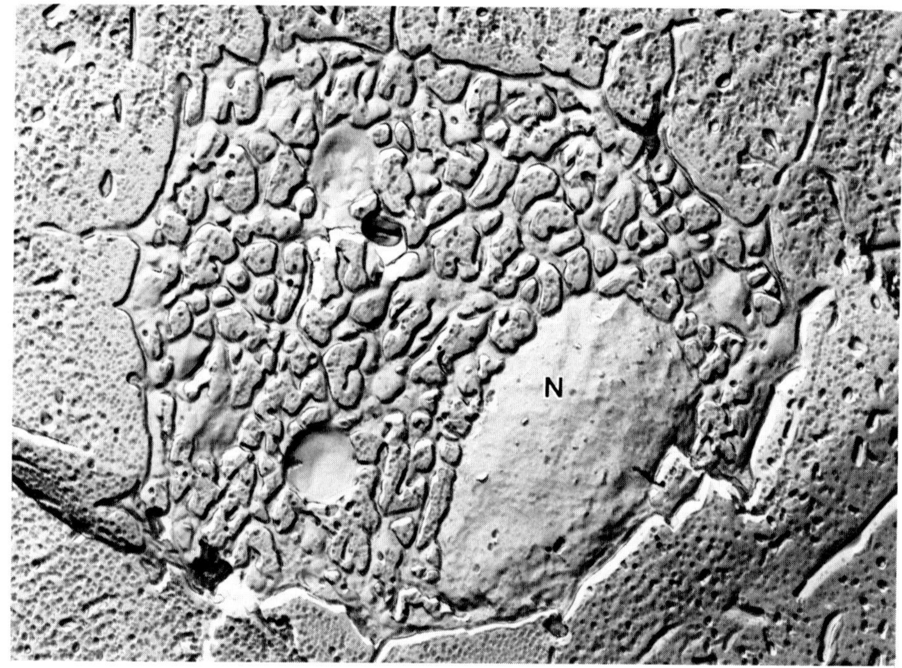

FIGURE 7. Electron micrograph of a freeze-etched tumor cell subsequently rewarmed to −17 from −27°C. (Magnification × 7500.) (From Shimada, K. and Asahina, E., *Contrib. Inst. Low Temp. Sci. Hokkaido Univ. Ser. B*, 30, 65, 1972. With permission.)

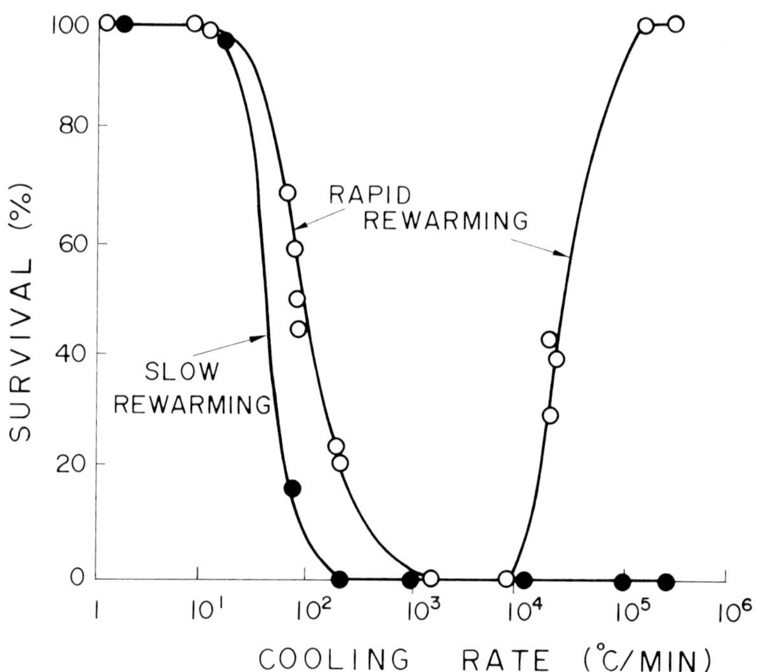

FIGURE 8. Survival of cortical cells from winter mulberry twigs cooled to −75°C at rates indicated on the abscissa without prefreezing and rewarmed either rapidly or slowly. Rewarming was done rapidly by immersion in water at 30°C and slowly in air at 0°C. (From Sakai, A. and Yoshida, S., *Plant Physiol.*, 42, 1695, 1967. With permission.)

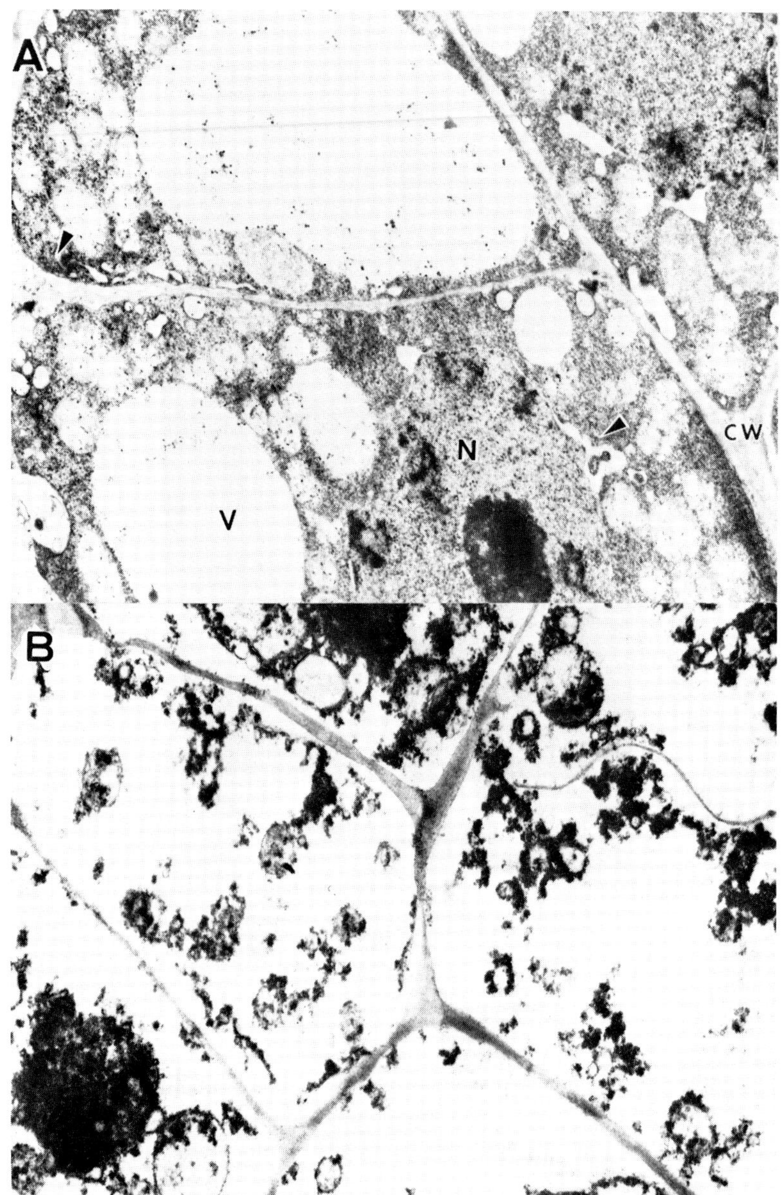

FIGURE 20.   The ultrastructure of cells in a carnation shoot apex immersed in LN in the presence of 10% DMSO from room temperature. Cooling rate: 5×10^{4}°C/min. A: immediately after rapid thawing, B: subsequent culturing for 24 hr.[18]

suspending medium is left in contact with the thawed cells when they are layered onto semisolid recovery medium.[21] Further investigations in this direction would certainly benefit cryopreservation. Withers[39,40] suggested that to date the most sound policy would appear to be to offer as little osmotic disturbance as possible in the days immediately following thawing and that thawed apices should be placed onto semisolid medium witbout washing.

Preconditioning, cryoprotectants, cooling and rewarming conditions which enabled shoot apices of herbaceous plants to survive freezing to the temperature of LN are listed in Table 2.

Table 2
SUCCESSFUL DEVELOPMENT OF APICAL MERISTEMS FROZEN TO THE TEMPERATURE OF LN

Species or cultivar	Pretreatment	Cryoprotectant	Cooling rate, prefreezing temperature (PF)	Thawing rate	Ref.
Carnation (stem) (*Dianthus caryophyllus*)	Hardening (4°C, 4—10 days)	DMSO 5—10%	50°C/min	Rapid	27
Carnation (stem) (*D. caryophyllus*)	None	DMSO 5—15%	0.5°C/min; PF: −40°C	Rapid	22
Strawberry (runner-tip) (*Fragaria x ananassa*)	None	DMSO 10—16% + 3% sucrose	Ultrarapid (5 × 10^{4}°C/min); PF: −15—−20°C	Rapid	24
(Mass propagation of plantlets)	Preculture 5% DMSO, 2 days	DMSO 5%	0.5—1.0°C/min; PF: −40°C	Rapid	26
Pea (seedlings) (*Pisum sativum*)	Preculture 5% DMSO, 2 days	DMSO 5%	0.5—1.0°C/min; PF: −40°C	Rapid	20
Pea (seedlings) (*P. sativum*)	Preculture 5% DMSO, 2 days; hardening at 0°C, 10 days	DMSO 5%	0.5—1.0°C/min; PF: −40°C	Rapid	18
Wild potato (*Solanum goniocalyx*)	None	DMSO 5%	Ultrarapid	Rapid	28
Tomato (seedlings) (*Lycopersicon esculentum*)	None	DMSO 15%	22—55°C/min	Rapid	29
Asparagus (stem) (*Asparagus officinalis*)	Preculture 4% DMSO for 2 days	DMSO 8—16%	0.5°C/min; PF: −40°C	Rapid	41
Malus species	None	None (water)	Step-by-step freezing; PF: −15 to −30°C	Rapid	38

Table 3
EFFECTS OF PLANT HORMONES IN CULTURE MEDIUM ON ROOT DEVELOPMENT OF CULTURED SHOOT APEX AFTER FREEZING IN LN[18]

Material	Root development (%)	
	Auxin + cytokinin	Auxin[a]
Carnation	47	76
Pea	0	50

[a]   Culture media for carnation and pea shoot apices containing NAA + kinetin and NAA + BA, respectively. When they reached 2 cm in length, they were transferred to their own medium without cytokinin.

Kartha et al.[30] reported that root formation on shoots produced by culturing pea meristems on B5 medium supplemented with benzyladenine (BA) alone or in combination with 1-naphthalene acetic acid (NAA) with induced by reculturing the shoots on half-strength B5 medium supplemented with NAA.[30] Frozen and rewarmed shoot apices were cultured on agar medium supplemented with both auxin and cytokinin.[18] Only 50% of carnation shoot apices, and none from pea, produced rooting within 30 days after thawing. To increase the rooting rate in culture, those apices that produced shoots without roots were recultured on agar medium supplemented with auxin only. As a result, about 80% of carnation and 50% of pea shoot apices formed roots within 30 days of reculturing (Table 3). Not all cells in the shoot apices frozen to the temperature of LN survived. Most of the cells in the area of subapical tissue appeared to be injured. In carnation shoot apices, most of the surviving cells are located in both areas, apical meristem and primordial leaf tissue.[18] Seibert and Wetherbee[27] showed in frozen carnation shoot apices that the cells in the meristematic region of the shoot apices remained viable after freezing and thawing, while those in the leaf primordia did not. In frozen pea-shoot apices, Haskins and Kartha,[31] however, observed that the undifferentiated cells in the area of the meristematic "dome" were less likely to survive freeze-preservation procedures than were cells elsewhere in the isolated meristems and that most of surviving cells are located in primordial leaf tissue and lateral areas of the "dome". The same result was also observed in potato-shoot apices by Grout and Henshaw.[32] Thus, it is evident that in the shoot apices of pea and potato, the undifferentiated cells in the area of the meristematic "dome" are more sensitive to freezing than are cells elsewhere in the isolated meristems. Withers[33] demonstrated that the survival of frozen-rewarmed carrot embryos was influenced by factors such as the developmental stages of the embryo at the time of sampling and the duration of in vitro exposure to plant hormones. The regrowth of frozen-rewarmed embryos upon return to culture proceeded by secondary embryo formation, the initial embryos not resuming their individual development.[33] Thus, to more fully explain differences in the sensitivity to freezing between the different area of an isolated shoot apex and of different species, further studies are required.

We have observed that several shoots developed from a single isolated and cultured carnation shoot apex following freezing and rewarming. Haskins and Kartha[31] observed that in pea shoot apices frozen to the temperature of LN, the surviving cells or group of cell have the capacity for active cell division and structural organization into shoot meristems without the callus-mediated process or organogenesis. The same fact was also observed in

Table 4
**SURVIVAL OF THE SHOOT APICES OF APPLE CV. FUJI
AFTER IMMERSION IN LN FOR 10 MIN FOLLOWING
PREFREEZING TO DIFFERENT TEMPERATURES AND
SUBSEQUENT REWARMING IN WATER AT 37° OR 2°C,
RESPECTIVELY**

Freezing solution and method of rewarming	Shoot survival (no. survival/total) at various prefreezing temperatures (°C)						
	−10	−20	−30	−40	−50	−70	−40[a]
Distilled water (0.5 mℓ); rewarmed in water at 37°C	7/8	6/8	7/8	6/8	4/6	4/4	3/3
Distilled water (0.5 mℓ); rewarmed in water at 2°C	5/6	6/6	4/5	5/6	6/6	4/4	—

[a] Apices were stored in LN for 24 hr.

the potato shoot apices by Grout and Henshaw.[32] Thus, isolated shoot apices are the most suitable explants for cultures as they have greater genetic stability than either callus or suspension cultures.

The mechanism of action of DMSO as a cryoprotectant is at present poorly understood. It may be considered that there are at least three locations where DMSO may act to avoid freezing injury: on the exterior solution, on the membrane itself, and on the interior solution. As to the membrane effects of DMSO, Lyman et al.[34] suggested that DMSO may exert a membrane effect in causing a decrease in the fluidity of the cell membrane at high temperatures. However, in spin-labeled carrot protoplasts, we have observed little or no decrease in the lower phase transition temperature near −20°C by ESR. Thus, to elucidate the action mechanism of DMSO, further studies will be needed.

III. CRYOPRESERVATION OF APICAL MERISTEMS OF HARDY FRUIT TREES

Sakai and Nishiyama[35] demonstrated that when winter buds from apple shoots immersed in LN for 2 hr after prefreezing to −40°C were grafted onto 2-year-old seedlings, most of the buds developed normally and continued their growth, and apple foliar buds immersed in LN for 23 months still remained alive. Winter buds of other hardy deciduous trees such as pea, raspberry, gooseberry, and currant also survived immersion in LN after prefreezing to −40°C and slow rewarming.[35]

During the past decade, there has been increased interest in the micropropagation of woody species[36,37] and now, micropropagation of apple and several other fruit trees and root stocks has become feasible. We have tested the cryopreservation of apical meristem using winter foliar buds of apple cultivars, root stocks, and *Malus* species.[38] Shoot apices about 1-mm long with 4 or 5 leaf primordia isolated from 21 apple cultivars including Fuji and 15 other *Malus* species and varieties all survived slow freezing to −40°C without DMSO. Most of the apple shoot apices suspended in water of 0.05 mℓ remained alive through immersion in LN following prefreezing between −10 to −70°C and subsequent slow or rapid rewarming (Table 4). Apices not prefrozen or apices prefrozen at −5°C were all killed by subsequent immersion in LN. Thus, in very hardy apices, DMSO is unnecessary to maintain viability after immersion in LN. After thawing, these apices, when transplanted onto paper medium and cultured, grew normally (Figure 21). These results proved that the shoot apices of hardy winter buds may be good material for the cryopreservation of germplasm of hardy fruit

FIGURE 21.   Shoot initiation from apple apices cultured in the
light after being frozen to $-196°C$ and thawed rapidly. Apex 9
weeks after thawing.[38]

species. The conventional method of maintaining wild species and old and current cultivars
of hardy deciduous fruit trees for long periods in a nursery and orchard requires extensive
space and labor. Our routine method of freeze preservation of plant apical meristems is
shown in Figure 22. Freeze preservation of isolated shoot apices in LN is also a potentially
reliable means of preserving germplasm of hardy fruit trees.

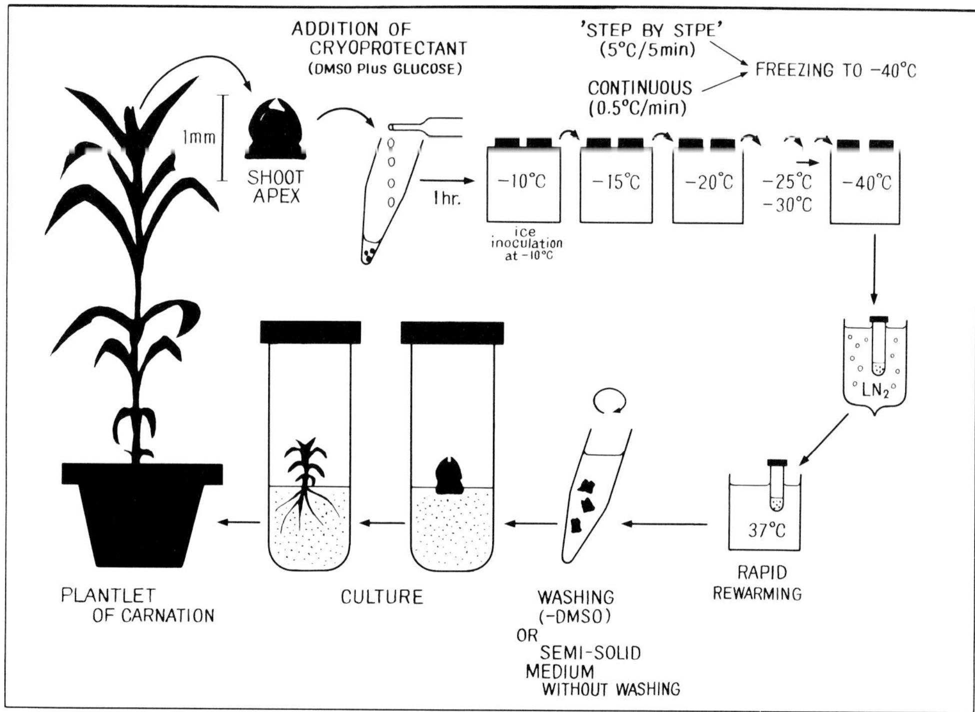

FIGURE 22.   Our routine method for freezing preservation of plant apical meristems. (From Sakai, A. and Uemura, M., *Plant Cold Hardiness and Freezing Stress*, Vol. 2, Academic Press, New York, 1982, 635. With permission.)

REFERENCES

1. **Sakai, A.,** Survival of plant tissue of super-low temperature, *Contrib. Inst. Low Temp. Sci.Haikkaido Univ. Ser. B*, 14, 17, 1956.

2. **Sakai, A.,** Survival of the twigs of woody plants at −196°C, *Nature (London)*, 185, 392, 1960.

3. **Sakai, A.,** Survival of plant tissue at super-low temperatures. III. Relation between effective prefreezing temperatures and degree of frost hardiness, *Plant Physiol.*, 40, 822, 1965.

4. **Sakai, A.,** Survival of woody plants in liquid helium, *Contrib. Low Temp. Sci. Hokkaido Univ. Ser. B*, 20, 212, 1962.

5. **Luyet, B. J.,** The vitrification of organic colloids and protoplasm, *Biodynamica*, 1(29), 1, 1937.

6. **Luyet, B. J. and Gehenio, P. M.,** The survival of moss virified in liquid air and its relation to water content, *Biodynamica*, 2(42), 1, 1938.

7. **Sakai, A.,** Survival of plant tissue at super-low temperature. IV. Cell survival and rapid cooling and rewarming, *Plant Physiol.*, 41, 1050, 1966.

8. **Sakai, A. and Yoshida, S.,** Survival of plant tissue at super-low temperature. VI. Effects of cooling and rewarming rates on survival, *Plant Physiol.*, 42, 1695, 1967.

9. **Mazur, P. and Schmidt, J.,** Interactions of cooling velocity, temperature, and warming velocity on the survival of frozen and thawed yeast, *Cryobiology*, 5, 1, 1968.

10. **Sakai, A. and Otsuka, K.,** Survival of plant tissue at super-low temperatures V. An electron microscope study of ice in cortial cells cooled rapidly, *Plant Physiol.*, 42, 1680, 1967.

11. **Shimada, K. and Asahina, E.,** Visualization of intracellular ice crystals formed in very rapidly frozen cells at −27°C, *Cryobiology*, 12, 209, 1975.

12. **Shimada, K. and Asahina, E.,** Innocuous ice crystallization within ascites tumor cells of rat by rapid freezing, *Contrib. Inst. Low Temp. Sci. Hokkaido Univ. Ser. B*, 30, 65, 1972.

13. **Mazur, P.,** Physical-chemical basis of injury from intracellular freezing in yeast, in *Cellular Injury and Resistance in Living Organisms*, Asahina, E., Ed., Institute of Low Temp. Sci., Sapporo, Japan, 1967, 171.

readily available in a genetically uniform state than potato seedlings. A stepwise series of meristem tips was employed, starting with the ungerminated seed — which it was correctly anticipated would survive exposure to LN without any special cryopreservation procedure — and progressing through stages of increasing hydration as the radicle and then the shoot-tip emerged. As might be expected, it became increasingly more difficult to achieve successful cryopreservation of the isolated meristem tips, but the most important result to emerge from these studies with tomato seedlings was an indication that faster cooling rates were more successful than those hitherto advocated for use with suspension cultures. Shoot-tips treated with 10 to 15% DMSO survived either direct immersion in LN (8% survival) or cooling first in LN vapor then plunging into LN (20% survival). Cooling slowly at a rate of 2°C min^{-1} to -16°C before taking them down to -196°C gave no survival at all.[18] These results were obtained at about the same time that Seibert[19] published the first report of the successful cryopreservation of an isolated shoot-tip; approximately 30% survival was obtained with a rapid-cooling technique in which LN was poured over explants of *Dianthus caryophyllus* which had been treated with 10% DMSO.

C. The Ultrarapid Freezing Method

The successes with tomato and *D. caryophyllus* suggested that it might be more profitable with potato also to explore rapid cooling procedures, and the investigations were continued with *Solanum goniocalyx*, a diploid species known to be particularly amenable to shoot-tip culture. The initial experiments employing moderately fast cooling rates (50 to 100°C min^{-1}) were unsuccessful, but direct immersion in LN (the ultrarapid freezing method described later) gave 20% survival of shoot-tips which had been protected with 10% DMSO; survival was noted in the form of both shoot regrowth and callus proliferation.[20] At about the same time, Bajaj[21] reported a maximum survival value of 26% among tuber sprout and axillary bud shoot-tips of Indian potato cultivars also frozen ultrarapidly and treated with a mixed cryoprotectant solution: 5% DMSO + 5% glycerol + 5% sucrose.

In the early unsuccessful experiments involving slow and moderately fast cooling rates, the meristem tips were generally immersed in a small volume of medium (up to 2 cm^3) contained in a plastic vial during the freezing and thawing sequences. As it was difficult to achieve a uniform rate of cooling throughout the specimen because of the thickness of the container and the volume of the medium, it was decided to eliminate both of these factors, and each meristem tip was simply supportd on the tip of a hypodermic needle which was plunged directly into LN, producing a cooling rate in excess of 1000°C min^{-1}.

In these first successful experiments the meristem tips (the apical dome plus 1 to 2 leaf primordia) were cultured for about 3 days on filter paper soaked in medium to allow recovery from the trauma of excision. Cryoprotection consisted of exposure to 10% DMSO for 1 hr at room temperature immediately prior to freezing in LN. The meristems were left in LN, frozen to the tip of the hypodermic needle, for periods up to several days before the needle was dipped into medium at a temperature of 35°C to bring about rapid thawing. After thawing and rinsing, the meristem tips were placed onto filter paper wicks in test tubes of culture medium (post-thaw culture medium) and incubated at 20°C with a 16-hr photoperiod for the first 5 days at 500 lx, and thereafter at 4000 lx. Survival involved either regrowth of the shoot apex or callus proliferation although sometimes it was difficult to be certain whether the original apical meristem resumed growth or whether it was replaced by another meristem rapidly regenerated from the callus. Transmission and scanning electron microscope studies[22] indicated that both types of regrowth probably occurred. These studies also showed that, generally, some of the cells in the meristem tip were irreparably damaged during the cryopreservation process and it could be concluded that the ability of the original meristem to maintain its organization and resume growth would depend on the proportion of cells damaged and their location; callus formation presumably indicated an intermediate level of damage leading to loss of organization.

Table 1
VARIABLES OPERATING AT EACH STEP OF
THE ULTRARAPID FREEZING PROCEDURE

Excision	Size of explant
	Age/physiology of plant
	Plant growth conditions
	Plant genotype
Prefreeze culture	Culture medium components
	Incubation conditions
	Duration of culture period
Cryoprotection	Type of cryoprotectant
	Mixtures of cryoprotectants
	Concentration of cryoprotectant
	Duration of protectant treatment
	Temperature of treatment
Ultrarapid freezing	Duration of period at $-196°C$
Rapid thawing	Temperature of thawing medium
Post-thaw culture	Culture medium components
	Incubation conditions
Regrowth	Length of ''lag'' phase before regrowth
	Morphological response (callus, shoots, roots)

Although these experiments were important in so far as they gave the first indications that it might be possible to devise a cryopreservation procedure for potato-shoot meristems, the survival rates were not high enough for germplasm storage purposes and it was also important to demonstrate whether similar procedures could be applied to other genotypes. To this end, investigations using this ultrarapid freezing protocol, and involving several potato genotypes, considered many of the variables operating at each step (Table 1), and some of these results are presented below and elsewhere.[23-26] In each of several genotypes tested, there was a definite requirement for a short culture period before freezing (Table 2). The optimum length of time of between one and three days was thought to allow the shoot-tips to recover from the trauma of excision without permitting so much growth that the size of the explant adversely affected the response to freezing.

Although DMSO was used routinely as a cryoprotectant, since it gave good results with *S. goniocalyx* and was widely used in other laboratories, a small experiment with *S. stenotomum* indicated that 10% glycerol alone or in combination with 10% DMSO could increase the survival level and did not diminish the incidence of shoot regrowth (Table 3). Survival during the post-thaw culture phase was more affected by light than by temperature (Table 4), and a low level of 500 lx was most favorable. However, shoot formation was observed in all of the conditions tested, including complete darkness.

Many experiments were concerned with manipulating the post-thaw culture medium in an attempt to encourage regrowth of single shoots without a callus phase and to improve survival. While some survival was observed if the whole procedure was carried out using hormone-free medium, the inclusion of auxin and cytokinin in the post-thaw culture medium doubled the amount of regrowth (Table 5). The optimum ratio of growth hormones provided in this final medium varied according to the genotype (Tables 6 and 7), but none of these median or any others of the many tested favored survival without callus proliferation. Table 6 illustrates how media giving vigorous shoot growth in control and nonfrozen but cryoprotected shoot-tips nevertheless allowed callus proliferation in certain of the thawed survivors.

A particularly noticeable feature of these results was their extreme variability even though, as far as possible the same basic procedure was always used. There was also some indication of a seasonal component in this fluctuation of response, possibly to be expected in view of

Table 2
**THE INFLUENCE OF THE PREFREEZE INCUBATION PERIOD ON THE
SURVIVAL OF SHOOT TIPS FROM SEVERAL POTATO GENOTYPES,
AFTER ULTRARAPID FREEZING IN LIQUID NITROGEN**

| | % Survival of genotypes (total frozen) | | | |
| | | | S. tuberosum ssp. tuberosum | |
Incubation period (hr)	*S. goniocalyx*[a]	*S. tuberosum* ssp. *andigena*[b]	cv. Desirée[c]	cv. Majestic[c]
0	0(15)	0(27)	0(20)	6.7(17)
18.5		24.0(25)		
24.0	50.0(18)		33.3(20)	0(18)
42.0		15.0(20)		
48.0	50.0(16)		7.1(20)	20.0(20)
65.5		17.2(29)		
72.5	35.0(17)		0(18)	6.2(19)
96.0	23.0(22)	29.7(27)	0(20)	0(19)
117.0		28.6(21)		
118.0	5.0(20)			
192.0		15.4(26)		

[a]   Cryoprotectant 10% DMSO; culture medium MS + 3% sucrose + 1 mg ℓ^{-1} BA (scored after 6 weeks).
[b]   Cryoprotectant 10% DMSO; culture medium MS + 3% sucrose + 1 mg ℓ^{-1} BA (scored after 22 days).
[c]   Cryoprotectant 5% DMSO; culture medium MS + 3% sucrose + 0.25 mg ℓ^{-1} NAA + 2.5 mg ℓ^{-1} BA (scored after 3 weeks).

Table 3
**SURVIVAL OF *SOLANUM STENOTOMUM* SHOOT TIPS
AFTER ULTRARAPID FREEZING IN LIQUID NITROGEN:
THE EFFECT OF DIFFERENT CRYOPROTECTANTS**

| | | | Number of survivors with | |
Cryoprotectant[a]	% Survival	Number of replicates	Callus	Shoots
DMSO	30.0	10	3	2
DMSO + glycerol	60.0	15	4	5
Glycerol	60.0	10	6	1
DMSO + glucose	14.0	14	2	0

Note: Medium: MS + 1 mg ℓ^{-1} BA + 3% sucrose.
Prefreeze incubation period 48 hr. (Scored after 5 months.)

[a]   Each cryoprotectant was used at a concentration of 10%, and applied in the culture medium.

the fact that environmental conditions can affect the size of the shoot-tip[27] and also the size and water content of its constituent cells.

Although the experiments already described have demonstrated that cryopreservation methods can be employed with potato meristem tips, it is obvious that they would require considerable improvement before they could be used routinely for germplasm storage with this species. High survival rates have been achieved, but the considerable variability in results according to experimental conditions and to the genotype is quite unacceptable. These particular experiments have been entirely concerned with attempts to improve the ultrarapid cooling method because of the relative simplicity of the procedure from the technical point

Table 4
EFFECT OF POST-THAW CONDITIONS ON THE SURVIVAL OF SHOOT TIPS OF *SOLANUM GONIOCALYX* AFTER ULTRARAPID FREEZING IN LIQUID NITROGEN[26]

Temperature (°C)	Light intensity (lx)	% Survival	Total	Number of survivors having	
				Callus	Shoots
20	0	36.4	22	2	6
20	500	57.7	26	14	1
20	4000	41.7	24	5	5
25	0	38.1	21	2	6
25	500	56.0	25	7	7
25	4000	42.3	26	2	9

Note: Cryoprotectant: 10% DMSO. Culture medium: MS + 3% sucrose + 1 mg ℓ^{-1} BA. (Scored after 27 days.)

Table 5
SURVIVAL OF *SOLANUM TUBEROSUM* SPP. *TUBEROSUM* CV. DESIRÉE SHOOT TIPS AFTER ULTRARAPID FREEZING IN LIQUID NITROGEN: THE INFLUENCE OF THE POST-THAW CULTURE MEDIUM

First medium[a]	Second medium[b]	% Survival[c]	Regrowth response
MSH	MSH	53.3	Callus
MSH	MSO	13.0	Callus
MSO	MSO	13.0	Callus + shoot
MSO	MSH	26.6	Callus

Note: MSH = MS + 3% sucrose + 0.25 mg ℓ^{-1} NAA + 2.5 mg ℓ^{-1} BA. MSO = MS + 3% sucrose. (Scored after 5 weeks.)

[a] Used for prefreeze incubation, cryoprotectant treatment (with 5% DMSO), and rinsing after freezing.
[b] Used for post-thaw culture.
[c] 15 Replicates per treatment.

of view. It is apparent, however, that the advantages of simplicity are accompanied by a lack of scope for adjustments in the basic procedure, which probably implies that the physiological state of the explant is particularly critical.

D. Alternative Cryopreservation Procedures in Potato

More recently, important progress has been made by Towill,[28] who, using *S. etuberosum* as a model system, has obtained a very high level of survival (71% average) following a two-step procedure: slow cooling (0.3°C/min) to −40°C, followed by immersion in LN. His work agreed with the results already described in stating a requirement for a 2-day prefreeze incubation period, and in the use of 10% DMSO as cryoprotectant. He notes that *S. etuberosum* was chosen partly for its amenability to shoot-tip culture, and both frozen-thawed and control shoot-tips gave rise to multiple shoots on culture. This work has also

Table 6
THE INFLUENCE OF THE POST-THAW CULTURE MEDIUM ON DESIRÉE SHOOT TIPS FROZEN IN LIQUID NITROGEN AND THEIR NONFROZEN CONTROLS

Medium code[a]	AC (%)	Zeatin (mg ℓ^{-1})	Zeatin riboside (mg ℓ^{-1})	BA (mg ℓ^{-1})	Kinetin (mg ℓ^{-1})	2iP (mg ℓ^{-1})	CM (%)	CH (g ℓ^{-1})	% Survival	Growth response F	NF	C
1	—	—	—	—	—	—	3	—	0	—	s	s
2	—	—	—	—	—	—	6	—	20	ca	s	s
3	—	—	—	—	—	—	10	—	0	—	s, ca	s, ca
4	0.3	—	—	—	—	—	10	—	40	ca	s	s
5	—	—	—	—	—	0.1	—	—	40	ca	ca	s, ca
6	—	—	—	—	—	0.5	—	—	40	ca	s.ca	s, ca
7	—	—	—	—	—	2.0	—	—	20	ca	ca	ca
8	—	—	—	—	0.02	—	—	—	0	—	s, ca	s, ca
9	—	—	—	—	0.2	—	—	—	100	ca	s, ca	s, ca
10	—	—	—	—	2.0	—	—	—	20	ca	ca	ca
11	—	1.0	—	—	—	—	—	—	50	ca	ca	ca
12	—	—	1.0	—	—	—	—	—	60	s.ca[c]	s[c]	s[c]
13[b]	—	—	—	1.0	—	—	—	—	100	ca	ca	ca
14[b]	0.3	—	—	—	—	0.1	—	—	20	ca	s	s
15[b]	—	—	—	—	—	0.1	—	1.0	80	s, ca	ca	s, ca

Note: AC = Activated charcoal; BA = N^6-benzyl adenine; 2iP = N^6-(2-isopentenyl)-adenine; CM = coconut milk; CH = casein hydrolysate. Prefreeze incubation = MS + 3% sucrose for 2 days. Cryoprotectant = 5% DMSO in MS + 3% sucrose (scored after 6 weeks). F = shoot tips treated with cryoprotectant, frozen ultrarapidly then thawed and cultured (5 replicates/medium); NF = shoot tips treated with cryoprotect and and cultured (2 replicates/medium); C = cultured shoot tips (5 replicates/medium); ca = callus; s = shoots.

[a]　All media contained MS + 3% sucrose + 0.05 mg ℓ^{-1} NAA + 1.0 mg ℓ^{-1} GA.
[b]　Same as footnote a, but with half strength MS.
[c]　All shoots vigorous on this medium.

Table 7
SURVIVAL OF SHOOT TIPS FROM SEVERAL POTATO GENOTYPES AFTER ULTRARAPID FREEZING IN LIQUID NITROGEN: THE INFLUENCE OF THE POST-THAW CULTURE MEDIUM

	Hormone content of medium[c] (mg ℓ^{-1})				Survival (%)	Total	Regrowth response
Genotype	GA	NAA	BA	2iP			
S. goniocalyx[a]	—	—	1.0	—	54	287	Callus, shoots
S. tuberosum ssp. *andigena*	—	—	1.0	—	35.6	121	Callus, shoots
S. tuberosum ssp. *tuberosum*	—	—	1.0	—	1.7	119	Callus
cv. Dr. McIntosh[a]	0.2	0.5	—	—	31.7	41	Callus, roots
cv. Majestic[b]	0.5	0.2	—	—	20.0	10	Callus
	0.2	0.5	—	—	11.1	9	Callus
	1.0	0.05	—	1.0	10.0	10	Callus
	—	0.25	2.5	—	25.0	12	Callus
	—	0.25	2.5	—[d]	28.5	14	Callus
cv. Desirée[b]	0.2	0.5	—	—	0	6	—
	—	0.1	2.0	—	20.0	5	Callus, shoots
	—	0.25	2.5	—	60.0	15	Callus, shoots
	—	0.25	2.5	—[d]	60.0	15	Callus, shoots
cv. Pentland Crown[b]	0.5	0.2	—	—	0	10	—
	0.2	0.5	—	—	11.1	9	Callus
	—	0.1	2.0	—	0	10	

[a] Cryoprotectant 10% DMSO.
[b] Cryoprotectant: 5% DMSO.
[c] Basal medium: MS + 3% sucrose.
[d] Plus 10% coconut milk.

been extended to several cultivars of *S. tuberosum*[29] (Chapter 6), and the technique has been shown to be especially effective when attention was paid to the use of a callus-inducing medium for the regrowth phase. Thus the effects of widespread damage in the shoot-tip were overcome by encouraging the proliferation of any surviving cells, but at the expense of disrupting the shoot-tip. Fortunately, shoots will readily regenerate from the callus, but it could pose a problem if high survival after cryopreservation can only be obtained by introducing a callus phase.

III. CONCLUSIONS

Having demonstrated that the ultrarapid freezing method does work with a range of potato genotypes, but frequently to an unacceptably low level of survival, it is recognized that many problems are waiting to be resolved. One of the most pressing questions concerns genetic stability: cells irreversibly damaged during ultralow freezing cannot contribute to regrowth, and hence the freezing process exerts a selection pressure on the cells from which the regenerated plants are derived. These difficulties may be compounded when a callus phase intervenes between thawing and plant regeneration.

Few reports of long-term storage of shoot-tips at $-196°C$ exist; Bajaj[30] noted regrowth of *S. tuberosum* shoot-tips thawed after storage in LN for several months, but gave no details of their morphology. Grout and Henshaw[20] stored shoot-tips of *S. goniocalyx* for up to 3 weeks in LN before thawing, and Stamp,[26] working with the same material and storing in

LN for 12 days, obtained a 50% survival rate (37% as callus alone and 13% as callus plus the original shoot). Kartha and colleagues[31] reported 95% survival among strawberry meristems stored for 1 week in LN, but this declined to between 50 to 65% during storage over the next 1 to 7 weeks (see Chapter 6, Section III. B for explanation). Earlier work from the same laboratory[32] with pea meristems stored in LN reported a smaller decrease in survival over a 26-week period: 68 to 61% of the explants were able to redifferentiate to whole plants after thawing.

The results of Towill and those of other workers with different species indicate that a two-step cooling procedure might combine some of the advantages of the slow and rapid cooling methods, with the initial slow-cooling stage presumably allowing some degree of protective dehydration and the rapid-cooling stage limiting the size of any ice crystals that are eventually formed. Quite clearly, this method allows more scope than the rapid-cooling method for the fine adjustments that might be required for the different accessions in a germplasm collection and for their different physiological states. However, results obtained by Kartha and associates[31] with meristems from strawberry shoot-tip cultures show just how precise this adjustment might need to be, since the recovery rates at initial cooling rates of 0.56, 0.84, and 0.95°C min^{-1} were 62, 95, and 33%, respectively (see Chapter 6).

The techniques that have already been described are perhaps reaching the state of refinement at which their use could be contemplated with a small, specialized germplasm collection, but undoubtedly much more standardized procedures would be required for use in the larger, more diverse collections. As far as potato is concerned, the present techniques, which frequently involve a callus phase after cryopreservation, could only be used with caution until evidence is obtained of acceptable levels of trueness-to-type among the plants regenerated (see Chapter 6, Section III. B).

Further, it would be a considerable advantage if the techniques could be applied to meristem tips isolated from shoot-tip cultures as well as to those taken directly from the plant. This would simplify the task of producing meristem tips in a uniform physiological state for cryopreservation purposes and it would shorten the recycling procedure that would be required from time to time when the cryopreserved stocks needed to be replenished. Unfortunately, there have been no reports of the successful cryopreservation of such meristem tips from potato, although there is evidence of success with other species[31] (see Chapter 6, Section III. A).

It remains to be seen whether improvements in the techniques can be made by careful attention to the cooling sequences or by improved cryoprotection, or whether somewhat different approaches will be required. The "colligative" approach[33] in which very high cryoprotectant levels are employed to avoid the problem of ice-crystal formation, has not yet been explored thoroughly with plant material, and at some time in the future, improved regeneration techniques and a greater understanding of genetic stability might lead to the use of other, more convenient, morphogenetically competent cells or tissues.

ACKNOWLEDGMENTS

The authors are grateful for financial support from the U.K. Overseas Development Administration for the collaborative project with the International Potato Centre, Peru.

REFERENCES

1. **Morel, G.,** Meristem culture techniques for the long-term storage of cultivated plants, in *Crop Genetic Resources for Today and Tomorrow*, Frankel, O. and Hawkes, J. G., Eds., Cambridge University Press, London, 1975, 327.

2. **D'Amato, F.,** The Problem of genetic stability in plant tissue and cell cultures, in *Crop Genetic Resources for Today and Tomorrow*, Frankel, O. and Hawkes, J. G., Eds., Cambridge University Press, London, 1975, 333.

3. **Henshaw, G. G.,** Technical aspects of tissue culture storage for genetic conservation, in *Crop Genetic Resources for Today and Tomorrow*, Frankel, O. and Hawkes, J. G., Eds., Cambridge University Press, London, 1975, 349.

4. **Wang, P. J. and Hu, C. Y.,** Regeneration of virus-free plants through *in vitro* cultures, *Adv. Biochem. Eng.*, 18, 61, 1980.

5. **Westcott, R. J., Henshaw, G. G., Grout, B. W. W., and Roca, W. M.,** Tissue culture methods and germplasm storage in potato. *Acta Hortic.*, 78, 45, 1977.

6. **Westcott, R. J., Henshaw, G. G., and Roca, W. M.,** Tissue culture storage of potato germplasm: culture initiation and plant regeneration, *Plant Sci. Lett.*, 9, 309, 1977.

7. **Murashige, T. and Skoog, F.,** A revised medium for rapid growth and bioassays with tobacco tissue cultures, *Physiol. Plant.*, 15, 473, 1962.

8. **Denton, I. R., Westcott, R. J., and Ford-Lloyd, B. V.,** Phenotypic variation of *Solanum tuberosum* L. cv. Dr. McIntosh regenerated directly from shoot-tip culture, *Potato Res.*, 20, 131, 1977.

9. **Westcott, R. J.,** Tissue culture storage of potato germplasm. 1. Minimal growth storage, *Potato Res.*, 24, 331, 1981.

10. **Westcott, R. J.,** Tissue culture storage of potato germplasm. 2. Use of growth retardants, *Potato Res.*, 24, 343, 1981.

11. **Roca, W. M., Bryan, J. E., and Roca, M. R.,** Tissue culture for the international transfer of potato genetic resources, *Am. Potato, J.*, 56, 1, 1979.

12. **Sakai, A. and Noshiro, M.,** Some factors contributing to the survival of crop seeds cooled to the temperature of liquid nitrogen, in *Crop Genetic Resources for Today and Tomorrow*, Frankel, O. and Hawkes, J. G., Eds., Cambridge University Press, London, 1975, 317.

13. **Quatrano, R. S.,** Freeze-preservation of cultured flax cells utilizing DMSO, *Plant Physiol.*, 43, 2057, 1968.

14. **Latta, R.,** Preservation of suspension cultures of plant cells by freezing, *Can. J. Bot.*, 49, 1253, 1971.

15. **Nag, K. K. and Street, H. E.,** Carrot embryogenesis from frozen cultured cells, *Nature (London)*, 245, 270, 1973.

16. **Dougall, D. K. and Wetherell, D. F.,** Storage of wild carrot cultures in the frozen state, *Cryobiology*, 11, 410, 1974.

17. **Sakai, A.,** Survival of plant tissue at super-low temperatures. III. Relation between effective pre-freezing temperatures and the degree of frost-hardiness, *Plant Physiol.*, 40, 882, 1965.

18. **Grout, B. W. W., Westcott, R. J., and Henshaw, G. G.,** Survival of shoot meristems of tomato seedlings frozen in liquid nitrogen, *Cryobiology*, 15, 478, 1978.

19. **Seibert, M.,** Shoot initiation from carnation shoot apices frozen to −196°C, *Science*, 191, 1178, 1976.

20. **Grout, B. W. W. and Henshaw, G. G.,** Freeze-preservation of potato shoot-tip cultures, *Ann. Bot.*, 42, 1227, 1978.

21. **Bajaj, Y. P. S.,** Tuberization in potato plants regenerated from freeze-preserved meristems, *Crop Impr.*, 5, 137, 1978.

22. **Grout, B. W. W. and Henshaw, G. G.,** Structural observations on the growth of potato shoot-tip cultures after thawing from liquid nitrogen, *Ann. Bot.*, 46, 243, 1980.

23. **Henshaw, G. G., O'Hara, J. F., and Westcott, R. J.,** Tissue culture methods for the storage and utilization of potato germplasm, in *Tissue Culture Methods for Plant Pathologists*, Ingram, D. S. and Helgeson, J. P., Eds., Blackwell Scientific, Oxford, 1980, 71.

24. **O'Hara, J. F. and Henshaw, G. G.,** Cryopreservation and strategies for plant germplasm conservation, *Cryo-Lett.*, 1, 261, 1980.

25. **Henshaw, G. G., Stamp, J. A., and Westcott, R. J.,** Tissue Cultures and Germplasm Storage, in *Plant Cell Cultures: Results and Perspectives*, Sala, F., Parisi, B., Cella, R., and Ciferri, O., Eds., Elsevier, Amsterdam, 1980, 277.

26. **Stamp, J. A.,** Freeze preservation of shoot-tips of potato varieties in liquid nitrogen, Master's thesis, University of Birmingham, Birmingham, England, 1978.

27. **Steward, F. C., Moreno, U., and Roca, W. M.,** Growth, form and composition of potato plants as affected by environment, *Ann. Bot.*, 48(Suppl. 2), 1, 1981.

28. **Towill, L. E.,** *Solanum etuberosum:* a model for studying the cryobiology of shoot-tips in the tuber-bearing *Solanum* species, *Plant Sci. Lett.*, 20, 315, 1981.

I. INTRODUCTION

Pollen biology is important in manipulating germplasm for breeding purposes. Historically, the need to retain pollen for breeding and genetics has kindled many studies into the physiology of pollen for prolonging viability. There is a tremendous number of crop plants for which pollen storage strategies are desirable. These include vegetables, agronomic grains, forages, fruit crops, landscaping plants, and plants for fiber. Within a species there exists numerous genotypes, some of cultivar status, others as breeding lines, for which storage is necessary. When evaluating storage techniques the goal is to identify a method which gives high percentages of survival for all genotypes of a species or for all materials which are used for crop improvement.

The development of methods for the successful storage of pollen requires information on floral biology, including shedding times, variation due to genotype and environmental influences, pollen collection procedures, viability tests, "receptivity" of the female, and incompatibility reactions. These topics are not within the scope of this chapter. Books by Heslop-Harrison[1] and Stanley and Linskens[2] provide access to some biochemical and physiological literature. Topics on pollination biology for specific crops are summarized in Fehr and Hadley[3] (field crops), Janick and Moore[4] (fruit crops), Franklin,[5] and Snyder and Clausen[6] (forest species).

The purpose of this chapter is to describe the methodology and factors necessary for successful cryopreservation of pollen, and to summarize the literature dealing with low-temperature pollen storage. Cryopreservation usually refers to the use of very low temperatures for preservation of biological materials, and within this context it is the viability of pollen that is preserved. Although cryogenic temperatures are usually considered below $-80°C$, lack of studies at these temperatures required expanding the coverage to include temperatures between 0 and $-80°C$ and freeze- and vacuum-drying. Once procedures are established so that injury is minimized during exposure to low temperatures, cryopreservation will provide long-term storage during which, in theory, the viability of the pollen does not change. Shorter periods of storage are adequate for some purposes and can be accomplished at higher subzero temperatures.

Pollen storage is desirable for several purposes. The most prevalent is the need to hybridize plants which flower at different times. This may require storage for a few days, which would be of value for crops with short-lived pollen (e.g., Graminae), or for about a duration of a year for woody species when a late-flowering strain is used to pollinate an earlier flowering strain. Prolonged storage is also desirable when dealing with materials which flower erratically. Often, the need for storage is based upon convenience or economy of space so that both members of a cross do not have to be grown at the same time.

Pollen storage can be of value in supplementing germplasm preservation strategies. Seed or clonal storage would provide the main preservation system, but pollen banks may be desirable.[7] Pollen storage for germplasm of forest, landscaping, and fruit trees could be efficient, economical, and space-saving.[8] The matter must be thoroughly explored, however, to determine strategies for selection of appropriate materials, replacement of pollen that either declines in viability or is used up, and the addition of pollen from new lines to the bank. These concerns apply to pollen banks for all species and not just woody plants. As with other propagules, pollen destined for international germplasm exchange requires safeguards and tests to minimize introduction and spread of disease.[9] Pollen-borne viruses and mixtures of pollen and fungal spores are not uncommon occurrences.

Pollen is suitable for shipment. The global concern for crop improvement and for germplasm collection requires transport of materials to many countries. The distribution of pollen, rather than seed, speeds the crossing of desired strains. This is evident for many woody lines which require several years to flower from the seedling stage.

Cryogenic storage is needed for germplasm preservation since materials may not be used for many years. As emphasized elsewhere in this volume, at temperatures below approximately $-130°C$, low molecular kinetic energies, very slow diffusion characteristics, and lack of liquid water preclude most types of chemical reactions; thus, no viability decline during storage should occur. The major problem in all hydrated biological systems is devising procedures to expose the material to low temperatures without having a loss in function due to ice crystallization. Pollen from many species can withstand dessication after which it is easily preserved at very low temperatures. The viability of cryostored pollen often approximates that of untreated, fresh pollen. Thus, one has the desirable feature that cryogenic storage may not exert selective pressure on the pollen. Under less stringent storage conditions, the surviving pollen may not be a random sample of the initial batch placed in storage. Data to support pollen selection at usual storage temperatures are sparse, however.

Although emphasis is on the application of pollen storage for genetic uses, storage is also needed to retain a batch of pollen in a constant condition over time for physiological and biochemical studies. The study of many processes demands a supply of biological materials in an unchanged state over a period of time. In addition, there is increasing development and use of pollen germination and pollen-tube elongation procedures for the detection of biological activity (e.g., toxicity, mutagenicity) of environmental pollutants.[10,11] Storage capabilities would allow a supply of uniform pollen for these studies. Some storage methods for biochemical and physiological purposes must be constructed to retain viability, others need only be concerned with retention of chemical activity in an unchanged state over time. Storage at cryogenic temperatures is desirable for these purposes.

Several minor reasons for pollen storage are cited and include the use in apiary culture and in the study of allergens.[2] Storage methods for these two uses do not need to retain viability; unaltered chemical composition is the goal. Emphasis on the use of microspores and, to a lesser extent, pollen for generation of haploids also provides a justification for development of storage procedures.[12] Cryopreservation methods using whole anthers may be feasible,[13] but have not been extensively examined.

II. VIABILITY

The first question that must be answered by any preservation problem is how pollen viability will be measured. The approach is usually based on the eventual use of the pollen. Most studies have been concerned with storage for pollinations; relatively few cite reasons for basic physiology studies. Thus, many search for a rapid method to estimate viability which can be applied to a large number of samples. The method chosen is then examined for its correlation with fruit or seed-set data. Depending upon the viability assay, the correlation may or may not be high. If low correlations are found, other rapid tests are examined. Correlation between two assays for a given species does not assure correlation in other species, although certain viability assays are inherently more predictive because they are based on a physiological characteristic (e.g., germination).[2]

Pollen viability assays fall into three general groups based upon (1) pollen staining/ fluorescence, (2) pollen germination, and (3) fruit or seed set. A few comments will suffice to indicate the potentials and limitations of each group. Stanley and Linskens[2] provide a thorough analysis of the various viability tests.

Those assays involving staining or fluorescence are diverse and include starch, acetocarmine, peroxidase, and fluorochromasia measurements. The general consensus is that staining of pollen with visible dyes is useful only where an imprecise estimate of viability is needed.[2] The starch viability assay uses iodine-potassium iodide (I-KI) to stain pollen containing starch. Weak correspondence of the I-KI test with the fluorescein diacetate assay was found using 2.5-year-old pollen from several plum cultivars stored at $-20°C$.[14] A sequential test of individual pollen grains with fluorescein diacetate and then with iodine vapor demonstrated this lack of correlation. Thick-walled pollen grains show little staining with I-KI.[15] Viable cotton pollen does not stain with iodine.[16] Other examples exist where nonviable grains become stained with I-KI and where unstained grains readily germinate.[2]

Acetocarmine is used to stain chromosomes and nuclei and has been used for viability estimates in pollen from a number of species.[17] Technical considerations have been reported.[18,19] Correlations between carmine staining and in vitro germination or seed set with potato pollen are low,[20] but the method is relatively simple and quick and is used where a rough estimate of "good" pollen is needed.

The peroxidase assay is based upon a pollen peroxidase oxidizing externally supplied benzidine to produce a dark blue color. The specific relationship of viability to coloration must be determined for each species. "Viable" potato and tomato pollens remain uncolored but "viable" sugar cane and sweet potato pollens turn blue.[21] These observations may relate to the presence of cell-wall bound peroxidases in pollen from some species.[22] Rigorous correlations of the peroxidase assay with other viability assays are not common. In the case of potato, the peroxidase test significantly correlated with the acetocarmine assays but not with in vitro germination of seed-set assays.[20]

The fluorochromasia and tetrazolium assays are better estimates of viability than simple stains because they are based on a membrane permeability barrier. The tetrazolium assay, for which several tetrazolium salts may be used, is dependent on uptake of the compound and its subsequent reduction to produce a colored product. Membrane integrity is necessary to prevent loss of the concentrated product. The selection of an appropriate tetrazolium salt is crucial. Triphenyl-tetrazolium chloride (TTC) has provided an "accurate" test of viability for several diverse species, although with others the test has not been "accurate".[2] Fresh pollen stained with TTC from several apple, peach, pear, and grape cultivars did not show a clear correlation with in vitro germination.[23] Pollen-sterile lines also gave high percentage of TTC-stained pollen. Werner and Chang[24] found staining with 3(4-5-dimethylthiazolyl)2,5-diphenyl tetrazolium bromide (MTT) correlated with in vitro germination of stored peach pollen, whereas nitro-blue tetrazolium (as well as aniline blue, iodine/potassium iodide, and propiono-carmine) did not. Among the 12 tetrazolium salts tested, MTT also gave the best correlation with in vitro germination for pollen from 9 varieties of plum.[25] A consideration for all studies which compare a viability test to in vitro germination is whether the in vitro culturing conditions and medium are optimum for pollen germination. Such optima are rarely reported in the literature; indeed, many studies utilize only a single medium.[26]

The fluorochromasia viability test involves uptake of fluorescein diacetate and subsequent cleavage by an esterase to produce fluorescein, the fluorescing molecule.[27] An intact membrane is required so that fluorescein is retained and concentrated within the cell. Use of this method for measuring pollen viability is recent and there are insufficient data from diverse species to predict general usefulness. Shivanna and Heslop-Harrison[28] have reported significant correlations between fluorochomasia and in vitro germination for the following nine species: *Carex nigra, C. ovalis, Cytisus battandieri, Digitalis purpurea, Eleocharis palustris, Iris pseudoacorus, Lonicera pericylmenum, Plantago lanceolata,* and *Secale cereale.*

Testing the ability of pollen to germinate on an artificial media (in vitro germination test) has long been used as a gauge of pollen viability and is often assumed to correlate well with in vivo germination. This approach must first elucidate a near-optimum growth medium that allows germination and growth with a minimum of pollen-tube bursting. Tests are necessary to show that a given medium promotes approximately maximum germination in pollen from a range of genotypes for a species. Adequate media now exist for binuclate pollen from many species; however, trinuclate pollen from many species still does not adequately germinate even when factors important for other pollen systems are varied. Snyder and Clausen[6] list media which are useful for many woody species. The major components of germination media are sucrose (10 to 40%), boron (50 to 200 ppm), calcium, and magnesium.

The importance of the first two are well described by Visser.[29] Sucrose serves two functions in the medium, that of being an energy source and of providing a suitable osmoticum. Current evidence suggests that external sucrose is metabolized by germinating pollen grains.[30] The osmotic role of sucrose for germinating pollen has long been recognized and it suffices to mention that other sugars or sugar alcohols can substitute. Boron is essential for pollen germination.[2] Germination in the absence of added boron is usually explained by adequate boron stores within the grain.[30,31] Pollen longevities described in the literature, especially from the early part of the century, should be viewed with caution if boron was not in the germination medium.

Many researchers have obtained high germination using the medium described by Brewbaker and Kwack.[26] This medium gave good germination for 86 species from 79 genera (39 families), and was devised while elucidating the nature of the population (or density) effect for pollen germination. Higher percentages are often observed when the pollen density is greater within the medium. The density effect can be attenuated by the inclusion of calcium in the germination medium. However, even with calcium, low pollen densities may exhibit reduced germination.[32]

The physical systems for in vitro germination include hanging drops, aerated solutions, and agar-solidified media.[2] When doing time studies, especially with multiple samples, it is convenient to fix samples so that they can be examined later. Simple microtechnique fixatives (e.g., Carnoy's solution) or storage at 4°C have proved adequate. Stanley and Linskens[2] discuss problems and suggestions for each system.

In vivo germination involves placing pollen on the stigma and measuring either pollen growth within the style, or, more commonly, the seed set that occurs. This assay is the most definitive since it answers whether the pollen can function under natural conditions. As always, crossing systems must be devised to avoid problems with incompatibility and genotypic systems diversity. The assay does not provide a quick answer — styles must be fixed, stained, and examined; or seeds must develop, be extracted, and then counted. The assay is best suited to a small number of samples. Quantitative expressions relating in vitro pollen germination to seed set are often difficult to obtain.[2] In some cases this may be due to the number of ovules available for fertilization and the number of pollen grains which can be borne on a receptive stigma.

Both in vitro and in vivo germination assays are not suited to rapid testing of multiple samples. Since information for crossing does not usually require exact estimates of germination, but only identification of samples with high germination potential, the development of other rapid viability tests is still desired. Techniques often using in analyzing freezing injury in leaf or crown samples, such as absorbance, conductivity, or amino-acid measurement of leachates, have been applied to pollen.[33,34] Such tests, however, are only useful for species which shed large amounts of pollen. For devising methodology to optimize storage conditions it is advantageous to use the best quantitative estimate of survival possible and, thus, with current technology, in vitro germination is preferred.

III. LOW-TEMPERATURE POLLEN STORAGE

There have been few comprehensive reports on the influence of low temperature on pollen survival and storability. Most data are descriptive, citing survival under a given set of conditions but without systematic variation of these conditions nor do they report on the physiological significance of the findings. Investigations have utilized a range of species but it is difficult to extrapolate whether all pollen from a genus or family behaves similarly in storage. As a general observation, members of the Ginkgoaceae, Pinaceae, Palmae, Saxifragaceae, Rosaceae, Leguminosae, Anacadiaceae, Vitaceae, and Primulaceae families possess pollen with fairly long lifetimes under conventional storage systems. Members of

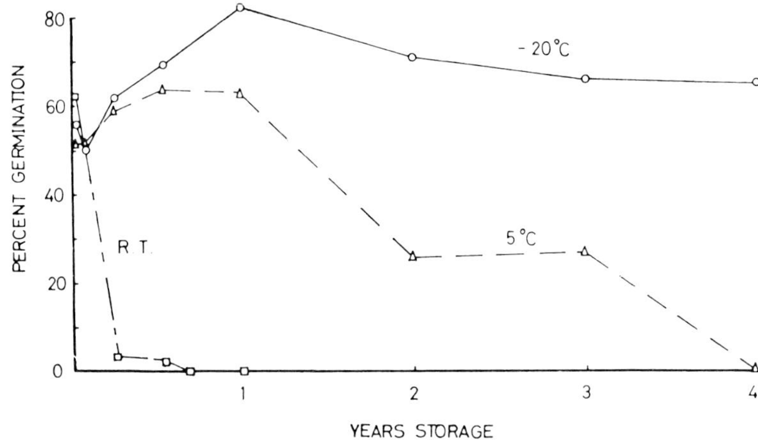

FIGURE 1.   Germination of freeze-dried peach pollen stored under vacuum at −20, 5°C, and room temperature (R.T.). (From Kozaki, I., *Gene Conservation,* Matsuo, T., Ed., JIBP Syntheses, Vol. 5, 1975, 89. With permission.)

the Liliaceae, Amaryllidaceae, Salicaceae, Ranunculaceae, Cruciferae, Rutaceae, Solanaceae, and Scrophulariaceae families possess pollen with intermediate longevities. Members of the Alismataceae, Graminaceae, Cyeraceae, Commelinaceae, and Junceae families are characterized by pollens with short lifetimes.[7]

A. Factors Affecting Storage

The major factors that determine pollen longevity in storage are temperature and moisture content.[2] Other factors, such as the composition and pressure of the gas phase over the pollen, are known to affect longevity but are rarely manipulated for optimum storage except for freeze- or vacuum-dried pollen.

It is well known that pollen longevity can be increased by lowering the storage temperature. Longevities have been tabulated for many species.[2,29,35,36] Data are usually obtained for storage temperatures above 0°C and using only a few sampling times; thus, maximum longevities often are not determined. The increased longevity of peach pollen stored at lower temperatures illustrates the trend observed for many species (Figure 1).[37] The rate of germination loss at a given temperature probably varies with species. Storage at about −20°C has increased longevities of some fruit- and forest-tree pollens to several years.[2,6,29] However, potato pollen remains viable for only about a year.[32]

The other major factor influencing longevity is moisture content. There is ample information to demonstrate that pollen longevity is affected by humidity during collection, pretreatment, and storage.[2,29] Pollen collection techniques sometimes involve short periods of drying to allow anther dehiscence. Drying the pollen before storage is often necessary to further lower the moisture content. Storage humidities are controlled so that pollen will not become rehydrated during the storage period.

Pollen from different species can be classified into two groups based upon nuclear number at anthesis. Binucleate pollens usually have a thicker exine, withstand considerable desiccation, and have greater longevities. Trinucleate pollens have a thinner exine, are sensitive to desiccation, and have short longevities. The majority of angiosperm families are composed of species bearing binucleate pollen.[38] Families with species bearing trinucleate pollen include the Graminaeae, Umbelliferae, Cruciferae, Araceae, Caryophyllaceae, and Chenopodiacieae. Some families have genera with binucleate pollen and other genera with trinucleate pollen, but species within a genus are only of one type. Storage and physiological studies on trinucleate pollen from families other than the Graminaeae are rare.

Most binucleate pollens store longer at lower relative humidities (rh). Visser[29] observed that at $+2°C$ binucleate pollen from apple, azalea, pea, rhododendron, and tomato survived best at 10% rh. Some other species store better at rh between 10 and 30%. Storage at 0% rh, however, greatly decreases longevity and has been observed for pollen from many, but not all, species.[2,29] This suggests that drying pollen below a certain moisture content may lead to loss of viability. Caution must be observed, however, in interpreting germination data after extensive drying. Often a subsequent slow rehydration (humidification) is then necessary to obtain high percentages of in vitro germination.[29,31,39,40] Visser et al.[39] found that hybrid-tea-rose pollen stored at 1°C and 0% rh for 1 year gave 27% germination without prior rehydration but 51% germination after only 30 min of rehydration. Pollen from the tuber-bearing *Solanum* species exhibits a similar phenomenon.[41] *S. polyadenium* pollen stored at 4°C for 18 days over anhydrous $CaCl_2$ gave 3% germination if tested without rehydration, but gave 59% germination if rehydrated at 4°C for 1 day (100% rh) prior to testing. Vacuum- and freeze-dried pollens also require rehydration prior to germination testing.[2,42-44] Rehydration may not be necessary for seed-set tests, possibly because rehydration on the stigma surface may be sufficiently slow.

Trinucleate pollens store longer at higher rh but the longevities vary considerably. Pollen from species of the Graminae survived better at 4°C than at warmer temperatures but longevities were still only about 2 to 5 days.[45,46] Trinucleate pollen from cabbage was stored for 25 days at 4°C without loss of viability.[47] Trinucleate pollens also differ in their ability to withstand desiccation. *Dieffenbachia maculata* pollen stored best at 5°C and 90% rh (2 to 5 days' longevity),[48] whereas *Spathiphyllum floribundum* pollen stored best at 7°C and 65% rh (24 weeks' longevity).[49] Both are members of the Araceae family. Although wheat pollen required high humidities to retain viability,[45] maize pollen can be dried considerably without loss of viability. The rate at which drying occurs is related to exine thickness but is not the reason for sensitivity per se. Studies with recalcitrant seeds suggest that desiccation sensitivity is related to membrane injury and such may be the case with trinucleate pollen.[50]

The kinetics of water movement are important in understanding how certain pollen samples survive low temperature treatments. Drying of pollen prior to storage is usually reported by citing time of exposure and a given rh. Occasionally, moisture contents of the pollen before and after exposure are given, but with no indication if moisture content equilibrium had been reached. Atmospheres with different rh can be produced by using saturated salt solutions or sulfuric acid solutions; however, these solutions must be carefully prepared and used.[51]

Pollen moisture content may be expressed either as a percentage of wet or dry weight. An equilibrium moisture content (EMC) curve for *Pinus ponderosa* pollen stored at different rh (Figure 2) illustrates that high atmospheric rh is necessary to greatly increase the water content of pollen.[52] These data are useful in selecting appropriate rh to control pollen moisture content. The time to reach equilibration increases with the difference between atmospheric rh and initial pollen moisture content. Dried pollen (ca. 0% moisture) took about 200 hr in 100% atmospheric rh to equilibrate, giving an 86% EMC (dry-weight basis); whereas only a 10-hr period was required to lower pollen with a 21% moisture content to an EMC of 9%. Studies with pollen from *Abies* and *Pseudotsuga* produced an EMC curve similar to that from *Pinus,* but that from *Cedrus* was distinctly different. Japanese pear pollens had moisture contents of 100, 85, and 30% when stored for 48 hr at 20°C over rh of 100, 95, and 88%, respectively.[53] Although not stated as equilibrium times, these results approximate the data in Figure 2.

Data for diverse species are lacking, but many pollens probably hydrate approximately as pollen from *P. ponderosa*. The observation that some pollen survives storage at 0% rh may mean these pollens possess higher percentages of bound water. The EMC and equilibrium times might be considerably different for these pollens. Critical studies require construction of an EMC curve for exact comparisons. There are presently no detailed studies

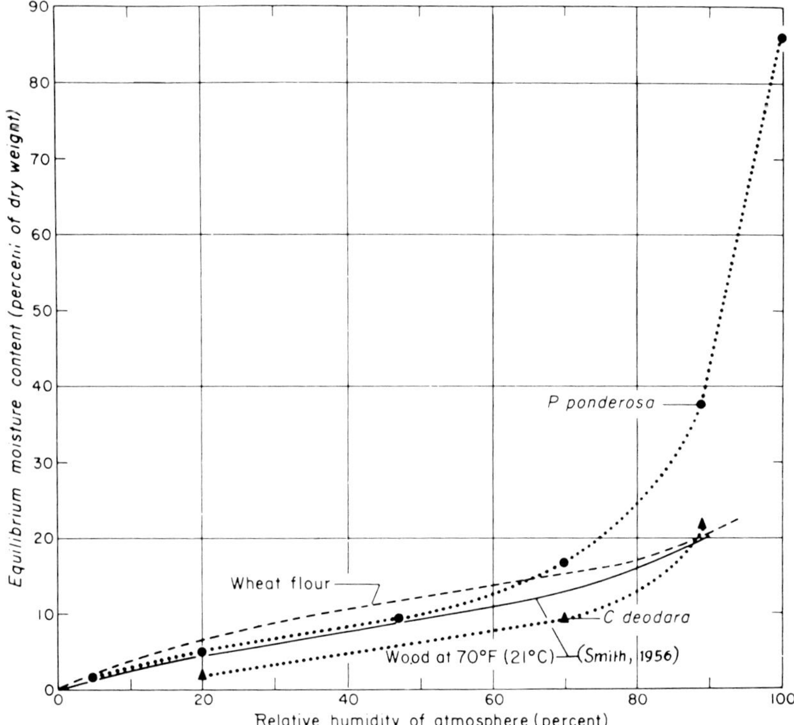

FIGURE 2.   Moisture content of pollen from *Pinus ponderosa, Cedrus deodara,* and wheat flour and wood after equilibration in atmospheres of known relative humidity. (From Lanner, R. M., *Silvae Genet.,* 11, 14, 1962. With permission.)

interrelating the kinetics of drying, rehydrating, and germination. Some researchers have observed that germination of pollen that was dried and rehydrated is delayed.[29]

The difference between germinability as measured by in vitro germination tests and fertility as measured by seed set needs to be emphasized in evaluating pollen viability studies. Many examples exist where pollen with low germination gave almost normal seed set and where pollen with high germination gave very low seed set.[2] The usual situation is that pollen with good germination gives good seed set and, thus, that the observation of germination can be taken as a measure of viability.

B. Storage at −10 to −34°C

Many species have been successfully stored at temperatures between −10 and −34°C (Table 1). Most of these diverse species have binucleate pollen and most reports stress the necessity for drying prior to low-temperature exposure. All the data come from a direct exposure of the pollen to the storage temperature. The method and rate of warming of pollen from these temperatures are not usually stated. Differing warming rates from −20°C did not notably influence in vitro germination of dried potato pollen.[124] This is not surprising since the small amount of water within the pollen probably is not crystallized at −20°C.

Differential thermal analyses of pollen with different moisture contents exposed to subzero temperatures are needed to determine at what temperature water within the pollen will freeze and how freezing affects survival.

Exotherm data from seed studies may be relevant. No exotherms for lettuce seeds with moisture levels less than 16% were detected during cooling.[34] Seeds with moisture contents less than 13% were not injured at −196°C, but at between 13 and 16% moisture content

Table 1

SURVIVAL OF POLLEN FROM SEVERAL SPECIES AFTER STORAGE AT −10 TO −34°C

Species	Storage temperature (°C)	Duration of storage	Germination[a] %B	%A	Comments	Ref.
Allium cepa	−18	198 Days	100	38		54
Amaryllis sp.	−11	214 Days	—	64		55
Asparagus officinalis	−20	60 Weeks	82	85		56
Brassica campestris	−20	50 Weeks	—	—	Normal seed set in 2 of 3 cultivars	57
B. oleracea	−20	78 Weeks	—	—	Normal seed set	57
B. napus	−20	50 Weeks	—	—	Normal seed set	57
Chamaecyparis obtusa	−20	364 Days	—	58		58
Citrus grandis	−20	3 Years	69	48	Normal seed set	59
C. hassaku	−20	3 Years	60	46	Normal seed set	59
C. natsudaidai	−20	3 Years	93	80	Normal seed set	59
C. reticulata	−18	57 Weeks	35	12		60
Coffea arabica	−18	26 Months	ca. 73	60	Pollen stored 6 months set seed	61
Corylus spp.	−18	12 Months	—	54		62
Cucumis melo	−18	30 Days	98	98		63
Juglans regia	−19	20 Days, 1 Year	—	—	Normal fruit set	64
J. nigra	−30	2 Years	59	56		65
Lilium auratum	−20	194 Days	—	66		66
L. auratum ''hybrid''	−20	194 Days	—	55		66
L. henryii	−20	194 Days	—	74		66
L. regale	''Deep freeze''	3 Years	100	3		67
Lycopersicon esculentum	−20	1132 Days	47	63	Normal seed set	29
Medicago sativa	−18	183 Days	—	—	Moderate seed set	68
M. sativa	−18	34 Days	88	73		63
Olea europaea	−18	367 Days	39	29		63
Phoenix lactylifera	−13	1 Year	—	—	Slightly reduced seed set	69
P. reclinata	−13	1 Year	—	—	Slightly reduced seed set	69

Table 1 (continued)

SURVIVAL OF POLLEN FROM SEVERAL SPECIES AFTER STORAGE AT −10 TO −34°C

Species	Storage temperature (°C)	Duration of storage	Germination[a]		Comments	Ref.
			%B	%A		
P. sylvestris	−13	1 Year	—	—	Slightly reduced seed set	69
Picea glauca	−20	9 Months	69	68		70
Pinus attenuata	−23	1 Year	80	50		71
P. contorta var. *latifolia*	−23	1 Year	60	60		71
P. contorta var. *murrayana*	−20	5 Years	70	57	Reduced seed set	72
P. coulteri	−23	1 Year	60	20		71
P. flexilis	−23	1 Year	20	30		71
P. jeffreyi	−20	5 Years	70	63	Reduced seed set	72
P. lambertiana	−23	1 Year	—	50		72
P. monticola	−20	5 Years	—	22	Reduced seed set	72
P. nigra	−23	1 Year	80	50		71
P. ponderosa	−20	5 Years	80	50	Reduced seed set	72
P. rigida	−23	1 Year	70	40		71
P. sabiniana	−20	5 Years	80	58	Reduced seed set	72
Pisum sativum	−5	3 Months	—	—	Good seed set, 15 months poor seed set	73
Poncirus trifoliata	−18	57 Weeks	23	7		60
Populus alba var. *pyramidalis*	−18	—	—	—	Remains viable for several years	74
P. deltoides	−18	—	—	—	Remains viable for several years	74
P. nigra	−18	—	—	—	Remains viable for several years	74
Prunus amygdalis	−18	801 Days	66	30		63
P. armeniaca	−18	440 Days	—	8		63
P. avium (cerasus)	−18	408 Days	47	40		63
P. brigantia	−18	402 Days	—	58		63
P. cerasifera	−18	435 Days	69	60		63

Species	Temp. (°C)	Duration	B	A	Notes	Ref.
P. communis	−18	801 Days	66	30		63
P. domestica	−20	2.5 Years	—	56	In 8 other cultivars, germination from 7 to 56%	14
P. domesticus (insititia)	−18	435 Days	69	60		63
P. persica (peach)	−18	425 Days	—	71		63
P. persica (nectarine)	−18	410 Days	—	76		63
P. salicina	−18	441 Days	43	38		63
Pyrus communis	−18	408 Days	85	74		63
P. domesticus	−18	385 Days	92	78		63
P. malus	−20	673 Days	76	63	Normal seed set	29
Rhododendron catawbiense	−20	662 Days	30	25		29
Ricinus communis	−18	1 Year	—	—	Normal seed set	75
Rosa sp. (hybrid tea)	−24	38—40 Weeks	64	36	Pollen 2 years old gave seed set	39
Rubus idaeus	−13	6 Months	—	—	Pyrene and seed set at 4 months	76
Solanum brevidens	−20	9 Months	100	<1	Survival at 4 months was 24%	77
S. stoloniferum	−20	9 Months	100	42		77
S. tuberosum Gp. Phureja	−20	10 Months	—	—	Normal seed set	78
S. tuberosum Gp. Tuberosum	−24	359 Days	—	—	Normal seed set	79
S. tuberosum Gp. Tuberosum	−34	1 Year	—	—	Reduced seed set	80
S. tuberosum Gp. Tuberosum	−20	11 Months	—	—	Normal seed set	81
S. tuberosum Gp. Tuberosum	−20	9 Months	100	29		77
Trifolium pratense	−15	26 Weeks	100	<2		82
Vitus vinifera	−12	4 Years	34	21	High berry set	83

[a] B: percentage in vitro germination before storage; A: percentage in vitro germination after storage.

the freezing tolerance dropped to $-40°C$. When seed moisture was 51%, supercooling occurred and allowed survival down to $-16°C$ (see also Chapter 10).

The longevity of pollen stored between -10 and $-34°C$ is not well defined. Pollen from 5 species of *Pinus* stored at $-20°C$ for 1 year gave normal seed set.[72] Seed set after 2 years of storage was reduced, yet some still occurred after a 5-year period. *Lilium regali* pollen retained control levels of germination for 1 year whereupon it decreased to 30% at 2 years and 3% at 3 years.[67] It appears from the limited data with air dried pollen that -10 to $-34°C$ storage will allow retention of high percentages of survival for about 1 to 3 years for many species. Reasons for decreased viability are not understood, but probably are related to respiration activities during storage.[2] Factors related to mineral nutrition and disease status of the stock plant, and collection techniques, also influence pollen storability.[2] Once fruit-tree pollen is removed from $-20°C$, the kinetics of viability decline at room temperature are similar to that of freshly collected pollen.[29] This has not been adequately studied for many species.

In vitro germination of pollen stored at low temperatures for short periods of time often is higher than that of the fresh pollen.[2,29,65] Exposure of pine pollen to $-20°C$ for a few days increased seed yields over that obtained with nonfrozen pollen.[72] No explanation has been experimentally verified, but these effects could be due to after-ripening processes occuring after shedding or the freezing process causing release of some needed nutrient.[2]

Genotypic differences in the ability of pollen to survive low temperature have been reported. Eight clones of black walnut gave varying percentages of in vitro germination after 2-yearstorage at $-30°C$ and at $-196°C$.[65] Pollen moisture contents ranged between 8 and 15% on a dry-weight basis and might account for the variation in survival observed. Differences in storability of pollen at $-20°C$ from different trees of *Pinus* species also have been observed.[72]

Thus, storage at $-20°C$ is sufficient to retain viability for crossing many species, but is not sufficient for purposes of germplasm preservation or physiological/biochemical studies.

C. Storage at Cryogenic Temperatures

Two of the earliest reports that pollen could survive very low temperatures were by Knowlton[85] who observed that Antirrinum pollen germinated after exposure to $-180°C$ and Bredemann et al.[86] who found *Lupinus* pollen viable after three months' storage in liquid air. Knowlton observed that a low pollen moisture content was necessary for survival at $-180°C$ and "the rate of the fall in temperature had no effect" on germination. Germination of *Lycopersicon, Pyrus*, and *Rhododendron* pollens was unaffected by exposure to $-180°C$.[29] Pollen from 25 species of the tuber-bearing *Solanum* species exhibited high percentages of germination after exposure to $-196°C$.[41] A listing of species whose pollen survived exposure and storage at temperatures lower than $-80°C$ is presented in Table 2.

Ichikawa and Shidei[98] stored pollen from 33 species at $-196°C$ and found that many retained the ability to germinate after storage (Table 3). Some pollens appeared to have decreased in germination with time of storage at $-196°C$, but considerable variation in germination occurred from test to test. Studies with *Cryptomeria, Pinus*, and *Larix* pollen, however, showed no effect of storage on seed set. This author also has observed unexplicable variations in germination with pollen from several *Solanum* species stored at $-196°C$. Several other reports cite unpredictability of germination assays from either freeze-dried or low-temperature stored materials.[2,43] Storage below $-130°C$ is believed to allow almost indefinite longevities but no long-term data are available for pollen. A need is apparent to devise and implement such studies as well as to determine why erratic viability is observed in some situations.

The method of exposing pollen to the cryogenic temperature is usually by direct immersion of a small vessel containing the dried pollen into baths of the desired temperature. Cooling

Table 2
SURVIVAL OF POLLEN FROM SEVERAL SPECIES AFTER STORAGE AT CRYOGENIC TEMPERATURES

Species	Storage temperature (°C)	Duration of storage	Germination %B	Germination %A	Comments	Ref.
Antirrhinum majus	−180	30 Min	—	60		85
Avena sativa	−192	1 Day	16	0	Gave + TTB[b] stain, no seed set	87
Cedrus deodara	−196	Brief	ca. 90	ca. 90		89
Chamaecyparis obtusa	−70	364 Days	—	63		58
Diospyros kaki	−80	1 Year	48	43	Normal fruit and seed set	89
	−196	1 Year	48	42	Normal seed set	89
Glycine max	−192	21 Days	65	24	Seed set	87
Gossypium hirsutum	−192	10 Days	8	1	No seed set	87
Juglans nigra	−196	2 Years	84	72		66
Larix leptolepis	−196	Brief	99	97		90
Lilium longiflorum	−196	60 Days	74	17		91
Lupinus polyphyllus	−190	93 days	78	78		86
Lycopersicon esculentum	−190	1062 Days	47	35		29
Medicago sativus	−192	84 Days	87	63		87
Persea americana	−196	1 Year	—	—	Gave normal ovule penetration	92
Phoenix dactylifera	−196	Brief	56	53		93
P. reclinata	−196	Brief	18	16		93
P. sylvestris	−196	Brief	17	17		93
Pseudotsuga menziesii	−196		—	—	Viability unchanged	94
Pyrus communis	−190	1062 Days	64	50		29
P. mauls	ca. −70	7 Months	—	83	Bee-collected	95
P. malus	−190	673 Days	76	68		29
Rhododendron catawbiense	−190	662 Days	30	25		29
Secale cereale	−192	7 Days	34	1	Seed set	87
Selanum brevidens	−196	9 Months	100	100		77
S. stoloniferum	−196	9 Months	100	98		77
S. tuberosum	−196	9 Months	100	100		77
S. spp.	−196	Min—hr	—	—	Of 25 species, all retained high % germination after exposure	41
Sorghum bicolor	−192	1 Day	—	—	Positive TTB[b] stain after treatment	87
Trifolium pratense	−196	26 Weeks	100	62	Seed set	82
T. aestivum	−192	1 Day	—	—	Positive TTB[b] stain after exposure, no seed set	87
Zea mays	−76	363 Days	59	48	Seed set	96, 97
	−196	40 Days	66	25		91

[a]   B: Percentage in vitro germination before storage; A: percentage in vitro germination after storage.
[b]   TTB: triphenyltetrazolium bromide viability test.

rates are not often stated and probably vary considerably because of the use of different vessel sizes. Glass ampules (1 cc), when immersed in liquid nitrogen, cool at about 100 to 200°C/min. Cryobiological parameters have not been explored to any extent to determine whether higher percentages of pollen survival are possible or to evaluate causes of injury.

Reduction in pollen moisture content below a threshold level prior to low-temperature exposure is critical for obtaining survival. This value differs widely for diverse species.

Table 3
SURVIVAL (GERMINATION RATE) OF DEEP-FROZEN STORED POLLEN

Species	% Water content	Storage period (months)				Storage period (years)						Control
		7—9	12—18	19—23	24—27	3	4	5	6	7	8	
Cedrus deodara	20.0[a]			48.8[b]	+/−							20.4
	12.7[c]			55.9	+							28.6
	12.7[d]			83.5	+							28.6
	14.0							87.4				
Pinus rigida	15.0		92.5									49.4
P. taeda			80.7									63.1
P. Luchuensis	11.0		96.9									82.3
P. banksiana	13.0		77.5									92.1
P. pinasta	13.0		99.5									100.0
P. denciflora	17.0		95.0									95.0
P. massoniana			87.7									28.4
P. thunbergii	20.4				76.6	28.1				+/−		
	20.4					19.4				+/−		
	20.4					0				0		
	18.0			23.5						76.3		
	10.0			32.6						79.1		
Keteleeria davidiana	13.0							0				60.6
Cryptomeria japonica	26.5		0				0					
	18.0		50.9				0			0		
	10.0		76.9				52.7			15.0		
	11.0							45.8				75.9
Thuja orientalis	38.6		0	0			0					
	15.6		80.2	43.4			48.2					
	10.8		92.0	72.6			63.4					
Juniperus sp.	41.6		36.8							0		
Cunninghamia lanceolata	14.0		78.7					30.0				78.0
Torreya nucifera	19.0	90.0							98.2			
Maghiolia grandiflora	26.0		85.0	25.0						32.4		
	17.0		62.6	36.7						67.8		
	8.0		46.3	79.4						26.3		

Michelia fuscata	16.0	84.4					85.0	
	12.7		94.4					
Liriodendron tulipifera	22.0		34.0	37.8	51.7		0	
	24.0	19.0					40.2	
Idesia polycarpa	23.0		95.0	76.5		67.8		97.8
	18.0		65.9	82.6		41.0		98.0
	17.0		95.0					98.0
	12.0		85.1	76.0		24.2		37.7
Gleditsia japonica	Dry	22.7	65.0		98.7		56.2	
Philadelphus satsumi	Dry	90.0			98.0		98.5	
Stewartia monadelpha	19.0	53.4			99.0		59.1	
Quercus dentata	34.3		0	0		0		
	12.0		51.9	48.3		69.4		
	9.7		91.6	74.4		21.4		
Cudrania tricuspidata	25.0	18.6					10.0	
Punica granatum	18.9	68.0					49.3	
	16.0	69.7					39.1	
Sapium sebiferum	16.0	11.2					0	
Quercus glauca	28.5	3.4						
Salix gracilistyla	19.0		26.5		78.0			
Platycarya storobilacea	16.0	19.9						
Aescula turbinata		39.6						
Phoenix sp.		82.0						
Datura sp.		82.0						

[a]   Slowly cooled.
[b]   Percentage germination after storage period at $-196°C$.
[c]   Rapidly cooled.
[d]   Directly cooled.

From Ichikawa, S. and Shidei, T., *Bull. Kyoto University For.*, 44, 47, 67, 1971. With permission.

Table 4
FREEZING TEMPERATURE AND SURVIVAL RATE
OF LARCH POLLEN AT DIFFERENT WATER
CONTENTS

Temperature (°C)[a]	Pollen water content (%)					
	11	30	40	50	60	76
0	94.1[b]	99.0	98.0	—	—	66.7
−5	—	—	—	—	—	—
−10	98.3	99.0	—	92.4	52.5	56.0
−15	—	—	—	30.4	—	—
−20	98.0	99.0	98.0	85.1	85.3	27.3
−25	—	—	—	75.8	88.9	—
−30	98.1	93.6	68.2	—	40.6	55.4
−35	97.0	92.7	72.2	2.5	38.2	2.0
−40	94.0	88.5	34.0	8.3	9.9	+
−45	96.9	96.2	36.2	0	0	16.0
−50	94.4	97.2	20.2	0	0	0
−55	—	—	—	0	—	—
−60	—	—	+	0	0	0
−65	—	—	—	0	—	—
−70	—	—	—	0	0	0
−75	—	—	—	0	—	—
−80	72.5	97.4	0	0	0	0
−196	98.0	97.0	0	0	0	0

[a]   Temperature to which pollen was rapidly cooled.
[b]   Percentage in vitro germination.

From Ichikawa, S., Kaji, K., and Kubota, Y., *Bull. Hokkaido For. Exp. Stn.*, 8, 11, 1970. With permission.

Pollen from many *Solanum* species survived − 196°C exposure when the pollen moisture content was less than about 35 to 40%.[41] Hop pollen survived liquid nitrogen treatment when moisutre content was reduced to about 10%.[125] The relationship between pollen moisture content and survival at different low temperatures for rapidly cooled pollen was determined for several forest trees. For larch pollen a moisture content less than 30% allowed survival at − 196°C (Table 4).[90] If moisture content was increased above 30%, injury occurred at higher temperatures. When the moisture content was 40%, pollen could not survive below about − 30°C. Similar patterns have been observed for pollen from other tree species.[90,98-100] When moisture content of *Larix* pollen was high (ca. 70%), a prefreezing technique similar to that described by Sakai for tissues[102] gave high percentages of survival at − 196°C. Exposure to prefreezing temperatures from − 15 to − 30°C for between 1 and 5 hr was effective in allowing pollen from several angiosperm and gymnosperm species to survive subsequent fast cooling to either − 80 or − 196°C.[99,100] The optimum prefreezing temperature and time of exposure differed among the species.

Ice formation was observed by X-ray diffraction in Douglas fir and Western white pine pollens that had greater than 36% moisture and were cooled to − 25°C.[102] Loss of viability occurred with ice formation. Free water as determined by nuclear magnetic resonance (NMR) spin echo at room temperature only occurred in *Tradescantia paludosa* pollen containing above 15% moisture.[103] These two observations further suggest that dried pollen possesses little freezable water and, hence, can survive low-temperature exposure.

Cooling and warming rates are not important when pollen has been dried sufficiently. *S. brevidens* pollen survived equally well when slowly or rapidly cooled to − 196°C and stored for 9 months.[77] *Cedrus deodara* pollen, with a moisture content below 20% and immersed in liquid nitrogen, survived equally well regardless of warming rate.[98] Intracellular ice formation is implicated as the injurying event when germination is lost after rapid cooling. This occurs in pollens with moisture levels above some species-specific value. *Pinus thunbergii* pollen, with a moisture content of about 30%, was rapidly cooled to $-196°C$ and then slowly rewarmed; it did not survive, but rapid warming under these conditions allowed survival.[99] Pollen with lower moisture content survived slow rewarming from $-196°C$. Extensive variations in cooling rates usually are not evaluated, but rate variation alone probably would not improve survival at any given temperature. Conventional cryobiological studies with cells in aqueous solutions provide well-defined survival curves describing injuries due to intracellular ice formation and solution effects.[104] Since pollen is cooled without any added solution, there is no opportunity for seeding with ice to avoid supercooling nor any extracellular ice to which cellular water can diffuse during slow cooling.

Numerous attempts to expose trinucleate pollen to temperatures below freezing have occurred. Tetrazolium bromide viability staining was positive for pollen from corn, sorghum, oats, rye, and wheat treated at $-180°C$.[87] Of corn, rye, and wheat pollen, only rye resulted in successful seed set. This limited success suggests that some trinucleate grains can survive low-temperature exposure. As contrasted to binucleate pollens, trinucleate pollen from different species may show considerable variation in tolerating freeze or desiccation stress. Sugar cane pollen did not retain viability at any subzero temperature and was very sensitive to desiccation injury.[105]

Maize pollen has been successfully cryopreserved after drying.[96,97] The water content of freshly collected maize pollen depends upon environmental conditions. Pollen collected at noon contained 25 to 35% moisture, whereas that collected in the morning contained 45 to 60%. In this study, high in vitro germinations and seed sets were observed with $-196°C$-treated pollen if pollen moisture levels were less than 25%. Surprisingly, maize pollen survived drying to moisture levels of 2.5%, and after $-196°C$ exposure, gave good germination and seed set. Pollen dried to moisture contents between 10 and 26% did not always survive $-196°C$. The highest percentages of seed set occurred with pollen which lost between 65 to 75% of the original water content. Injury in trinucleate pollen, therefore, may not be due solely to drying below a critical moisture level but may also be related to the initial moisture content of the pollen. Nath and Anderson[91] observed survival of maize pollen after rapid cooling to $-196°C$, but survival decreased considerably with days of storage at $-196°C$. Changes in isozyme staining after electrophoresis were more prevalent in lily and maize pollen cooled at 100°C/min and 150°C/min compared to 200°C/min.[106,107]

There are few reports on the use of cryoprotectants for obtaining survival of pollen during low-temperature treatment. Exposure of lily pollen to 1% dimethylsulfoxide (DMSO) within the germination medium had little effect on the percent of germination or pollen-tube growth.[108] Pollen-tube initiation was inhibited at 5 and 10% DMSO. The inhibition was reversible. Respiration and starch metabolism also were altered during exposure to 5% DMSO. The limited evidence suggests that pollen responds similarly to other cells that are exposed to DMSO. Viable lily pollen was obtained after $-20°C$ exposure in a medium containing 5% DMSO.[109] Pollen from several cultivars of date palm, *Phoenix dactylifera* and *P. reclinata*, gave some survival when slowly frozen to $-196°C$ after exposure to a solution containing polyethylene glycol, glucose, and DMSO. Survival was best, however, after direct exposure to $-196°C$ without cryoprotectant treatment.[93] *Cedrus deodora* pollen slow-cooled to -40 or $-50°C$ in 2 M solutions of sucrose, glycerol, sodium chloride, or DMSO gave only a slight increase in germination. Here, too, the highest percentages of germination occurred by drying the pollen to less than 20% moisture and directly immersing it in liquid nitrogen. Cryoprotectants alone or in combination have not been successful in preserving sugar cane pollen at temperatures below zero.[126]

The dearth of information for use of cryoprotectants with binucleate pollen is probably because of the success using simple dehydration combined with rapid exposure to low temperatures. Species possessing trinucleate pollen are difficult to preserve by these simple techinques. Studies using cryoprotectants should provide a logical system to attain protection, but the lack of publications suggests that encouraging results have not been obtained.

D. Preservation by Freeze- or Vacuum-Drying

Pollen from many species can be preserved by freeze- or vacuum-drying (Table 5). In freeze-drying, the pollen is fast-cooled to a subzero temperature and then subjected to a vacuum to remove subliming water vapor. Vacuum-drying does not use the initial freezing step; the pollen is directly exposed to a vacuum which lowers the sample temperature through evaporative cooling. King[43] has presented an excellent review and summary of his extensive studies.

To develop a useful procedure, optima for the initial pollen moisture content, cooling method, duration and extent of drying, and rehydration must be determined. The moisture content of pollen prior to freeze-drying may affect survival. Japanese pear pollen with initial moisture levels between 17 and 31% exhibited much lower germination after freeze-drying than did pollen with 4 to 8% moisture.[53] Western white pine pollen with about 40% moisture did not survive freeze-drying, whereas that with about 18% did; air-drying of the pine pollen to about 8 to 10% prior to the freeze-drying process increased the length of time that the sample could be freeze-dried and still retain viability.[118] Air drying (8% moisture) combined with prechilling (36 days, 0°C) was the best condition for treatment of Douglas fir pollen prior to freeze-drying.[94] Cold treatment of Western white pine pollen prior to freeze-drying also enhanced survival.[118] Whitehead,[111] however, did not observe an effect of the initial water content on viability of coconut pollen after freeze-drying.

Since vacuum-drying has been as effective as freeze-drying for maintaining pollen viability,[43,110,111] there appears to be no advantage to freezing the sample prior to the drying process. Length of the drying portion of the process influences subsequent survival; survival is highest after short periods of drying (from one to a few hours).[42] The bulk of the moisture is removed quickly during drying (Figure 3). King[42] found with 12 species that the optimum length of drying was between 0.5 and 3 hr; however, *Ipomoea batatas* pollen gave high percentages of in vitro germination even after 30 hr of drying.

Continued drying may lower the moisture content below a critical value necessary for retention of viability. Although data for various species are limited, this value is 1 to 2% for peach[53] and pine[118] pollen and about 3.5% for coconut pollen.[111] *Citrus grandis* pollen began to decrease in germination after only 5 min of drying during which the moisture content dropped from 60 to 50%.[59]

As discussed earlier, when pollen is dried to a low moisture level, as in exposure to drying agents and the vacuum- or freeze-drying process, rehydration prior to germination testing is essential.[29,37] Freeze-dried coconut pollen did not germinate at all unless rehydrated prior to testing.[112] Two hours at 100% rh and 23°C were sufficient to yield maximum germination. The rehydration time that produced maximum germination for oil palm pollen increased with increasing length of the vacuum-drying step.[44] Rehydration in 80 to 100% rh for several hours often is best, although longer durations may drastically decrease viability. Longer periods of rehydration are possible at lower temperatures. Examination of the equilibrium moisture content curve for *P. ponderosa* illustrates that this pollen does not greatly increase in water content unless exposed to rh above about 70%.[49]

Longevity of pollen after freeze- or vacuum-drying is greater at lower temperatures. Freeze-dried pollen from *Picea*, *Pinus*, and *Pseudotsuga* exhibited higher percentages of germination after 4 years when stored at −10°C compared to −1° or +17°C.[113] Douglas fir pollen stored for 1 and 2 years retained high percentages of germination at both −18 and 3°C

Table 5
SURVIVAL OF POLLEN FROM SEVERAL SPECIES AFTER FREEZE- OR VACUUM-DRYING AND SUBSEQUENT STORAGE

Species	Treatment[a]	Storage temperature	Duration of storage	Germination[b]		Comments	Ref.
				%B	%A		
Allium cepa	F	Room	23 Months	—	—	Seed set, 60%	43
A. cepa	F	−18°C	198 Days	100	60		55
Amarylis sp.	F	Room	34 Months	68	22		43
Antirrhinium majus	F	Room	13 Months	60	40		43
Asparagus officinalis	F	−1°C	6 Months	68	60		56
Beta vulgaris	V	+5°C	400 Days	45	25		110
Betula verrucosa	V	+5°C	900 Days	58	20		110
Carya illinoensis	F	Room	38 Months	7	30		43
Citrus sp.	F	Room	38 Months	48	46		43
Citrus sp.	F	−196°C	2 Years	100	80		59
Cocos nucifera	F	Room	3 Months	60	49		111
C. nucifera	F	Room	6 Months	—	28		112
Elacis guineensis	V	−10°C	1 Year	89	79		44
Ficus carica	F	Low	354 Days	10	22		42
Fragaria grandiflora	F	Room	13 Months	15	29		43
Gladiolus sp.	F	Room	12 Months	48	21		43
Larix sp.	F	−10°C	48 Months	—	—	Normal seed set	113
Lilium auratum (hybrid)	F	−20°C	194 Days	—	8		66
L. henryi	F	−20°C	194 Days	—	18		66
L. longiflorum	F	+5°C	180 Days	70	~5		91
Liquidambar styraciflua	F	Room	6 Months	56	49		43
Lycopersican esculentum	F	Room	24 Months	—	—	Reduced seed set	43
Magnolia sp.	F	Room	4 Months	—	57		43
Malus domesticus	F	Room	121	23	30		42
Medicago sativa	V	−21°C	11 Years	—	—	Reduced seed set	114
Nicotiana glutinosa	V	+5°C	900 Days	65	28		110
Picea sp.	F	−10°C	24 Months	95	51	Normal seed set	113
P. abies	V	+4°C	5 Years	—	84		115

Table 5 (continued)
**SURVIVAL OF POLLEN FROM SEVERAL SPECIES AFTER FREEZE- OR VACUUM-DRYING AND
SUBSEQUENT STORAGE**

Species	Treatment[a]	Storage temperature	Duration of storage	Germination[b]		Comments	Ref.
				%B	%A		
P. glauca	V	+4°C	5 Years	—	72		115
P. glauca	V	−20°C	9 Months	69	48		70
Pisum sativum	V	−25°C	1 Year	—	—	Reduced seed set	116
P. sativum	V	+5°C	700 Days	60	25		110
Pinus banksiana	V	+4°C	5 Years	—	88		115
P. caribaea	F	Room	34	84	81		43
P. cambra	V	+5°C	12 Years	78	72		117
P. griffithii	V	+5°C	5 Years	68	53		117
P. koraiensis	V	+5°C	7 Years	70	45	Germination at 8 years, 0%	116
P. monticola	F	+3°C	6 Weeks	99	72		118
P. nigra	V	−10°C	48 Months	80	45	Normal seed set	113
P. nigra	V	+5°C	900 Days	90	45		110
P. peuce	V	+5°C	5 Years	42	50	Germination at 6 years, 0%	117
P. strobus	V	+5°C	8 Years	67	60	Germination at 9 years, 0%	117
P. sylvestris	V	−10°C	48 Months	88	60	Normal seed set	113
P. taeda	F	Room	86 Days	80	79		42
P. taeda	F	Room	379 Days	—	32		42
Populus hybrids	V	−18°C	4—5 Years	—	—	Reduced seed set	119
Prunus avium	F	Room	27 Months	10	28		43
P. persica	F	−20°C	9 Years	73	82		53
P. persica	F	−20°C	4 Years	63	65		37
P. persica	F	Room	21 Months	—	47		43
P. salicina	F	Room	39 Months	56	32		43
Pseudotsuga sp.	F	−10°C	48 Months	86	58		110
P. manziesii	F	−18°C	2 Years	—	88		94

Pyrus communis	F	Room	17 Months	49	37		43
P. serotina	F	−20°C	6 Years	60	75		53
P. serotina		−196°C	3 Years	60	84		53
Ricinus communis	F	−20°C	1 Year	—	—	Seed set	76
Secale cereale	V	+5°C	120 Days	25	18	At 160 days, 0%	110
Solanum tuberosum	F	Room	24 Months	—	—	Seed set	43
S. tuberosum	V	+5°C	400 Days	48	30		110
Trifolium pratense	V	+5°C	400 Days	48	33		110
T. pratense	F	Room	—	—	—		82
Tulipa sp.	V	+5°C	400 Days	72	58		110
Vaccinium augustifolium	F	−20°C	4 Months	—	4	At 144 hr, 15%	120
V. corymbosum	F	Room	6 Months	—	60		43
Zea mays	F	Room	12	—	—	Seed set	43
Z. mays	F	22°C	135 Days	70	~5		91
Z. mays	V	+5°C	300 Days,	25	18	At 400 Days, 0%	110

[a] Samples were either freeze-dried (F) or vacuum-dried (V) prior to storage.

[b] B: percentage germination before storage; A: percentage germination after storage.

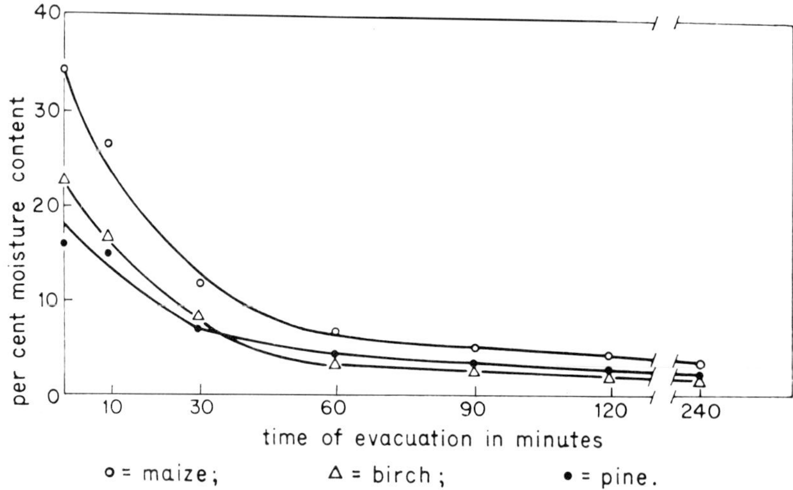

FIGURE 3.   Kinetics of moisture loss from pollen of maize, birch, and pine during vacuum-drying. (From Jensen, C. J., *Kobenhaver K. Vet. Landbohoejsk.*, 133, 1964. With permission.)

compared to 20°C.[94] Even at 20°C, some germination occurred after 2 years of storage. These pollen samples also retained the ability to set seed. Vacuum-dried pollen of jack pine, white pine, and Norway spruce retained viability after 5 years of storage at +4°C.[115] Vacuum-dried alfalfa pollen set seed after 11 years of storage at −21°C.[114] As with other storage procedures, considerable variation was found among different genotypes of a species.[113] Freeze- or vacuum-dried pollen stored at room temperature has greater longevity than fresh pollen. This longevity could allow shipment of some pollen to distant locations without insulated packaging.

Storage atmosphere can affect the longevity of freeze- or vacuum-dried pollen. Most studies with freeze- or vacuum-dried pollen have used either a nitrogen atmosphere or vacuum for storage. A vacuum or nitrogen atmosphere prolonged maize pollen viability as compared to air storage.[91] Vacuum storage was better than a nitrogen atmosphere storage for vacuum-dried pea pollen;[116] however, the reverse was true for freeze-dried *Asparagus officinalis* pollen.[56] The detrimental aspects of oxygen on dried organisms have long been known. Oxygen exposure can generate long-lived free radicals, but it has been difficult to determine whether these directly cause a loss in viability.[121]

The reasons for loss of viability during the freeze- or vacuum-drying process or during subsequent storage are not well defined. Most studies have not examined extensive variations in the procedure to optimize survival. Sample temperature during the drying process is often not recorded. Thus, it is difficult to determine why some pollen samples survive the process well and others do not. Drawing from the literature on freeze-drying fungal spores and cells, Mazur[122] analyzed freeze-drying and vacuum-drying techniques and emphasized the following points: (1) for retaining viability it is best not to subject the sample to temperatures where freezing damage alone might occur; (2) slow cooling to the temperature of choice is often preferable to rapid cooling; (3) drying should not be so severe as to remove "bound" water — to do so usually drastically lowers viability. The main advantage that freeze- or vacuum-drying has over freeze preservation is that samples can be stored at 0°C rather than ca. −75°C and still retain a high percent of survival over time.

Insufficient information exists for evaluating the best freezing temperature and cooling rate for pollen at different moisture levels. Prefreezing peach and Japanese pear pollen at −1, −20, and −40, or −196°C did not significantly alter germination after freeze-drying,[53] but this is expected if the pollen was initially quite dry. Lily and maize pollen have survived fast cooling better than slow cooling prior to the drying step,[91] but initial moisture contents were not stated so that reasons for injury are speculative. If the water content of pollen is high, the initial prefreezing step, especially when done with rapid cooling, may be injurious, probably because of intracellular ice formation.[118] Samples of Western white pine pollen with high moisture contents that were exposed to $-78°C$ and then freeze-dried showed a more rapid appearance of metabolites in the leachate compared to an air-dried sample similarly treated. This suggests that the initial freezing step produced disruptive intracellular ice.

Injury may also occur during the drying phase of a freeze- or vacuum-drying protocol.[123] Permeability and respiration changes were observed in freeze-dried lily pollen, but not in pollen that was only frozen to -65 to $-75°C$. Western white pine pollen that was air-dried and then subjected to freeze-drying also showed permeability changes during drying.[118] These observations implicate membrane alterations during extensive dehydration. The organization of membranes in dry cells is speculative. The usual phopholipid bilayer membrane structure present in a hydrated cell may be rearranged into a hexagonal array with the hydrophilic heads in a center channel and the hydrophobic tails forming the confluent phase.[50] This change in membrane structure occurs at 20 to 30% water content, so one can suppose that pollen membranes are altered during both drying and freeze-drying. If an analogy with seed inhibition is possible, the low viability of pollen which is rapidly hydrated (by being directly immersed in the medium for a germination test) may be due to extensive loss of metabolites to the medium, whereas the slow rehydration in a humid atmosphere allows the membranes to rearrange without extensive metabolite loss.

Thus, pollen can be preserved for fairly long periods of time after freeze- or vacuum-drying. Viability is often high if critical factors are evaluated and optimized. Unless pollen is stored at low temperatures, however, there is a distinct decrease in survival over periods of one to a few years. These results are not that different from simple storage at about $-20°C$. The added technical requirements for freeze- or vacuum-drying are probably not worth the effort for pollen from many species, but the method may be worthwhile if shipment of pollen is desired. Most success with freeze- or vacuum-drying has been with binucleate pollen grains. Studies with trinucleate pollen from diverse species are lacking, but the success observed with maize and rye suggest that further studies are warranted.

IV. SUMMARY

Pollen cryobiology has not been systematically or extensively studied. Information on low-temperature storage was obtained through empirical studies whose goals were to increase pollen longevity for breeding purposes. These studies gave practical procedures for storage of some pollens, but the physiology and biochemistry of pollen during storage, particularly with regard to viability decline, remain poorly understood.

Longevity of binucleate pollen can be increased through moisture reduction and low temperatures. If pollen moisture content is suitably adjusted, no damage occurs during cooling to temperatures lower than $-40°C$ and subsequent warming. This is usually true for pollen cooled to $-196°C$ as well, regardless of cooling or warming rates. Many pollens do not survive below $-40°C$ if the moisture content is greater than 20 to 30%. Drying below 1 to 5% moisture content can, of itself, decrease viability. Viable pollen can be obtained after freeze- and vacuum-drying procedures if initial moisture content, freezing procedure, and drying procedure are controlled. Longevities of freeze- and vacuum-dried pollen are increased at lower storage temperature.

Longevities at temperatures within the range -10 to $-43°C$ vary with species but typically are from 1 to 3 years. Storage between -80 and $-34°C$ have not been reported but would be expected to provide increased longevity now that suitable refrigeration systems are available. Storage data at temperatures below $-80°C$ are sparse but suggestive of greater longevities. Although samples stored at cryogenic temperature should not show viability loss, data in the literature have illustrated an occasional decline. Procedural problems, such as loss of refrigerant or uncontrolled warming and refreezing, need to be eliminated. Detailed studies of binucleate pollen longevity at different subzero temperatures, and particularly at cryogenis, are needed if such temperatures are to be used for germplasm preservation.

Temperatures below freezing have not been adequately examined for most trinucleate pollens. Experiments interrelating storage temperature, critical moisture content, and metabolism are necesary to define the mechanism of storage injury. Trinucleate pollens are desiccation-sensitive but species-specific limits need to be determined. For instance, maize pollen can survive low-temperature exposure, whereas all attempts with sugar cane pollen have failed. The potential for using cyroprotectants to increase survival in trinucleate pollen will only be realized if injury is adequately explained.

REFERENCES

1. **Heslop-Harrison, J.,** *Pollen: Development and Physiology,* Butterworths, London, 1971.
2. **Stanley, R. G. and Linskens, H. F.,** *Pollen: Biology, Biochemistry and Management,* Springer-Verlag, Basel, 1974.
3. **Fehr, W. R.and Hadley, H. H., Eds.,** *Hybridization of Crop Plants,* Am. Soc. Agron., Madison, Wis., 1980.
4. **Janick, J. and Moore, J. N.,** *Advances in Fruit Breeding,* Purdue University Press, West Lafayette, Ind., 1975.
5. **Franklin, E. C., Ed.,** Pollen Management Handbook, agriculture Handbook N-O. 587, U.S. Department of Agriculture, Washington, D.C., 1981.
6. **Snyder, E. B. and Clausen, K. E.,** Pollen handling, in Seeds of Woody Plants in the United States, Schopmeyer, C. S., Ed., Agriculture Handbook No. 450, U.S. Department of Agriculture, Washington, D.C., 1974.
7. **Harrington, J. F.,** Seed and pollen storage for conservation of plant gene resources, in *Genetic Resources in Plants — their Exploration and Conservation,* Frankel, O. H. and Bennett, E., Eds., Blackwell, Scientific, Oxford, 1970, 501.
8. **Schoenike, R. E. and Bey, C. F.,** Conserving genes through pollen storage, in Pollen Management Handbook, Franklin, E. C., Ed., U.S. Department of Agriculture Handbook No. 587, Washington, D.C., 1981, 72.
9. **Miller, T. and Belcher, E. W.,** Pollen contamination and quarantine restrictions, in Pollen Management Handbook, Franklin, E. D., Ed., U.S. Department of Agriculture Handbook No. 587, Washington, D.C., 1981, 70.
10. **Mascarenhas, J. P.,** Microspore and microgametophyte development in relation to biological activity of environmental pollutants, *Environ. Health Perspect.,* 37, 9, 1981.
11. **Pfahler, P. L.,** *In vitro* germination characteristics of maize pollen to detect biological activity of environmental pollutants, *Environ. Health Perspect.,* 37, 125, 1981.
12. **Kasha, K. J., Ed.,** *Haploids in Higher Plants: Advances and Potential,* University of Geulph, Guelph, Ontario, Canada, 1974.
13. **Bajaj, Y. P. S.,** Effect of super-low temperature on excised anthers and pollen-embryos of *Atropa, Nicotiana,* and *Petunia, Phytomorphology,* 28, 171, 1978.
14. **Jeffries, C. J.,** Sequential staining to assess viability and starch content in individual pollen grains, *Stain Technol.,* 52, 277, 1977.
15. **Alexander, M. P.,** Differential staining of aborted and nonaborted pollen, *Stain Technol.,* 44, 117, 1969.
16. **Sarvella, P.,** Vital-stain testing of pollen viability in cotton, *J. Hered.,* 55, 154, 1964.
17. **Belling, J.,** On counting chromosomes in pollen mother cells, *Am. Nat.,* 55, 573, 1921.
18. **Thomas, P. T.,** The acetocarmine method for fruit materials, *Stain Technol.,* 15, 167, 1940.
19. **Marks, G. E.,** A controllable carmine technic for plants and small chromosomes, *Stain Technol.,* 27, 333, 1952.
20. **Janssen, A. W. B. and Hermsen, J. G. Th.,** Estimating pollen fertility in *Solanum* species and haploids, *Euphytica,* 25, 577, 1976.
21. **King, J. R.,** The peroxidase reaction as an indicator of pollen viability, *Stain Technol.,* 35, 225, 1960.

22. **Dashek, W. V., Erickson, S. S., Hayward, D. M., Lindbeck, G., and Mills, R. R.,** Peroxidase in cytoplasm and cell wall of germinating lily pollen, *Bot. Gaz.,* 140, 261, 1979.
23. **Oberle, G. D. and Watson, R.,** The use of 2,3,5-triphenyl tetrazolium chloride in viability tests of fruit pollens, *Proc. Am. Soc. Hortic. Sci.,* 61, 299, 1953.
24. **Werner, D. J. and Chang, S.,** Stain testing viability in stored peach pollen, *HortScience,* 16, 522, 1981.
25. **Norton, J. D.,** Testing of plum pollen viability with tetrazolium salts, *Proc. Am. Soc. Hortic. Sci.,* 89, 132, 1966.
26. **Brewbaker, J. L. and Kwack, B. H.,** The essential role of calcium ion in pollen germination and pollen tube growth, *Am. J. Bot.,* 50, 859, 1963.
27. **Heslop-Harrison, J. and Heslop-Harrison, Y.,** Evaluation of pollen viability by enzymatically induced fluorescence; intracellular hydrolysis of fluorescein diacetate, *Stain Technol.,* 45, 115, 1970.
28. **Shivanna, K. R. and Heslop-Harrison, J.,** Membrane state and pollen viability, *Ann. Bot. (London),* 47, 759, 1981.
29. **Visser, T.,** Germination and storage of pollen, *Meded. Landbouwhogesch. Wageningen,* 55, 1, 1955.
30. **Johri, B. M. and Vasil, I. K.,** Physiology of pollen, *Bot. Rev.,* 27, 325, 1961.
31. **Linskens, H. F.,** Pollen physiology, *Ann. Rev. Plant Physiol.,* 15, 255, 1964.
32. **Towill, L. E.,** Preservation of potato pollen, in *The Potato,* Bajaj, Y. P. S., Ed., Springer-Verlag, Basel, in press.
33. **Ching, T. M. and Ching, K. K.,** Rapid viability tests and aging study of some coniferous pollen, *Can. J. For. Res.,* 6, 516, 1976.
34. **Foster, G. S. and Bridgewaer, F.,** Viability tests to evaluate pollen reliability in loblolly pine controlled pollinations, *For. Sci.,* 25, 270, 1979.
35. **Holman, R. M. and Brubaker, F.,** On the longevity of pollen, *Univ. Calif. Publ. Bot.,* 13, 179, 1926.
36. **Altman, P. L. and Dittmer, D. S.,** Life spans: pollen, in *Biology Data Book,* Vol. 1, 2nd ed., Fed. Am. Soc. Exp. Biol., Bethesda, Md., 1972, 242.
37. **Kozaki, I.,** Storage methods of pollen, in *Gene Conservation,* JIBP Syntheses, Vol. 5, Matsuo, T., Ed., 1975, 89.
38. **Brewbaker, J. L.,** Biology of the angisper pollen grain, *Indian J. Genet. Plant Breed.,* 19, 121, 1959.
39. **Visser, T., DeVries, D. P., Welles, G. W. H., and Scheurink, J. A. M.,** Hybrid tea-rose pollen. I. Germination and storage, *Euphytica,* 26, 721, 1977.
40. **Visser, T. and Oost, E. H.,** Pollen and pollination experiments. III. The viability of apple and pear pollen as affected by irradiation and storage, *Euphytica,* 30, 65, 1981.
41. **Towill, L. E.,** Liquid nitrogen preservation of pollen from tuber-bearing *Solanum* species, *HortScience,* 15, 177, 1981.
42. **King, J. R.,** The freeze-drying of pollens, *Econ. Bot.,* 15, 91, 1961.
43. **King, J. R.,** The storage of pollen — particularly by the freeze-drying method, *Bull. Torrey Bot. Club,* 92, 270, 1965.
44. **Hardon, J. J. and Davies, M. D.,** Effects of vacuum-drying on the viability of oil palm pollen, *Exp. Agric.,* 5, 59, 1969.
45. **Goss, J. A.,** Development, physiology, and biochemistry of corn and wheat pollen, *Bot. Rev.,* 34, 333, 1968.
46. **Manthriratna, M. A. P. P. and Hayward, M. D.,** Pollen development and variation in the genus *Lolium, Z. Pflanzenzucht.,* 69, 210, 1973.
47. **Chiang, M. S.,** Cabbage pollen germination and longevity, *Euphytica,* 23, 579, 1974.
48. **Henny, R. J.,** Germination of *Dieffenbachia maculata* 'Perfection' pollen after storage at different temperatures and relative humidity regimes, *HortScience,* 15, 191, 1980.
49. **Henny, R. J.,** Germination of *Spathiphyllum* and *Vriesea* pollen after storage at different temperatures and relative humidities, *HortScience,* 13, 596, 1978.
50. **Simon, E. W.,** Membranes in dry and imbibing seeds, in *Dry Biological Systems,* Crowe, J. H. and Clegg, J. S., Eds., Academic Press, New York, 1978, 205.
51. **Winston, P. W. and Bates, D. H.,** Saturated solutions for the control of humidity in biological research, *Ecology,* 41, 232, 1960.
52. **Lanner, R. M.,** Controlling the moisture content of conifer pollen, *Silvae Genet.,* 11, 114, 1962.
53. **Akihama, T., Omura, M., and Kozaki, I.,** Further investigation of freeze-drying for deciduous fruit tree pollen, in *Long Term Preservation of Favorable Germplasm in Arboreal Crops,* Akihama, T. and Nakajima, K., Eds., Fruit Tree Research Station M.A.F., Japan, 1978, 1.
54. **Kwan, S. C., Hanson, A. R., and Campbell, W. F.,** Storage conditions for *Allium cepa* L. pollen, *J. Am. Soc. Hortic. Sci.,* 94, 569, 1969.
55. **Pfeiffer, N. E.,** Longevity of pollen of *Lilium* and hybrid *Amaryllis, Contrib. Boyce Thompson Inst.,* 8, 141, 1936.
56. **Snope, A. J. and Ellison, J. H.,** Storage of asparagus pollen under various conditions of temperature, humidity, and pressure, *Proc. Am. Soc. Hortic. Sci.,* 83, 447, 1963.

57. **Cunningham, A.,** Pollen viability after low temperature storage in *Brassica campestris, B. oleracea,* and *B. napus, Cruciferae Newsl. (Eucrapia),* 6, 18, 1981.

58. **Yamamoto, C. and Saito, M.,** Long-term storage of *Chamaecyparis obtusa* pollen under various conditions of humidity and temperature — preliminary results, in *Long Term Preservation of Favorable Germ Plasm in Arboreal Crops,* Akihama, T. and Nahakima, K., Eds., Fruit Tree Research Station M.A.F., Japan, 1978, 13.

59. **Kobayashi, S., Ikeda, I., and Nakatani, M.,** Long-term storage of citrus pollen, in *Long Term Preservation of Favorable Germ Plasm in Arboreal Crops,* Akihama, T. and Nakajima, K., Eds., Fruit Tree Research Station M.A.F., Japan, 1978, 8.

60. **Sahar, N. and Spiegel-Roy, P.,** Citrus pollen storage, *HortiScience,* 15, 81, 1980.

61. **Walyaro, D. J. and Van Der Vossen, H. A. M.,** Pollen longevity and artificial cross pollination in *Coffea arabica* L., *Euphytica,* 26, 225, 1977.

62. **Zielinski, Q. B.,** Techniques for collecting, handling, germinating, and storing pollen of the filbert (*Corylus* ssp.), *Euphytica,* 17, 121, 1968.

63. **Griggs, W. H., Vansell, G. H., and Iwakiri, B. T.,** The storage of hand-collected and bee-collected pollen in a home freezer, *Proc. Am. Soc. Hortic. Sci.,* 62, 304, 1953.

64. **Griggs, W. H., Forde, H. I., Iwakiri, B. T., and Asay, R. N.,** Effect of subfreezing temperature on the viability of Persian walnut pollen, *HortScience,* 6, 235, 1971.

65. **Farmer, R. E., Jr. and Barnett, P. E.,** Low temperature storage of black walnut pollen, *Cryobiology,* 11, 366, 1974.

66. **Pfeiffer, N. E.,** Effect of lyophilization on the viability of *Lilium* pollen, *Contrib. Boyce Thompson Inst.,* 18, 153, 1955.

67. **Saxena, H. K. and Saini, J. P.,** Effect of storage on viability of Regal lily pollen grains, *Indian J. Plant Physiol.,* 22, 269, 1979.

68. **Hanson, C. H.,** Longevity of pollen and ovaries of alfalfa, *Crop Sci.,* 1, 114, 1961.

69. **Crawford, C. L.,** Effectiveness of date pollen following cold storage, *Proc. Am. Soc. Hortic. Sci.,* 35, 91, 1937.

70. **Feret, P. P. and Stairs, G. R.,** Vacuum-dried spruce pollen: pollen viability and seed germination, University of Wisconsin For. Res. Notes No. 153, 1970.

71. **Duffield, J. W. and Callaham, R. Z.,** Deep-freezing pine pollen, *Silvae Genet.,* 8, 22, 1959.

72. **Callaham, R. Z. and Steinhoff, R. J.,** Pine pollen frozen five years produces seed, *U.S. For. Serv. Res. Pap. NC,* 6, 94, 1966.

73. **Warnock, S. J. and Hagedorn, D. J.,** Germination and storage of pea (*Pisum sativum*) pollen, *Agron. J.,* 48, 347, 1955.

74. **Knox, R. B., Willing, R. R., and Pryor, L. D.,** Interspecific hybridization in poplars using recognition pollen, *Silvae Genet.,* 21, 65, 1972.

75. **Haaring, R., Zimmerman, L. H., and Smith, J. D.,** Short- and long-period storage of pollen of *Ricinus communis* L., *Crop Sci.,* 9, 17, 1969.

76. **Dale, A.,** Some consequences of pollen storage in the raspberry (*Rubus idaeus*), *Euphytica,* 26, 745, 1977.

77. **Weatherhead, M. A., Grout, B. W. W., and Henshaw, G. G.,** Advantage of storage of potato pollen in liquid nitrogen, *Potato Res.,* 21, 331, 1978.

78. **DeMaine, M. J.,** A simple technique for the short-term storage of pollen in a dihaploid-inducing potato clone, *J. Agric. Sci. Camb.,* 89, 511, 1977.

79. **Blomquist, A. W. and Lauer, F. I.,** A simplified technique for handling and storing potato pollen, *Am. Potato J.,* 39, 340, 1962.

80. **King, J. R.,** Irish potato pollen storage, *Am. Potato J.,* 32, 460, 1955.

81. **Howard, H. W.,** The storage of potato pollen, *Am. Potato J.,* 35, 676, 1958.

82. **Engelke, M. C. and Smith, R. R.,** Effect of storage on germination and viability of red clover pollen grains, *Agronomy Abstr.* Am. Soc. Agron., Madison, Wis., 1974, 52.

83. **Olmo, H. P.,** Storage of grape pollen, *Proc. Am. Soc. Hortic. Sci.,* 41, 219, 1942.

84. **Junttila, O. and Stushnoff, C.,** Freezing avoidance by deep supercooling in hydrated lettuce seeds, *Nature (London),* 260, 325, 1977.

85. **Knowlton, H. E.,** Studies in pollen with special reference to longevity, *Cornell Univ. Exp. Stn. Mem.,* 52, 747, 1922.

86. **Bredeman, G., Garber, K., Harteck, P., and Suhr, K. A.,** Die Temperaturabhangigkeit der Lebensdauer von Blutenpollen, *Naturwissenschatten,* 34, 279, 1947.

87. **Collins, F. C., Lertmongkol, V., and Jones, J. P.,** Pollen storage of certain agronomic species in liquid air, *Crop Sci.,* 13, 493, 1973.

88. **Otsuka, K.,** Survival of pollen cells at super-low temperatures, *Contrib. Low Temp. Sci. Ser. B,* 29, 107, 1971.

89. **Wakisaka, I.,** Ultra low temperature storage of pollens of Japanese persimmons (*Diospyros kaki* Linn f), *Engei Gakkai Zasshi,* 33, 291, 1964.

90. **Ichikawa, S., Kaji, K., and Kubota, Y.,** Studies on the storage of larch (*Larix leptolepis*) pollen at super-low temperatures, *Bull. Hokkaido For. Exp. Stn.*, 8, 11, 1970.

91. **Nath, J. and Anderson, J. O.,** Effect of freezing and freeze-drying on the viability and storage of *Lilium longiflorum* L. and *Zea mays* L. pollen, *Cryobiology*, 12, 81, 1975.

92. **Sedgley, M.,** Storage of avocado pollen, *Euphytica*, 30, 595, 1981.

93. **Tisserat, B., Nelson, M. D., Ulrich, J. M., and Finkle, B. J.,** Survival of *Phoenix* pollen grains under cryogenic conditions, *Crop Sci.*, in press.

94. **Livingston, G. K. and Ching, K. K.,** The longevity and fertility of freeze-dried Douglas-fir pollen, *Silvae Genet.*, 16, 98, 1967.

95. **Griggs, W. H., Vansell, H., and Reinhardt, J. F.,** The germinating ability of quick-frozen, bee-collected apple pollen stored in a dry ice container, *J. Econ. Entomol.*, 43, 549, 1950.

96. **Barnabas, B. and Rajki, E.,** Storage of maize (*Zea mays* L.) pollen at $-196°C$ in liquid nitrogen, *Euphytica*, 25, 747, 1976.

97. **Barnabas, B. and Rajki, E.,** Fertility of deep-frozen maize (*Zea mays* L.) pollen, *Ann. Bot. (London)*, 48, 861, 1981.

98. **Ichikawa, S. and Shidei, T.,** Fundamental studies on deep-freezing storage of tree pollen. III, *Bull. Kyoto Univ. For.*, 44, 47, 1972.

99. **Ichikawa, S. and Shidei, T.,** Fundamental studies on deep-freezing storage of tree pollen, *Bull. Kyoto Univ. For.*, 42, 51, 1972.

100. **Ichikawa, S. and Shidei, T.,** Fundamental studies on deep-freezing on tree pollen. II, *Bull. Kyoto Univ. For.*, 43, 9, 1972.

101. **Sakai, A.,** Survival of plant tissue at superlow temperatures. III. Relation between effective prefreezing temperatures and the degree of frost hardiness, *Plant Physiol.*, 40, 882, 1965.

102. **Ching, T. M. and Slabaugh, W. H.,** X-ray diffraction analysis of ice crystals in coniferous pollen, *Cryobiology*, 2, 321, 1966.

103. **Nikolaev, G. M., Tainla, E. A., Tukeeva, M. I., Morozova, E. M., Zubarev, V. A., and Ermakov, I. P.,** Investigation of the water state in mature pollen grains of *Tradescantia paludosa*, *Fiziol. Rast. (Moscow)*, 29, 312, 1982.

104. **Mazur, P.,** Freezing and low-temperature storage of living cells, in *Basic Aspects of Freeze Preservation of Mouse Strains*, Muehlbock, O., Ed., Springer-Verlag, Basel, 1976, 1.

105. **Moore, P. H.,** Studies in sugarcane pollen. II. Pollen storage, *Phyton*, 34, 71, 1976.

106. **Anderson, J. O. and Nath, J.,** The effects of freeze-preservation on some pollen enzymes. I. Freeze-thaw stresses, *Cryobiology*, 12, 160, 1975.

107. **Anderson, J. O. and Nath, J.,** The effects of freeze-preservation on some pollen enzymes. II. Freezing and freeze-drying stresses, *Cryobiology*, 15, 469, 1978.

108. **Dickinson, D. B. and Cochran, D.,** Dimethyl sulfoxide: reversible inhibitor of pollen tube growth, *Plant Physiol.*, 43, 411, 1968.

109. **Skelnik, D. L. and Ascher, P. D.,** Dimethyl sulfoxide as a cryoprotectant for pollen, *HortScience*, 11, 303, 1976.

110. **Jensen, C. J.,** Pollen storage under vacuum, *Kobenhaver K. Vet. Landbohoejsk.*, 133, 1964.

111. **Whitehead, R. A.,** Freeze-drying and room temperature storage of coconut pollen, *Econ. Bot.*, 19, 267, 1965.

112. **Benard, G.,** Quelques aspects de la lyophilisation du pollen de cocotier, *Oleagineaux*, 10, 447, 1973.

113. **Dietze, V. W.,** Gefriertrocknung und Lagerung von Pollen verschiedener Waldbaumarten, *Silvae Genet.*, 22, 154, 1973.

114. **Hanson, C. H. and Campbell, T. A.,** Vacuum-dried pollen of alfalfa (*Medicago sativa* L.) viable after eleven years, *Crop Sci.*, 12, 874, 1972.

115. **Schoenike, R. E. and Stewart, D. M.,** Fifth year results of vacuum-drying storage and additives on the viability of some conifer pollens, *For. Sci.*, 9, 96, 1963.

116. **Layne, R. E. C. and Hagedorn, D. J.,** Effect of vacuum-drying, freeze-drying, and storage environment on the viability of pea, *Crop Sci.*, 3, 433, 1963.

117. **Ahlgren, C. E. and Ahlgren, I. F.,** Viability and fertility of vacuum dried pollen of 5-needle pine species, *For. Sci.*, 24, 100, 1978.

118. **Ching, T. M. and Ching, K. K.,** Freeze-drying pine pollen, *Plant Physiol.*, 39, 705, 1964.

119. **Herrman, Von S.,** Verfahren zur Konservierung und Erhaltung der Befruch-tungsfahigkeit von Wald-baumpollen uber mehrere Jahre, *Silvae Genet.*, 25, 223, 1976.

120. **Wood, G. W. and Barker, W. G.,** Preservation of blueberry pollen by the freeze-drying process, *Can. J. Plant Sci.*, 44, 387, 1964.

121. **Heckly, R. J.,** Effects of oxygen on dried organisms, in *Dry Biological Systems*, Crowe, J. H. and Clegg, J. S., Eds., Academic Press, New York, 1978, 257.

122. **Mazur, P.,** Survival of fungi after freezing and dessication, in *The Fungi*, Vol. 3, Ainsworth, G. C. and Sussman, A. S., Eds., Academic Press, New York, 1968, 325.

123. **Davies, M. D. and Dickinson, D. B.,** Effects of freeze-drying on the permeability and respiration of germinating lily pollen, *Plant Physiol.,* 24, 5, 1971.
124. **Towill, L. E.,** unpublished data.
125. **Haunold, A. and Stanwood, P.,** personal communication.
126. **Finkle, B. J.,** personal communication.

Chapter 10

CRYOPRESERVATION OF SEED GERMPLASM FOR GENETIC CONSERVATION

Phillip C. Stanwood

TABLE OF CONTENTS

I. INTRODUCTION

From the beginning of agriculture, man has used seed to preserve important plants that were needed for food, feed, fiber, and medicinal purposes. Coupled with the advancement and modernization of agriculture there has been an ever increasing dependence on plant diversity, easy access to that diversity, and the ability to use that diversity to improve the production and quality of crop plants. The collection and preservation of seed germplasm is crucial to the crop plant improvement process.[1] Consequently, a high priority exists to assemble, store, and preserve as much genetic variability as possible of the important crop plants and their relatives.

Overcoming physiological deterioration of a stored seed sample is one of the most important and difficult tasks in the preservation process. Large-scale mechanical refrigeration systems, which hold seed at temperatures down to $-20°C$ have greatly increased the storage life of a seed sample.[2] However, deterioration and loss of viability can still occur with increased time in storage, resulting in a loss of precious genetic material. Cryopreservation techniques (e.g., liquid nitrogen — LN2 at $-196°C$) provide the potential for "indefinite" preservation by reducing metabolism to such a low level[3-5] that all biochemical processes are significantly reduced and biological deterioration virtually stopped.

The longevity of seeds or the maintenance of seed viability is a balance between extrinsic and intrinsic deleterious factors and repair or protective mechanisms. Deleterious factors may include depletion of essential metabolites, denaturation of macromolecules, accumulation of toxic metabolites, attack by microorganisms and insects, and effects of ionizing radiation.[6-8] Repair and protective mechanisms are not well understood but could be as fundamental as repair of DNA molecules[9] or general enzyme repair and replacement activities.[10] Protective mechanisms such as hard seed coats can exclude oxygen and/or water from the embryonic tissue thus extending seed life. Depending upon the particular mechanism(s) involved and external factors such as storage temperature, seed moisture content, and oxygen availability,[11-12] the life of a seed may be shortened or extended. It has not been demonstrated that temperatures of -40, -70, or $-196°C$ are significantly better than $-20°C$, in terms of physiological preservation, although estimates from Roberts[8] and Harrington[12] would suggest that the colder the storage, the longer the potential storage time. The potential advantage of LN2 storage is that all metabolism would be essentially stopped, causing a stalemate between factors, thus, greatly extending storage life.

If all physiological factors associated with storage temperature and seed storage life were equal, the use of LN2 as a storage medium would be preferred over mechanical refrigeration systems because of the relative reliability and ease of use of LN2 storage systems. Liquid nitrogen storage vats are passive in nature and only require that LN2 be added to the reservoir on a periodic basis (e.g., 2 weeks to 4 months, depending on the construction of the vat and size of the LN2 reservoir). Pumps, mechanical devices, or electricity are not required for their operation. A reliable source of bulk LN2 is necessary and is generally available throughout the world.

The importance of preserving seed genetic material demands that the best and safest possible storage technique be employed. LN2 storage appears to be both practical and desirable for long-term preservation of numerous kinds of seed.

II. SEED RESPONSE TO LN2 EXPOSURE

The utilization of LN2 as a storage medium is predicted on the capability of seeds to survive LN2 exposure without significant damage to viability. Seeds fall into three general categories with regard to exposure to LN2 temperatures:

1.   Desiccation-tolerant LN2-tolerant

2. Desiccation-tolerant LN2-sensitive
3. Desiccation-sensitive LN2-sensitive

Seeds within a species tend to fall under one distinct category, however there may be external factors or handling procedures, such as rate of cooling/rewarming, which may cause a given kind of seed to fall into more than one category.

Roberts[8] describes two major categories of seed based on their response to dehydration: (1) desiccation tolerant or "orthodox" seeds which can withstand drying to 1 to 3% moisture without damage to viability; and (2) desiccation sensitive or "recalcitrant" seeds which are generally killed if dried below a critical value, usually between 12 and 35% moisture content. Lower seed moisture content is associated with an increase in storage life and, conversely, higher seed moisture content shortens the storage life of a sample. Because of the association between moisture content and storability, the terms orthodox and recalcitrant often have mixed meanings. In the content of cryopreservation, the terms desiccation-tolerant and/or sensitive are more precise and thus preferred.

A. Desiccation-Tolerant LN2-Tolerant Seed

Seeds of most common agricultural and horticultural species are tolerant to desiccation and exposure to LN2. Researchers have had considerable success in cooling such seeds to LN2 temperatures and rewarming them to ambient (20°C) without loss of viability. A listing of the many examples found in the literature[13-44] dating back to the late 1800s is given in Table 1. With over 155 species and 455 selections represented in the list, one is led to conclude that LN2 preservation of this category of seeds is quite feasible. In many cases this is true. However, one has to be aware that there are certain simple factors which must be accounted for or understood in order to achieve successful cryopreservation of such seeds, namely seed moisture content, cooling/rewarming rates, and potential physical damage to the seeds.

1. Seed Moisture Content

Seed moisture content is probably the most critical factor relating to successful cryopreservation. An optimum seed moisture content must be attained. If the seed moisture content is too high, instantaneous death is observed during the cooling/rewarming process. Too low a moisture content in seeds of some species such as sesame (see next section), can result in partial loss of viability of a sample depending upon the cool/rewarming protocol.

High moisture freezing limit (HMFL) is a seed moisture threshold that, if exceeded, will result in a decrease in viability of a seed sample during the LN2 cooling/rewarming process. This threshold is normally a relatively narrow seed moisture content range within a species but can vary between species. Examples of seed HMFL, which range from a low of 9.6% for sesame (*Sesamum indicum* L.), to a high of 28.5% for bean (*Phaseolus vulgaris* L.), are given in Table 2.[14] The HMFL phenomenon in clover (*Trifolium* sp.) and wheat (*Triticum aesitivum* L.) seeds is graphically illustrated in Figures 1A and 1B respectively. In these examples, the seeds were adjusted to different moisture levels before being exposed to LN2. Clover had a sharp drop in viability when the seed moisture content reached approximately 25%. A similar response was noted in wheat, except that a wider range for the HMFL was observed. The other species listed in Table 2 exhibited similar HMFL responses.

2. Cooling/Rewarming Rate

Much of the literature suggest that the cooling/rewarming rate is not a significant factor in the survival of most species of seed exposed to LN2. Notable exceptions to this are exemplified by experiments conducted with lettuce (*Lactuca sativa* L.) and sesame (*Sesamum indicum* L.) seeds. Roos[45] working with lettuce seeds with moisture contents near the HMFL

Table 1
SPECIES OF SEED REPORTED TO HAVE SURVIVED LN2 EXPOSURE

Species/selection	Moisture (%)	Duration	Germination (%)		Comments	Ref.
			Before	After		
Abies concolor (Gord. & Glend.) Lindl. ex Hildebr.	9.8	1 Y	59	50	−200°C/min cool., 30°C/min rewarm., 2 selections	13, 14
Abies procera Rehd. Sele. Cascade Collection	8.9	1 Y	37	25	−200°C/min cool., 30°C/min rewarm.	14
Abies procera Rehd.	10 1	3 Y	51	52	−200°C/min cool., 30°C/min rewarm., 4 selections	13, 14
Alcea rosa L. cv. Majorette	7.2	1 M	65	63	−200°C/min cool., 30°C/min rewarm.	13, 14
Allium cepa L. cv. Suttons A1	4.0	ca. 3.75 Y	N.R.	83.5	Exposed seeds showed no evidence of cytological abnormalities	15
Allium cepa L. cv. Silver-shinned	N.R.	1 M	N.R.	No change	Rapid cooling, rewarming N.R., liquid air treatment on seedlings	16
Allium cepa L. cv. Early Mild Bunching	8.3	2 M	80*	94	−200°C/min cool., 40°C H_2O (90 sec) rewarm.	17
Allium spp.	6.5	3 Y	90	92	−200°C/min cool., 30°C/min rewarm., 21 cultivars	13, 18, 14
Amaranthus tricolor L. cv. Early Splendor	6.1	1 Y	96	91	−200°C/min cool., 30°C/min rewarm.	14
Anemone coronavia L.	N.R.	ca. 3 M	N.R.	No change	Exposed seeds showed no evidence of cytological abnormalities	15
Anethum graveolens L. cv. Dill Stokes #158	8.6	600 D	76	85	−200°C/min cool., 40°C H_2O (90 sec) rewarm.	17
Antirrhinum majus L.	N.R	ca. 3 M	N.R.	No change	Exposed seeds showed no evidence of cytological abnormalities	15
Antirrhinum majus L. cv. Debutante	3.1	1 Y	94	92	−200°C/min cool., 30°C/min rewarm.	14
Antirrhinum majus L. cv. Floral Carpet	6.8	600 D	78	71	−200°C/min cool; 40°C H_2O (90 sec) rewarm.	17
Apium graveolens L.	10.2	3 Y	89	90	−200°C/min cool., 30°C/min rewarm., 2 cultivars	14
Apium graveolens L. cv. Florida 683	7.6	600 D	77*	71	−200°C/min cool; 40°C H_2O (90 sec) rewarm.	17
Aquilegia sp. cv. Early Splendor	6.1	1 Y	65	73	−200°C/min cool., 30°C/min rewarm.	14

Species						Ref.
Arabidopsis thaliana L. Heynh.	N.R.	N.R.	93	No change	−200°C/min cool., 37°C H$_2$O rewarm.	19
Arabidopsis thaliana L. Heynh.	N.R.	ca. 4 H	96	95	Liquid air/radiation experiments	20
Arachis hypogaea L. cv. Florunner	7.3	30 D	100	100	−200°C/min cool., 30°C/min rewarm.	21, 13
Arachia hypogaea L.	N.R.	24—108 H	N.R.	N.R.	Plants grown from single peanut cotyledons exposed to −195.8°C	22
Arachis hypogaea L. cv. Mammoth Jumbo	N.R.	48 H	50	10	Germination reported after 48 hr at 25°C.	23
Asparagus densiflorous (Kunth.) Jessop. cv. Sprengeri	7.8	1 Y	88	82	−200°C/min cool., 30°C/min rewarm.	14
Asparagus officinalis L.	N.R.	ca. 3 M	N.R.	No change	Exposed seeds showed no evidence of cytological abnormalities	15
Asparagus officinalis L. cv. Viking	8.6	600 D	82*	86	−200°C/min cool., 40°C H$_2$O (90 sec) rewarm.	17
Astragalus cicer L.	5.6	1 Y	85	95	−200°C/min cool., 30°C/min rewarm., 3 cultivars	13, 14
Aubrieta deltoidea (L.) DC. cv. Manon	6.5	1 Y	57	75	−200°C/min cool., 30°C/min rewarm.	14
Avena sativa L.	10—12	110 H	N.R.	N.R.	Slow cooling, slow rewarming, liquid air immersion.	24
Avena sativa L.	8.4	180 D	94	97	−200°C/min cool., 37°C/min rewarm.	21, 13
Avena sativa L. cv. Kenota	N.R.	60 D	N.R.	No change	Rapid cooling, rewarming N.R., liquid air treatment on seedlings	16
Avena sp.	N.R.	48 H	92	92	Rapid cooling, rewarming N.R., liquid air immersion	25
Begonia cucullata Willd. var. hooderi (A.D.C.) L.B. (Sm. + Schub.) cv. Glamour Picottee	N.R.	1 Y	74	82	−200°C/min cool., 30°C/min rewarm.	14
Begonia sp. cv. Nonstop Pink	N.R.	1 Y	69	72	−200°C/min cool., 30°C/min rewarm.	14
Bellis perennis L. cv. Super Enorma Mix	5.2	1 Y	73	89	−200°C/min cool., 30°C/min rewarm.,	14
Beta vulgaris L.	6.8	3 Y	88	90	−200°C/min cool., 30°C/min rewarm., 3 cultivars	14
Beta vulgaris L.	N.R.	30 D	N.R.	No change	Rapid cooling, slow rewarming, liquid air treatment on seedlings	16
Beta vulgaris L. cv. ``71/3-1-3-L10``	6.3	30 D	96	91	−200°C/min cool., 30°C/min rewarm.	13, 18
Beta vulgaris L.	7.3	3 Y	89	92	−200°C/min cool., 30°C/min rewarm., 8 cultivars	14
Brassica alba L.	N.R.	36 H	85	72	Rapid cooling, liquid air immersion	25
Brassica alba L.	N.R.	6 H	100	100	Liquid hydrogen immersion	26
Brassica campestis	N.R.	36 H	91	88	Rapid cooling, liquid air immersion	25

Table 1 (continued)
SPECIES OF SEED REPORTED TO HAVE SURVIVED LN2 EXPOSURE

Species/selection	Moisture (%)	Duration	Germination (%)		Comments	Ref.
			Before	After		
Brassica carinata A.Br.	N.R.	30 D	N.R.	No change	Rapid cooling, rewarming N.R., liquid air treatment on seedlings	16
Brassica napus L. cv. Swede	N.R.	24 H	72	88	Rapid cooling, liquid air immersion	27
Brassica oleracea L. var. botrytis	5.5	3 Y	95	97	$-200°C$/min cool., $30°C$/min rewarm., 7 cultivars	14
Brassica oleracea L. var. botrytis	4.8	3 Y	94	97	$-200°C$/min cool., $30°C$/min rewarm., 3 cultivars.	13, 14
Brassica oleracea L. var. capitata	7.2	3 Y	95	94	$-200°C$/min cool., $30°C$/min rewarm., 8 cultivars	14, 18
Brassica oleracea var. botrytis cv. Early Abundance	6.1	600 D	92*	100	$-200°C$/min cool., $40°C$ H_2O (90 sec) rewarm.	17
Brassica oleracea var. botrytis cv. Rapine	5.5	600 D	96*	92	$-200°C$/min cool., $40°C$ H_2O (90 sec) rewarm.	17
Brassica oleracea var. capitata cv. Early Greenball	6.2	600 D	96*	97	$-200°C$/min cool., $40°C$ H_2O (90 sec) rewarm.	17
Brassica oleracea var. gemmifera cv. Early Morn	5.7	600 D	82*	97	$-200°C$/min cool., $40°C$ H_2O (90 sec) rewarm.	17
Brassica oleracea var. napobrassica cv. The Laurentian	6.4	600 D	99*	100	$-200°C$/min cool., $40°C$ H_2O (90 sec) rewarm.	17
Browallia sp. cv. Blue Bells Improved	4.7	1 Y	93	94	$-200°C$/min cool., $30°C$/min rewarm.	14
Callestephus chinensis (L.) Nees cv. Reine Marguerite	6.8	600 D	57*	68	$-200°C$/min cool., $40°C$ H_2O (90 sec) rewarm.	17
Callestephus chinesis (L.) Nees cv. Grego Mix	5.7	1 Y	87	83	$-200°C$/min cool., $40°C$ H_2O (90 sec) rewarm.	17
Campanula sp. cv. Cup and Saucer Mix	6.3	1 Y	80	97	$-200°C$/min cool., $30°C$/min. rewarm.	14
Cannabis sativa L.	N.R.	36 H	24	7	Rapid cooling, liquid air immersion	25
Capsicum annum L. cv. Early Canada Bell	6.5	600 D	77*	87	$-200°C$/min cool., $40°C$ H_2O (90 sec) rewarm.	17

Species						
Capsicum annum L.	4.3	3 Y	98	98	−200°C/min cool., 30°C/min rewarm., 3 cultivars	14
Capsicum frutescens L. cv. "PI 163184"	6.2	180 D	93	93	−200°C/min cool., 30°C/min rewarm.	13, 18
Carthamus tinctorius L. cv. "PI 259996"	5.6	180 D	86	86	−200°C/min cool., 30°C/min rewarm.	13, 21
Catharanthus roseus (L.) G. Don cv. Little Delicata	5.7	30 D	83	78	−200°C/min cool., 30°C/min rewarm.	14
Celosia argentea L. cv. Dwarf Mixed Improved	8.1	1 Y	70	80	−200°C/min cool., 30°C/min rewarm.	14
Cereus giganteus Engelm. cv. Giant Saguaro	4.2	1 Y	90	97	−200°C/min cool., 30°C/min rewarm.	14
Chrysanthemum coccinem Willd. cv. Robinson's Single Mix	5.5	30 D	90	87	−200°C/min cool., 30°C/min rewarm.	14
Citrullus lanatus (Thunb.) Matsum and Nakai	5.6	3 Y	97	97	−200°C/min cool., 30°C/min rewarm., 5 cultivars	13, 14, 18
Citrullus vulgaris cv. Sugar Baby	N.R.	48 H	83	63	Germination reported after 48 hr at 25°C	23
Citrus limon (L.) Burm. F.	5.4	3 H	33	34	Rapid and slow cooling, DMSO 1 hr, rapid rewarming at 35°C in water	28
Cleome spinosa Jacq. cv. Rose Queen	3.1	1 Y	89	85	−200°C/min cool., 30°C/min rewarm.	14
Coleus bluemi Benth.	7.1	1 Y	89	95	−200°C/min cool., 30°C/min rewarm.	14
Convolvulus tricolor	10—12	110 H	N.R.	No change	Slow cooling, slow rewarming, liquid air immersion	24
Crambe abyssinica Hochst.	7.4	3 Y	66	79	−200°C/min cool., 30°C/min rewarm., 4 cultivars	13, 14
Cucumis melo L.	6.0	3 Y	96	95	−200°C/min cool., 30°C/min rewarm., 5 cultivars	13, 14, 18
Cucumis melo L.	4.8	3 Y	93	90	−200°C/min cool., 30°C/min rewarm., 2 cultivars	14
Cucumis melo L. cv. Pride of Iowa	N.R.	48 H	83	63	Germination reported after 48 hr at 25°C	23
Cucumis sativus cv. Ohio MR-17	N.R.	48 H	97	97	Germination reported after 48 hr at 25°C	23
Cucumis sativus L.	N.R.	600 D	94	94		29
Cucumis sativus L.	5—10	2 Y	Varies	No change	Rates N.R. 10 hybrids	30
Cucumis sativus L.	6.3	3 Y	97	96	−200°C/min cool., 30°C/min rewarm., 7 cultivars	14
Cucumis sativus L. cv. Early White Spine	N.R.	60 D	N.R.	No change	Rapid cooling, rewarming N.R., liquid air treatment on seedlings	16
Cucumis sativus L. cv. Perfection	5.1	180 D	95	92	−200°C/min cool., 30°C/min rewarm.	13, 18
Cucumis sativus L. cv. Straight 8	6.6	600 D	79*	87	−200°C/min cool., 40°C H_2O (90 sec) rewarm.	17
Cucurbita maxima Duch.	<13	6 M	82	79	−200°C/min cool., 30°C/min rewarm.	31

Table 1 (continued)
SPECIES OF SEED REPORTED TO HAVE SURVIVED LN2 EXPOSURE

Species/selection	Moisture (%)	Duration	Germination (%)		Comments	Ref.
			Before	After		
Cucurbita pepo cv. Table Queen Acorn	N.R.	48 H	73	67	Germination reported after 48 hr at 25°C	23
Cucurbita pepo L.	N.R.	6 H	96	100	Liquid hydrogen immersion	26
Cucurbita pepo L.	10—12	110 H	N.R.	No change	Slow cooling, slow rewarming, liquid air immersion	24
Cucurbita pepo L. cv. Sugar Pie	N.R.	48 H	100	97	Germination reported after 48 hr at 25°C	23
Cucurbita sp. cv. Wood's Earliest Prolific	6.7	180 D	82	79	−200°C/min cool., 30°C/min rewarm.	13, 18
Cucurbita spp. cv. Hubbard	N.R.	60 D	N.R.	No change	Rapid cooling, rewarming N.R., liquid air treatment on seedlings	16
Cyclamen persicum Mill. cv. Albadonna	9.0	1 Y	31*	51	−200°C/min cool., 30°C/min rewarm.	14
Cyclanthera explodens	10—12	110 H	N.R.	No change	Slow cooling, slow rewarming, liquid air immersion	24
Dactylis glomerata L.	6.0	180 D	90	92	−200°C/min cool., 30°C/min rewarm.	21
Dahlia pinnata Cav. cv. Early Bird	6.4	30 D	90	90	−200°C/min cool., 30°C/min rewarm.	14
Datura metel L.	<13	180 D	91	89	−200°C/min cool., 30°C/min rewarm.	31
Daucus carota L.	6.8	3 Y	81	86	−200°C/min cool., 30°C/min rewarm., 19 cultivars	14
Daucus carota L.	N.R.	36 H	65	59	Rapid cool., rewarm. N.R., liquid air	25
Daucus carota L. cv. Gold Pak Improved	7.2	600 D	76*	81	−200°C/min cool., 40°C H_2O (90 sec) rewarm.	17
Daucus carota L. subsp. *sativus* (Hoffm.) Arcang. cv. Long Imperator 58	6.1	7 D	87	82	−200°C/min. cool., 30°C/min rewarm.	13, 18
Delphinium elatum L. cv. Giant Pacific Galahad	5.8	1 Y	59	68	−200°C/min cool., 30°C/min rewarm.	14
Dianthus caryophyllus L. cv. Enfant de Nice	9.6	600 D	92*	90	−200°C/min cool., 40°C H_2O (90 sec) rewarm.	17
Dianthus chinensis L. cv. Baby Doll	9.3	1 Y	97	96	−200°C/min. cool., 30°C/min rewarm.	14
Dianthus sp. cv. Dwarf Mixture F, Hybrid	9.6	1 Y	91	86	−200°C/min cool., 30°C/min rewarm.	14
Digitalis purpurea L. cv. Giant Shirley Mix	3.8	600 D	88*	91	−200°C/min cool., 40°C H_2O (90 sec) rewarm.	17

Dimorphotheca sinuata D.C. cv. Glistening White	7.3	30 D	89	95	−200°C/min cool., 30°C/min rewarm.	14
Eleocharis coloradoensis	N.R.	14 D	59[a]	59	Damp dried to remove external water	32
Elymus giganteus Vahl cv. "PI 421133"	8.6	3 Y	13	5	−200°C/min cool., 30°C/min rewarm.	13, 14
Eragrostis curvula (Schrad.) Nees	8.2	3 Y	94	91	−200°C/min cool., 30°C/min rewarm., 3 cultivars	13, 14
Exacum affine Balf. f. cv. Tiddly Winks	38	1 Y	89	89	−200°C/min cool., 30°C/min rewarm.	14
Fagaria chiloensis (L.) Dutch. cv. Baron Solemacher	6.3	600 D	76*	80	−200°C/min cool., 40°C H_2O (90 sec) rewarm.	17
Fagopyrum esculentum Moench.	N.R.	60 D	N.R.	No change	Rapid cooling, rewarming N.R., liquid air treatment on seedlings	16
Fagopyrum esculentum Moench.	N.R.	2 H	N.R.	No change	Slow cooling, slow rewarming, liquid helium immersion	33
Festuca pratensis Huds.	>12	24 H	85	91	Rapid cooling, rewarming N.R., liquid air immersion	27
Festuca rubra L. subsp. *rubra* cv. "PI 552194"	9.7	180 D	94	94	−200°C/min cool., 30°C/min rewarm.	21
Festuca spp.	8.7	3 Y	90	90	−200°C/min cool., 30°C/min rewarm., 6 cultivars	13, 14, 2
Funkia sieboldna	10—12	110 H	N.R.	No change	Slow cooling, slow rewarming, liquid air immersion.	24
Gaillardia X *grandiflora* Van Houtte cv. Monarch Strain	6.9	30 D	80	80	−200°C/min cool., 30°C/min rewarm.,	14
Glycine max cv. Horel	6—14	ca. 1 H	N.R.	80	−1.7°C/sec cool., 0.5°C/sec rewarming	34
Glycine max cv. T202	N.R.	N.R.	80	No change	−200°C/min cool., 37°C H_2O rewarm.	19
Glycine max cv. Wayne	N.R.	48 H	7	7	Germination reported after 48 hr at 25°C	23
Glycine max L. (Merr.)	7.1	3 Y	84	57	−200°C/min cool., 30°C/min rewarm.	13, 14, 21
Gossypium hirsutum L.	<13	180 D	92	90	−200°C/min cool., 30°C/min rewarm.	31
Helianthus annuus L.	3.0	3 Y	92	89	−200°C/min cool., 30°C/min rewarm., 3 cultivars.	13, 14, 21
Helianthus annuus L.	N.R.	36 H	70	65	Rapid cooling, liquid air immersion	25
Helianthus annuus L.	10—12	110 H	N.R.	No change	Slow cooling, slow rewarming, liquid air immersion	24
Helianthus annuus L.	—	—	91	91		29
Helianthus annuus L. cv. Russian Mammoth	N.R.	60 D	N.R.	No change	Rapid cooling, rewarming N.R., liquid air treatment on seedlings	16
Heracleum villosum	10—12	110 H	N.R.	No change	Slow cooling, slow rewarming, liquid air immersion	24

Table 1 (continued)
SPECIES OF SEED REPORTED TO HAVE SURVIVED LN2 EXPOSURE

Species/selection	Moisture (%)	Duration	Germination (%)		Comments	Ref.
			Before	After		
Hordeum districhon	10—12	110 H	N.R.	No change	Slow cooling, slow rewarming, liquid air immersion	24
Hordeum sp.	>12	24 H	95	96	Rapid cool., rewarming, N.R., liquid air immersion	27
Hordeum vulgare L.	N.R.	2 H	N.R.	No change	Slow cool., slow rewarm., liquid helium immersion	33
Hordeum vulgare L.	N.R.	6 H	100	100	Liquid hydrogen immersion	26
Hordeum vulgare L.	8.7	3 Y	83	95	−200°C/min cool., 30°C/min rewarm., 3 cultivars	14
Hordeum vulgare L. cv. Sacramento	N.R.	30 D	N.R.	No change	Rapid cooling, rewarming N.R., liquid air treatment on seedlings	16
Hordeum vulgare L. cv. Hosomugl	8.9	180 D	100	99	−200°C/min cool., 30°C/min rewarm.	13, 21
Hordeum sp.	N.R.	48 H	100	96	Rapid cooling, rewarming N.R., liquid air immersion	25
Hypoestes phyllostachya Bak.	3.8	1 Y	95	97	−200°C/min cool., 30°C/min rewarm.	14
Iberis sempervirens L. cv. Sempervirens	5.8	30 D	67	72	−200°C/min cool., 30°C/min rewarm.	14
Impatiens balsamina L.	10—12	110 H	N.R.	No change	Slow cooling, slow rewarming, liquid air immersion	24
Impatiens balsamina L. cv. Scarlet Parade	7.3	1 Y	98	98	−200°C/min cool., 30°C/min rewarm.,	14
Impatiens wallerana Hook. f. cv. Elfin Crimson	4.0	1 Y	77*	76	−200°C/min cool., 30°C/min rewarm.	14
Lactuca sativa L.	N.R.	ca. 3 M	N.R.	No change	Exposed seeds showed no evidence of cytological abnormalities	15
Lactuca sativa L.	5—13	1/2 H	98[a]	98	20°C/hr cooling rate, 2°C for 2—4 hr, rewarming rate	35
Lactuca sativa L.	4.6	3 Y	97	98	−200°C/min cool., 30°C/min rewarm., 9 cultivars	13, 14, 18
Lactuca sativa L. cv. Minilake M.I.,	6.2	600 D	97*	100	−200°C/min cool, 40°C H_2O (90 sec) rewarm.	17

Species					
Lepidium sativum	N.R.	36 H	100	100	Rapid cooling, rewarming N.R., Liquid air immersion	25
Lespedeza stipulacea Maxim. cv. Korean Iowa 6	7.8	3 Y	86	93	−200°C/min cool., 30°C/min rewarm.	14,21
Limonium sinuatum (L.) Mill. cv. Apricot	9.6	30 D	86	80	−200°C/min. cool., 30°C/min rewarm.	14
Linum sp.	>12	24 H	98	87	Rapid cooling, rewarming N.R., liquid air immersion	27
Linum usitatissium L.	N.R.	2 H	N.R.	No change	Slow cooling, slow rewarming, liquid helium immersion	33
Linum usitatissium L.	5.0	180 D	41	34	−200°C/min cool., 30°C/min rewarm.	21
Linum usitatissium L.	6.7	600 D	59*	88	−200°C/min cool., 40°C H_2O (90 sec) rewarm.	17
Lobelia erinus L.	N.R.	36 H	51	13	Rapid cooling, liquid air immersion	25
Lobelia erinus L. cv. Sapphire	9.5	1 Y	90	88	−200°C/min cool., 30°C/min rewarm.	14
Lobularia maritima (L.) Desv. cv. New Carpet of Snow	5.6	1 Y	75	83	−200°C/min cool., 30°C/min rewarm.	14
Lolium multiflorum cv. Billion	11—21	ca. 1 H	80[a]	90	−50°C/sec cool., 25°C/sec rewarm.	34
Lotus corniculatus L. cv. Dawn	9.7	180 D	56	5	−200°C/min cool., 30°C/min rewarm.	21
Lotus corniculatus L. cv. Lot 1504	5.2	3 Y	72	72	−200°C/min cool., 30°C/min rewarm., 3 cultivars	13
Lotus sp.	Low	10 H	N.R.	No change	Liquid helium immersion	36
Lotus tetragonolobus L.	10—12	110 H	N.R.	No change	Slow cooling, slow rewarming, liquid air immersion	24
Lupinus sp.	Low	130 H	N.R.	No change	Liquid air immersion	37
Lycopersicon esculentum Mill. cv. Bush Beefsteak	8.7	600 D	61*	51	−200°C/min cool., 40°C H_2O (90 sec) rewarm.	17
Lycopersicon esculentum Mill. cv. Santa Clara	N.R.	30 D	N.R.	No change	Rapid cooling, slow rewarming, liquid air treatment on seedlings	16
Lycopersicon esculentum Mill.	7.2	3 Y	90	89	−200°C/min cool., 30°C/min rewarm., 6 cultivars	14
Lycopersicon esculentum Mill. cv. T 148	5.3	180 D	91	93	−200°C/min cool., 30°C/min rewarm.	13,18
Lycopersicon esculentum Mill.	40	48 H	98	86	Rapid cooling, 40°C H_2O (30 sec) rewarm., 15% DMSO used as a cryoprotectant	38
Manihot esculenta	6.34	3 H	80[a]	80	Rapid cooling, liquid air immersion	39
Matthiota Incana (L.) R.Br. Cv. White Christmas	6.2	1 Y	67	69	−200°C/min cool., 30°C/min rewarm.	14

Table 1 (continued)
SPECIES OF SEED REPORTED TO HAVE SURVIVED LN2 EXPOSURE

Species/selection	Moisture (%)	Duration	Germination (%)		Comments	Ref.
			Before	After		
Medicago sp.	Low	130 H	N.R.	No change	Liquid air immersion	36
Medicago sativa L.	6.5	3 Y	86	90	−200°C/min cool., 30°C/min rewarm., 4 cultivars	14
Medicago sativa L.	N.R.	9 M	95[a]	92	Liquid air immersion	40
Medicago sativa L. cv. Grimm	N.R.	60 D	N.R.	No change	Rapid cooling, rewarming N.R., liquid air treatment on seedlings	16
Medicago sativa L. cv. Dupuits	5—17	ca. 1 H	N.R.	80	−50°C/sec cool; 25°C/sec rewarm.	34
Medicago sativa L. subsp. *sativa* cv. Grimm	7.0	180 D	92	89	−200°C/min cool., 30°C/min rewarm.	13, 21
Melilotus indica L.	N. R.	30 D	N.R.	No change	Rapid cooling, rewarming N.R., liquid air treatment on seedlings	16
Melilotus spp.	N.R.	12 M	70[a]	74	Liquid air immersion	40
Moluccella laevis L.	5.5	1 Y	79*	83	−200°C/min cool., 30°C/min rewarm.	14
Nermesia strumosa Benth. cv. Hi Fi Mixture	4.2	1 Y	85	93	−200°C/min cool., 30°C/min rewarm.	14
Nicotiana alata Link and Otto (*N. affinis* T. Moore) cv. Sensation Mix	5.7	1 Y	95	92	−200°C/min cool., 30°C/min rewqrm.	14
Nicotiana sp.	Low	10 H	N.R.	No change	Liquid helium immersion	36
Nicotiana tabacum L.	N.R.	2 H	N.R.	No change	Slow cooling, slow rewarming, liquid helium immersion	33
Nicotiana tabacum L.	4.7	3 Y	95	92	−200°C/min cool., 30°C/min rewarm., 4 cultivars	14, 21
Nicotiana tabacum L. cv. Speight G-7	13.2	180 D	94	97	−200°C/min cool., 30°C/min rewarm.	13, 21
Nicotiana tabacum L. cv. Samsum	N.R.	N.R.	87	No change	−200°C/min cool., 37°C H_2O rewarm.	19
Nierembergia sp. cv. Regal Robe	5.6	30 D	72	68	−200°C/min cool., 30°C/min rewarm.	14
Ocimum basilicum L. cv. Stokes 159	6.5	600 D	57*	58	−200°C/min cool., 40°C H_2O (90 sec) rewarm.	17
Onobrychis vicifolia Scop.	7.3	3 Y	77	86	−200°C/min cool., 30 °C/min rewarm., 3 cultivars	13, 14, 21
Oryza sativa cv. Horyu	5—15	ca. 1 H	100[a]	80	−50°C/sec cool., 25°C/sec rewarm.	34

Species						
Oryza sativa cv. Starbonnet	N.R.	N.R.	90	No change	−200°C/min cool., 37°C H_2O rewarm.	19
Oryza sativa L.	9.7	3 Y	96	92	−200°C/min cool., 30°C/min rewarm., 2 cultivars	13, 14, 2
Oryza sativa L.	<18	2 H	95[a]	95[a]	Various rates of cooling and rewarming	41
Papaver orientale L. cv. Oriental Mix	5.4	1 Y	64	66	−200°C/min cool., 30°C/min rewarm.	14
Papaver somniferum L. cv. "Pl 414021"	17.5	180 D	91	92	−200°C/min cool., 30°C/min rewarm.	18
Papaver somniferum L.	10.1	3 Y	90	84	−200°C/min cool., 30°C/min rewarm.	14
Pastinaca sativa L.	N.R.	ca. 3 M	N.R.	No change	Exposed seeds showed no evidence of cytological abnormalities	15
Pelargonium zonale (L.) L'Her. ex. Alt. cv. F2 Florist Mix	7.3	30 D	20	25	−200 °C/min cool., 30°C/min rewarm.	14
Pennisetum americanum L. cv. Pearl Tift 239 DB	7.5	3 Y	88	89	−200°C/min cool., 30°C/min rewarm.	13,14,21
Petroselinum crispum (Mill.) Nym. ex. A.W. Hill	6.7	3 Y	86	77	−200°C/min cool., 30°C/min rewarm.	13, 14
Petroselinum crispum (Mill.) Nym. ex A.W. Hill	N.R.	36 H	28	1	Rapid cooling, rewarming N.R., liquid air immersion	25
Petunia sp.	Low	10 H	N.R.	No change	Liquid helium immersion	36
Petunia sp.	7.5	3 Y	92	90	−200°C/min cool., 30°C/min rewarm., 6 cultivars	14
Petunia X *hybrida* Vilm. cv. Magic Mix	5.4	1 Y	86	85	−200°C/min cool., 30°C/min. rewarm.	14
Petunia [X] *hybrida* Vilm. cv. Red Devil	4.8	600 D	55*	80	−200°C/min cool., 40°C H_2O (90 sec) rewarm.	17
Peucedanum sativum (*Pastinaca sativa* L.)	N.R.	36 H	45	18	Rapid cooling, rewarming N.R., liquid air immersion	25
Phaseolus lanatus L. cv. Cangreen	N.R.	48 H	80	67	Germination reported after 48 hr at 25°C	23
Phaseolus multiflourus cv. Haricot	N.R.	36 H	100	90	Rapid cooling, rewarming N.R., liquid air immersion	25
Phaseolus sp.	—	—	95	97		29
Phaseolus vulgaris L. cv. Spartan Arrow	7.0	30 D	100	100	−200°C/min cool., 30°C/min rewarm.	13, 18
Phaseolus vulgaris L. cv. Contender	N.R.	48 H	90	67	Germination reported after 48 hr at 25°C	23
Phaseolus vulgaris L.	7.1	3 Y	97	99	−200°C/min cool., 30°C/min rewarm., 3 cultivars	14
Phaseolus vulgaris L.	N.R.	ca. 3 M	N.R.	No change	Exposed seeds showed no evidence of cytological abnormalities	15
Phleum pratense L. cv. 1-776	12.3	180 D	82	86	−200°C/min cool., 30°C/min rewarm.	13, 21

Table 1 (continued)
SPECIES OF SEED REPORTED TO HAVE SURVIVED LN2 EXPOSURE

Species/selection	Moisture (%)	Duration	Germination (%) Before	Germination (%) After	Comments	Ref.
Phleum sp.	>12	24 H	70	69	Rapid cooling, rewarming N.R., liquid air immersion	27
Phlox drummondii Hook. cv. Twinkle	7.4	1 Y	85	91	−200°C/min cool., 30°C/min rewarm.	14
Picea sp. sele, Brewer	5.4	1 Y	62	65	−200°C/min cool., 30°C/min rewarm.	14
Pinus lambertiana Dougl.	6.6	1 Y	85	75	−200°C/min cool., 30°C/min rewarm., 4 cultivars	13, 14
Pinus ponderosa Dougl. ex. P. & C. Laws	4.7	3 Y	74	77	−200°C/min cool., 30°C/min rewarm.	13, 14
Pisum sativum	N.R.	36 H	95	75	Rapid cooling, rewarming N.R., liquid air immersion	25
Pisum sativum	27—40	1 M	100	94	−200°C/min cool., rapid rewarm. in tap water, seedlings (7—12 mm) used	42
Pisum sativum cv. Alaska Wilt Resistant	N.R.	48 H	30	3	Germination reported after 48 hr at 25°C	23
Pisum sativum L.	—	—	—	—		29
Pisum sativum L.	N.R.	ca. 3 M	N.R.	No change	Exposed seeds showed no evidence of cytological abnormalities	15
Pisum sativum L.	N.R.	6 H	100	100	Liquid hydrogen immersion.	26
Pisum sativum L. cv. Progress 9	7.2	30 D	91	96	−200°C/min cool., 30°C/min rewarm.	13, 18
Pisum sp.	>12	24 H	66	65	Rapid cooling, rewarming N.R., liquid air immersion	27
Pisum sp. cv. Canadian	N.R.	60 D	N.R.	No change	Rapid cooling, rewarming N.R., liquid air treatment on seedlings	16
Pisum sp. cv. American Wonder	N.R.	60 D	N.R.	No change	Rapid cooling, rewarming N.R., liquid air treatment on seedlings	16
Pisum sp.	10—12	110 H	N.R.	No change	Slow cooling, slow rewarming, liquid air immersion	24
Poa spp.	7.2	3 Y	86	90	−200°C/min cool., 30°C/min rewarm., 3 cultivars	13, 14, 21
Primula sinensis L.	N.R.	ca. 3 M	N.R.	No change	Exposed seeds showed no evidence of cytological abnormalities	15

Species					
Primula veris L. cv. Yellow Shades	8.2	30 D	41	45	−200°C/min cool., 30°C/min rewarm.	14
Pseudotsuga menziesii (Mirbel. Franco	6.8	3 Y	78	78	−200°C/min cool., 30°C/min rewarm., 4 selections	13, 14
Pyrus malus	N.R.	48 H	12	4	Rapid cooling, rewarming N.R., liquid air immersion	25
Raphanus sativus L.	5.1	3 Y	98	98	−200°C/min cool., 30°C/min. rewarm., 8 cultivars	13,14
Raphanus sativus L.	N.R.	36 H	97	88	Rapid cooling, rewarming N.R., liquid air immersion	25
Raphanus sativus L. cv. Sabine	N.R.	ca. 3 M	N.R.	No change	Exposed seeds showed no evidence of cytological abnormalities	15
Raphanus sp.	Low	130 H	N.R.	No change	Liquid air immersion	37
Raphanus sativus L. cv. Champion	6.2	600 D	87*	98	−200°C/min cool., 40°C H_2O (90 sec) rewarm.	17
Ricinus cambgiensis	N.R.	36 H	100	100	Rapid cooling, rewarming N.R., liquid air immersion	25
Ricinus communis L.	8.7	7 D	71	74	−200°C/min cool., 30°C/min rewarm.	13, 21
Saccharum spp. cv. Early Amber Sugarcane	N.R.	60 D	N.R.	No change	Rapid cooling, rewarming N.R. liquid air treatment on seedlings	16
Salpiglossis sinuata Rulz. & Pav. cv. Splash F1 Mixture	3.0	1 Y	79	70	−200°C/min cool., 30°C/min rewarm.	14
Salvia splendens F. Sellow ex. Roem. & Schult. cv. Regal Purple	5.5	1 Y	68	65	−200°C/min cool., 30°C/min rewarm.	14
Sanvitalia procumbens Lam. cv. Gold Braid	6.3	30 D	76	74	−200°C/min cool., 30°C/min rewarm.	14
Satureja hortensis L. cv. Dark Opal	7.3	1 Y	74	87	−200°C/min cool., 30°C/min rewarm.	14
Scabiosa caucasica Bleb. cv. House Mixture	6.2	30 D	17	18	−200°C/min cool., 30°C/min rewarm.	14
Schizanthus sp. cv. Ball Giant Flowered	6.7	1 Y	62	69	−200°C/min cool., 30°C/min rewarm.	14
Secale cereale L.	N.R.	48 H	96	90	Rapid cooling, rewarming N.R., liquid air immersion	25
Secale cereale L. cv. Caribou	8.9	180 D	81	87	−200 °C/min cool., 30°C/min rewarm.	21
Sedum sp. cv. Rock Garden Blend	8.3	1 Y	88	87	−200°C/min cool., 30°C/min rewarm.	14
Sesamum indicum L.	4.2	7 D	96	38	−200°C/min cool., 30°C/min rewarm.	13, 21
Setaria lutescens	5—16	5 M	N.R.	30—70	Poured LN2 on seeds until 3 cm deep, let it evaporate., thawed for 1 hr in 24°C H_2O; also cycling	43
Sinningla speclosa (Lodd.) Hiern. cv. Orchid Osera	N.R.	30 D	87	81	−200°C/min cool., 30°C/min rewarm.	14

Table 1 (continued)
SPECIES OF SEED REPORTED TO HAVE SURVIVED LN2 EXPOSURE

Species/selection	Moisture (%)	Duration	Germination (%)		Comments	Ref.
			Before	After		
Solanum melongena L. cv. Florida Market	6.2	180 D	95	92	−200°C/min cool., 30°C/min rewarm.	13, 18
Solanum pseudocapsicum L. cv. Ball Christmas Cherry	5.9	30 D	97	96	−200°C/min cool., 30°C/min rewarm.	14
Sorghum bicolor (L.) Moench.	9.3	3 Y	96	96	−200°C/min cool., 30°C/min rewarm., 7 cultivars	13, 14, 21
Sorghum bicolor (L.) Moench.	N.R.	30 D	N.R.	No change	Rapid cooling, rewarming N.R., liquid air treatment on seedlings	16
Sorghum bicolor (L.) Moench. cv. ``PI 329554''	12.1	7 D	84	80	−200°C/min cool., 30°C/min rewarm.	21
Sorghum sudanense (Piper)[Stapf]	<13	180 D	95	97	−200°C/min cool., 30°C/min rewarm.	31
Sorghum vulgare L.	N.R.	2 H	N.R.	No change	Slow cool., slow rewarm., liquid helium	33
Spinacia oleracea L.	N.R.	2 H	N.R.	No change	Slow cool., slow Rewarm., liquid helium	33
Spinacia oleracea L.	7.5	3 Y	95	97	−200°C/min cool., 30°C/min rewarm., 5 cultivars	13, 14
Spinacia oleracea L. cv. Stenning	N.R.	30 D	N.R.	No change	Rapid cooling, rewarming N.R., liquid air treatment on seedlings	16
Tagetes erecta L. cv. Dolly Blend	5.4	600 D	31*	62	−200°C/min cool., 40°C H₂O (90 sec) rewarm.	17
Tagetes patula L. cv. Harvest Moon	7.0	30 D	91	88	−200°C/min cool., 30°C/min rewarm.	14
Thuja plicata Donn ex. D. Don sele. Western Red	11.6	3 Y	63	62	−200°C/min cool., 30°C/min rewarm.	13, 14
Thunbergia sp. cv. Alta	7.3	30 D	84	80	−200°C/min cool., 30°C/min rewarm.	14
Toreni fournieri Linden ex. E. Fourn cv. Complete Blue	5.7	1 Y	80	88	−200°C/min cool., 30°C/min rewarm.	14
Trifolium alba	N.R.	N.R.	83	No change	−200°C/min cool., 37°C H₂O rewarm.	19
Trifolium pratense L.	>12	24 H	72	90	Rapid cooling, rewarming N.R., liquid air immersion	27
Trifolium pratense L. cv. Chesapeake	6.4	180 D	90	83	−200°C/min cool., 30°C/min rewarm.	21

Species						
Trifolium spp.	7.3	3 Y	85	89	−200°C/min cool., 30°C/min rewarm., 6 cultivars	14
Trifolium subterraneum L.	N.R.	N.R.	97	No change	−200°C/min cool., 37°C H$_2$O rewarm.	19
Trigonella faenum-groecum L.	10—12	110 H	N.R.	No change	Slow cooling, slow rewarming., liquid air immersion	24
Triticum aestivum L.	8.8	3 Y	87	83	−200°C/min cool., 30°C/min rewarm., 3 cultivars.	14
Triticum aestivum L. cv. Mukakomugl	5—16	ca. 1 H	80[a]	80	−50°C/sec cool., 0.5°C/sec rewarm.	34
Triticum aestivum L. cv. Prodox	9.0	180 D	94	90	−200°C/min cool., 30°C/min rewarm.	13, 21
Triticum sativum [Sativum Lam.]	N.R.	6 H	96	100	Liquid hydrogen immersion	26
Triticum sp.	N.R.	2 H	N.R.	No change	Slow cool., slow rewarm., liquid helium	33
Triticum sp.	—	—	98	99		29
Tricitum sp.	Low	130 H	N.R.	No change	Liquid air immersion	37
Triticum sp.	N.R.	N.R.	100	100	Rapid cooling, liquid air immersion	25
Triticum spp.	<17	2 H	90[a]	90[a]	Various rates of cooling & rewarming	41
Triticum vulgare	10.6	2 Min.	92	92	−200°C/min cool., 30°C/min rewarm.	44
Tropaeolum majus L.	<13	180 D	12	9	−200°C/min cool., 30°C/min rewarm.	31
Tropaeolum majus L. cv. Golden Jewel	8.2	30 D	86	85	−200°C/min cool., 30°C/min rewarm.	18
Tsuga heterophylla (Raf.) Sarg. Sele. Western	8.2	3 Y	70	78	−200°C/min cool., 30°C min rewarm.	13, 14
Ulmus americana L.	6.4	3 Y	90	83	−200°C/min cool., 30°C min rewarm.	13, 14, 21
Verbena X *hybrida* Voss cv. Ideal Florist Mix	6.2	1 Y	48	50	−200°C/min cool., 30°C min rewarm.	14
Vicia benghalensis L.	N.R.	30 D	N.R.	No change	Rapid cooling, rewarming N.R., liquid air treatment on seedlings	16
Vicia faba	N.R.	48 H	3	0	Germination reported after 48 hr at 25°C	23
Vicia sativa L. subsp. *sativa*	N.R.	2 H	N.R.	No change	Slow cool., slow rewarm., liquid·helium	33
Vicia sativa L. subsp. *sativa* cv. Purple	8.3	180 D	94	90	−200°C/min cool., 30°C/min rewarm.	21
Vicia sp.	9.7	3 Y	94	93	−200°C/min cool., 30°C/min rewarm., 3 cultivars	14
Vigna sp.	Low	130 H	N.R.	No change	Liquid air immersion	37
Vigna ungulculata L.	8.1	180 D	93	96	−200°C/min cool., 30°C/min rewarm.	21
Viola cornuta L. cv. Jersey Gem	6.0	1 Y	62*	67	−200°C/min cool., 30°C/min rewarm.	14
Viola tricolor L. cv. Lake of Thun	6.5	1 Y	78	76	−200°C/min cool., 30°C/min rewarm.	14
Zea mays L.	N.R.	ca. 3 M	N.R.	No change	Exposed seeds showed no evidence of cytological abnormalities	15
Zea mays L. cv. Prior	N.R.	N.R.	100	No change	−200°C/min cool., 37°C H$_2$O rewarm.	19
Zea mays L.	8.9	3 Y	98	93	−200°C/min cool., 30°C/min rewarm.	13, 14, 21

Table 1 (continued)
SPECIES OF SEED REPORTED TO HAVE SURVIVED LN2 EXPOSURE

Species/selection	Moisture (%)	Duration	Germination (%)		Comments	Ref.
			Before	After		
Zea mays L.	N.R.	48 H	90	90	Rapid cooling, rewarming N.R., liquid air immersion	25
Zea mays L. subsp. *mays* cv. Golden Bantum	N.R.	60 D	N.R.	No change	Rapid cooling, rewarming N.R., liquid air treatment on seedlings	18
Zea mays L. subsp. *mexicana* cv. Teosinite	8.1	30 D	99	99	$-200°C$/min cool., $30°C$/min rewarm.	21
Zinnia elegans Jacq.	5.9	3 Y	89	90	$-200°C$/min cool., $30°C$/min rewarm., 4 cultivars	13, 14

Note: N.R., not reported; *, control; Y, years; M, months; D, days; H, hours.

[a]   Estimated from highest germination reported.

Table 2
SEED HIGH MOISTURE FREEZING LIMITS (HMFL) OF SELECTED CROP SPECIES

| | | Seed survival | | |
| | | Moisture below HMFL | Moisture above HMFL | |
Species	HMFL (% M)	Germination (%)	Lethal Freezing[a] (°C)	Germination (%)
Barley	20.8 (1.2)	98	−12; −13	18
Bean	27.2 (1.2)	99	−25	84
Cabbage	13.8 (0.3)	90	−28	0
Carrot	21.7 (1.6)	83	−25	0
Cauliflower	14.2 (1.1)	97	−25	0
Clover	25.6 (0.5)	95	−15	2
Cucumber	16.4 (0.9)	98	−23; −26	1
Fescue	23.0 (3.8)	98	−25	2
Onion	24.7 (0.8)	70	−18; −22	0
Pepper	18.6 (1.2)	99	−22; −25	0
Radish	16.8 (0.9)	99	−25	4
Sesame	9.3 (1.6)	97	−18; −26	0
Tomato	18.5 (1.6)	93	−20; −25	0
Wheat	26.8 (4.7)	96	−7	25

Note: Lethal freezing temperature determined by differential thermal anaylsis; Numbers in parentheses are standard deviations.

of 19% (or critical moisture level — CM, as defined by Roos), found that seed cooled at 200°C/min had nearly 100% survival. However, viability was significantly reduced or lost if the seeds were cooled at 22.2°C/min or slower. Sesame seeds at very low moisture contents had the opposite response. In Figure 2, sesame seed was adjusted to a series of moisture levels and subjected to a range of LN2 cooling rates.[13] At slow cooling rates (1 to 30°C/min), 100% survival was observed regardless of the seed moisture content. However, at 200°C/min cooling (comparable to exposing the seed directly to LN2), sample viability dropped as seed moisture level was reduced.

Determination of the cause of injury to the lettuce or sesame seed was not clarified, but it is clear, at least in some species, that cooling rate has the potential of reducing viability, depending on the relative seed moisture level. Cooling/rewarming rate and seed moisture content interactions must be understood for each species of seed that is being considered for LN2 preservation. If the seed moisture content is kept within a reasonable moisture range, slow or fast cooling/rewarming rates should not be a significant survival factor for seeds of most species.

3. Physical Damage to the Seed

Cooling and rewarming seeds through extremes of temperature impose great physical stress upon them. If the seed matrices are capable of expanding and contracting with such cooling/rewarming stresses, physical damage probably will not occur. This appears to be the case with most desiccation-tolerant LN2-tolerant seeds. Seeds of some species however, have problems with LN2 causing physical damage. For example, bean (*Phaseolus vulgaris* L.) seeds (Table 3) were adjusted to three moisture levels and exposed to LN2 and LN2 vapor then evaluated for physical damage.[14] When the seeds were directly immersed into LN2, regardless of seed moisture content, 9 to 14% of the seeds showed some type of

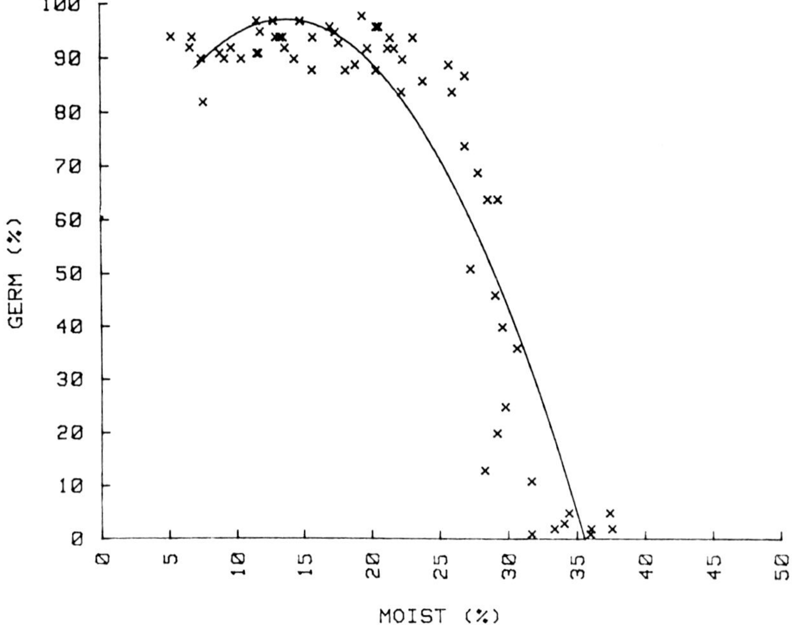

FIGURE 1A.   Clover seed — high moisture freezing limit (HMFL).

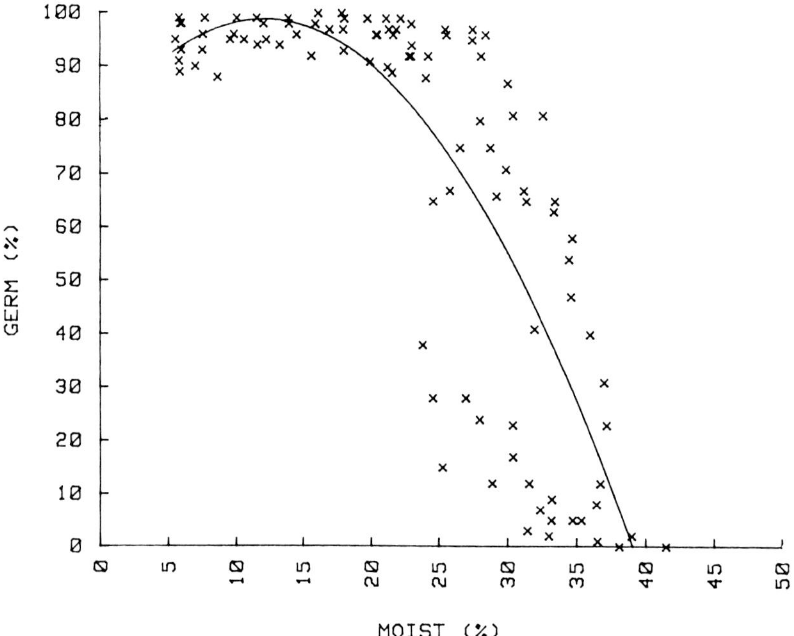

FIGURE 1B.   Wheat seed — high moisture freezing limit (HMFL).

breakage when rewarmed. The most predominant type of breakage was longitudinal splitting of seeds into halves (cotyledons) leaving the embryo axis attached to one cotyledon. If the seed coats were clipped before the seeds were exposed to LN2, no breakage was observed. A reduction in breakage was noted when the seeds were exposed to LN2 and then transferred to the LN2 vapor phase for a period of time before rewarming to room temperature (23°C).

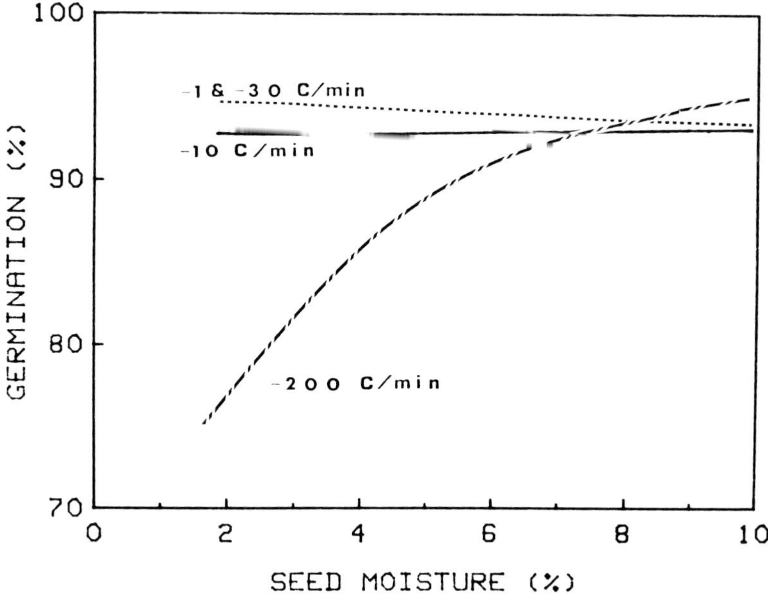

FIGURE 2.   Sesame seed — germination as influenced by moisture level and rate of cooling to liquid nitrogen.

No breakage was noted when the seeds were exposed only to LN2 vapor (approximately −150°C). These results suggest that a majority of the breakage occurred in the rewarming phase, probably at the lower temperature (−150 to −196°C). It is important to note that viability was not affected because the embryo axis developed normally during germination following all the LN2 exposure treatments, except that one cotyledon was missing.

Seeds of other species that have shown some type of physical (breakage) damage are: (1) flax (*Linum usitatissimum* L.) (20 to 30% breakage, some loss of viability, no procedure found to overcome damage); (2) radish (*Raphanus sativus*) and some soybean (*Glycine max*) selections (5 to 15% breakage, no loss of germination); and (3) alfalfa (*Medicago sativa* L.) (no estimate of percent breakage, no loss of viability).[14] Other types of physical damage to seeds exposed to LN2 have not been reported.

B. Desiccation-Tolerant LN2-Sensitive Seed

This category is exemplified by seeds of many fruit and nut crops such as *Prunus* sp., *Juglans* sp., *Corylus* sp. and *Coffee* sp. Most can be dried to moisture contents less than 10%,[46] but cannot withstand storage temperatures lower than −40°C.[14] Seeds stored at lower temperatures show significantly reduced viability or in many cases all the seeds were killed, depending on the species. Seeds that fall into this category are also noted for their short storage life of usually less than 5 years.

Seeds in the desiccation-tolerant LN2-sensitive category usually contain high amounts of storage lipids (oils, etc.), as high as 60 to 70% in some species.[47] Whether high oil content has any bearing on seed sensitivity to LN2 is unclear. For instance, filbert (*Corylus* sp.) seeds can be reduced to 6 to 8% moisture content and subjected to 0 to −20°C without loss of viability. Differential thermal analysis (DTA)[35,48] to −40°C (Figure 3) showed no exotherms that could be associated with the freezing of water and the viability of the filbert seeds were not affected.[14] However, as the temperature was lowered below −40°C an exotherm developed which appeared to coincide with loss of seed viability. Oils extracted from the filbert seeds exhibited a similar DTA profile. Although not studied further and confirmed, these preliminary results indicate that solidification of the oils occurred at tem-

Table 3
BREAKAGE OF BEAN SEED WHEN
EXPOSED TO LIQUID NITROGEN

LN2 seed exposure treatment	Seed breakage (%)[a] Seed moisture (%)		
	4	12	19
Direct immersion, LN2	9 x	14 x	11 x
Direct immersion, LN2 Seed coats clipped	0 z	0 z	0 y
LN2 vapor	0 z	0 z	0 y
LN2 vapor then LN2	7 xy	9 y	4 y
Direct immersion, LN2 then LN2 vapor	2 yz	2 z	1 y

[a] Seed breakage (%) with the same letter (x,y,z) within a given seed moisture are not significantly different at the 5% level of probability according to the LSD test.

peratures below $-40°C$ and may be related to seed damage associated with LN2 cooling of this species.

More extensive research is needed in this area to further characterize the behavior of desiccation-tolerant LN2-sensitive seeds during the LN2 cooling/rewarming cycle. Research is important because such seeds represent a large number of important species which are now being preserved only in the vegetative form.[49] Development of suitable cryopreservation techniques could open the door to improved long-term preservation of such germplasm which might lead to improved methods of breeding and propagation.

C. Desiccation-Sensitive LN2-Sensitive Seed

Desiccation-sensitive seeds are the most difficult to preserve. Chin and Roberts[50] provide a listing of such seeds. In addition to being sensitive to drying, they are generally shortlived and sensitive to extremes of temperature.[51,52] Confounding the situation is the fact that seed of many species in this group, which have been classed as desiccation-sensitive, may in fact tolerate dehydration. For example, lemon (*Citrus limon* L.) seeds, once thought to be killed by drying, have been shown to withstand desiccation to 1.2% moisture and exposure to LN2 without significant loss of viability.[28,53] A number of species (mostly tropical) that had been thought to be sensitive to desiccation but never fully tested and confirmed are listed in Table 4. The data in Table 4 show that most seeds survived both desiccation and LN2 exposure. Exceptions were silver maple (*Acer saccharinum*) and Areca palm (*Areca* sp.) seeds which appeared to be both desiccation- and LN2-(freeze) sensitive.[54] More extensive experimentation is needed to determine variations within species.

Several attempts have been made to develop long-term preservation techniques for desiccation-sensitive seeds with minimal success.[55,56] Most approaches have been directed toward reducing general metabolism through chemical and biophysical means such as reduced oxygen availability or enhancing general maintenance-type metabolism.[6,7] Keeping seed in a physiologically balanced state to extend storage life appears more appropriate for short- to medium-term storage, because it relies on maintaining tissue in a high metabolic state. This is biologically difficult and requires a considerable amount of time, resources, and an understanding of deleterious and maintenance metabolism. Assuming that this approach is more appropriate for short-term applications, development of cryogenic techniques would probably have the greatest potential for providing efficient and effective long-term preservation of desiccation-sensitive seeds.

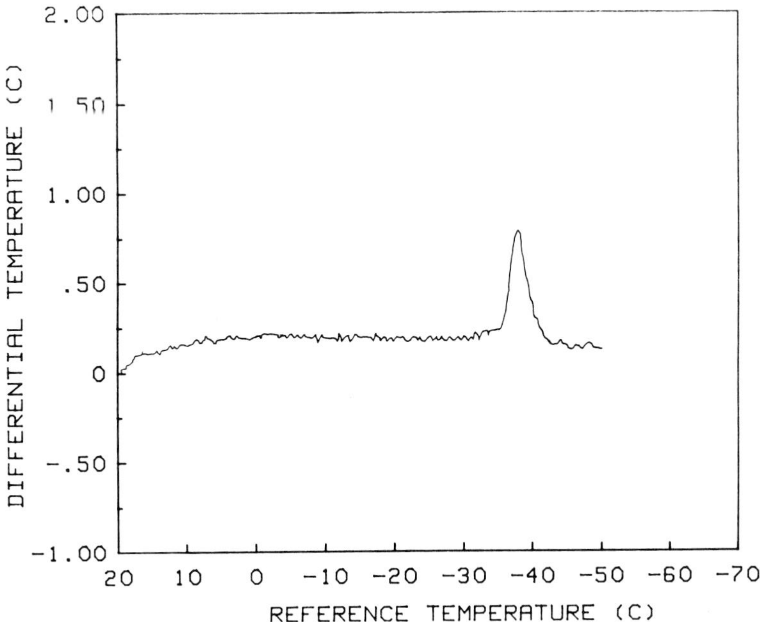

FIGURE 3.   Filbert seed — differential thermal analysis (DTA) profile showing oil exotherm at approximately −35°C.

Successful manipulation of seed moisture content appears to be the key to cryopreservation of desiccation-sensitive seed. Similar to vegetative materials discussed in other chapters, there is a paradox in that seed moisture must be removed to inhibit or stop freezing damage but removal of water is often fatal.

Cryopreservation of blood cells faced a similar obstacle of moisture removal but was overcome by using the "colligative" cryoprotection approach.[56] In that technique, cell water is replaced with a cryoprotectant such as glycerol. The level of substitution must be high enough to afford freezing protection but cannot exceed osmotic and toxic limits. Withers[57] described a technique for infiltrating proline into vegetative tissue to accomplish cryoprotection (see Chapter 12). Some preliminary work was done to determine the movement of DMSO, glycerol, proline, and PEG 4000 into seed of *Anthurium* sp., a desiccation-sensitive flower species.[58] In that work, it was suggested that L-proline infiltration at 23°C was higher and more consistent than the other cryoprotectants studied. Proline infiltration was not enhanced by the addition of DMSO. Conversely, DMSO infiltration was enhanced when combined with PEG 4000, glycerol, or proline. Studies are being continued using high concentrations of proline (4.5 M) in an attempt to achieve successful cryoprotection.[14]

Hydrated tomato (*Lycopericon esculentum* L.) seeds (simulating desiccation-sensitive seed) were infiltrated with DMSO (up to 15% concentration v/v) and exposed to LN2 without loss of viability.[38] It was not clear from the data if the hydrated tomato seeds had been osmotically dehydrated back down to or below the HMFL or if direct cryoprotection was afforded by the DMSO. Successful cryopreservation of truly desiccation-sensitive seed has yet to be achieved.

III. STRATEGY OF IMPLEMENTATION

Routine cryopreservation of seeds depends upon technical feasibility and preservation procedures and the relationship between cost and benefits. Costs and benefits are influenced by laboratory and/or location factors such as objectives, types of seeds being preserved, and

Table 4
DEHYDRATION AND FREEZING CHARACTERISTICS OF SEED
FROM SEVERAL SPECIES

Species	Tolerance to desiccation	Lowest moisture content survived[a]	Unfreezable water content[b]	Survival in LN2
		(% Fresh weight)		(% Control)
Strawberry guava	Tolerant	12	17	91
Passion fruit	Tolerant	2	15	97
Ceara rubber[c]	Tolerant	10	ND	102
Dwarf schefflera[c]	Tolerant	11	ND	55
Common guava	Tolerant	6	ND	98
Papaya	Tolerant	10	ND	100
Apple of sodom	Tolerant	9	18	128
Prickly poppy[c]	Tolerant	10	ND	107
Seamberry[c]	Tolerant	12	ND	97
Coffee	Tolerant	8	24	0
Areca palm	Sensitive	43	42	0
Silver maple	Sensitive	40	32	0

Note: ND no difference.

[a] Based on results of TTC reduction tests for areca palm embryos and germination tests for all other seeds.

[b] The moisture content below which only unfreezable water was found in DTA studies.

[c] Seeds received from Hawaii at low moisture contents, 10—12% that were not dehydrated to lower moisture contents.

available resources, therefore they will not be discussed here. A useful starting point for those interested in making such comparisons might be the cost analysis that was presented for the National Seed Storage Laboratory, Fort Collins, Colorado.[13] Technical feasibility and preservation techniques are addressed under the following topics: (1) preliminary evaluation; (2) cryopreservation equipment and seed containers; and (3) viability monitoring procedures.

A. Preliminary Evaluation
1. Desiccation-Tolerant LN2-Tolerant Seed

Utilization of cryogenic storage should be relatively straightforward for desiccation-tolerant LN2-tolerant seeds following initial screening tests to determine or establish optimum seed moisture levels and LN2 cooling/rewarming rates on a species by species basis. Care must be taken to examine enough selections to adequately determine potential variation within a species. Individual sample screening can be accomplished using 10 to 100 seeds and a short LN2 exposure time (24 hr). If the test subsample is not damaged, the whole sample could be placed into LN2 storage with a high degree of assurance that the sample would be safely preserved. Samples that do not pass the initial screening test should be placed in more conventional storage until further research can be conducted to determine the appropriate conditions for LN2 storage.[59]

Intraspecies variation to LN2 damage is probably not very great, consequently selection by selection examination may not be required.

Most desiccation-tolerant LN2-tolerant seeds tested showed complete survival when moisture content was between 5 and 10% and the cooling rate was kept to 1 to 30°C/min.

2. Desiccation-Tolerant LN2-Sensitive and Desiccation-Sensitive LN2-Sensitive Seed

Routine cryopreservation of desiccation-tolerant LN2-sensitive and desiccation-sensitive

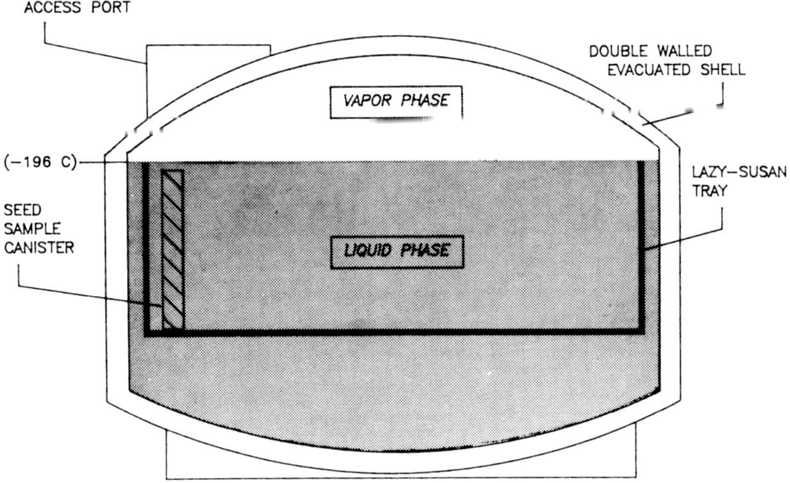

FIGURE 4A. Liquid nitrogen storage tank — liquid phase storage of seed.

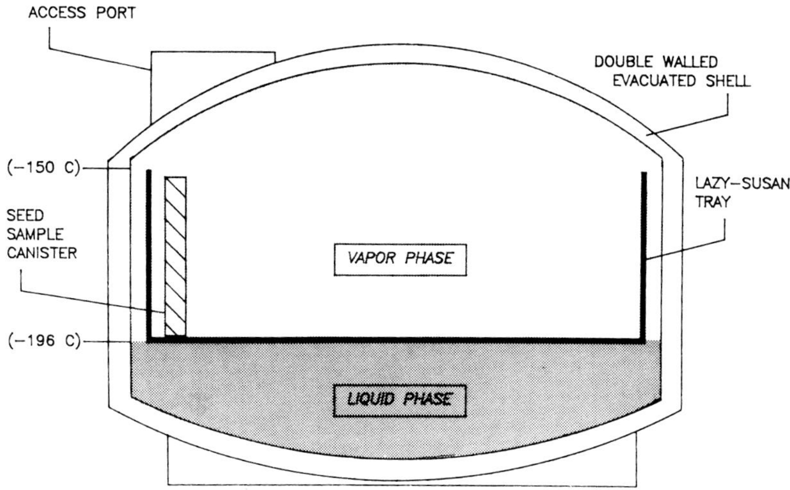

FIGURE 4B. Liquid nitrogen storage tank — vapor phase storage of seed.

seed is not possible at this time. Further research must be conducted on such seeds to delineate pre- and post-LN2 exposure treatments and cooling/rewarming procedures. Research on such species will be slow and difficult because specific procedures will have to be developed for each species of seed.

B. Cryopreservation Equipment and Seed Containers

Successful cryopreservation depends upon the use of good equipment and a flexible, efficient inventory system. A good LN2 seed storage tank should be a selfcontained, vacuum-insulated vessel with the LN2 reservoir and seed samples in the same cavity (Figures 4A and 4B.). This assures a consistent storage temperature and reduces the dependence on external controls and filling devices. Typically such LN2 tanks can maintain operating temperatures for extended periods of time (30 to 120 days) without addition of LN2. Most seeds can withstand direct immersion into LN2, however, there are indications (discussed earlier) that seed stored in the vapor above the LN2 may avoid some types of physical damage. The temperature in the "vapor phase" (Figure 4B.) should be sufficiently low to obtain the cryogenic storage effect. Seeds cooled in the vapor phase cool slower, at rates less than 30°C/min compared to 200°C/min in the liquid phase, which may be more desirable. Additionally, there is less concern for personnel safety with vapor phase storage. Consequently, LN2 vapor storage would probably be the most desirable method of preserving seed germplasm.

Most LN2 storage tanks have access from the top, therefore a vertical inventory scheme is necessary to accommodate easy access to the seeds being cryopreserved. Small seeded species can be easily stored in 1-, 2-, and 5-mℓ plastic cryovials which in turn snap onto canes (generally 6 vials/cane). A 650-ℓ LN2 tank can hold up to 50,000 1-mℓ vials. Larger seeded species such as bean, can be easily stored in heat sealable bags or similar seed storage containers. Large diameter (2 to 3 cm), heat sealable plastic tubes have been suggested as suitable containers for the larger seeded species. Glass containers should be avoided for safety reasons. Seed should be stored in sealed containers to prevent cross contamination by microoranisms and to reduce the potential for slow, long-term extreme desiccation of the seeds.

C. Viability Monitoring Procedures

Theoretically, if a seed sample can be cooled to LN2 temperatures and rewarmed without damage to viability, the time the sample is held in the LN2 (days, weeks, years, centuries) is insignificant. However, because extended LN2 storage of seed has never been done, prudence dictates that viability monitoring must be done at specified intervals. A yearly sampling of the accessions of a given species in LN2 storage would provide a reasonably reliable estimate of seed viability for the species. This would permit individual selections to be tested on a periodic basis. An acceptable sampling level would be 1 to 2% of the total number of accessions of a species. One of the major advantages expected of cryopreservation is that of being able to reduce the frequency of viability testing of individual seed samples. Although 1 to 2% of the samples of a species would not completely eliminate testing, it would greatly reduce the rate of loss of stored seeds through a viability monitoring program.

IV. CONCLUSIONS

Cryopreservation of seed germplasm at or near the temperature of LN2 has the potential of significantly reducing deterioration of seed to such a low level that essentially "indefinite" preservation can be achieved. Most common agronomic and horticultural species have desiccation-tolerant LN2-tolerant seed that can be preserved in LN2 without major problems. For best results, moisture content should be adjusted to the optimum for each species, which is commonly 5 to 10%, and a cooling rate of 30°C/min or slower should be used.

Desiccation-tolerant LN2-sensitive and desiccation-sensitive seed cannot be preserved in LN2 at this time. Research on such seeds must be continued in order to develop procedures by which they can be safely and routinely cryopreserved.

ACKNOWLEDGMENTS

The author would like to thank all the individuals who helped in the preparation of this manuscript, especially Kathi Johnson, Sue Sherman, and Karin Conniff. Data in Table 4 provided by Mike Becwar.

REFERENCES

1. **Anon.**, *Conservation of Germplasm Resources an Imperative*, National Academy of Sciences, Washington, D.C., 1978.
2. **Anon.**, *Report of IBPGR Working Group on Engineering, Design and Cost Aspects of Long-Term Seed Storage Facilities*, International Board for Plant Genetic Resources, FAO, Rome, 1976.
3. **Ashwood-Smith, M. J. and Farrant, J.**, *Low Temperature Preservation in Medicine and Biology*, Pitman Medicine, London, 1980.
4. **Bajaj, Y. P. S.**, Gene preservation through freeze-storage of plant cells, tissue and organ culture, *Acta Hortic.*, 63, 75, 1976.
5. **Bajaj, Y. P. S.**, Establishment of germplasm banks through freeze-storage of plant tissue culture and their implications in agriculture, in *Plant Cell and Tissue Culture: Principles and Applications*, Sharp, W. M., Larsen, P. O., Paddock, E. F. and Raghavan, V., Eds., Ohio University Press, Columbus, 1979, 745.
6. **Osborn, D. J.**, Nucleic acids and seed germination, in *The Physiology and Biochemistry of Seed Dormancy and Germination*, Khan, A. A., Ed., North-Holland, Amsterdam, 1977, 319.
7. **Osborn, D. J.**, Senescence in seeds, in *Senescence in Plants*, Thimann, K. U., Ed., CRC Press, Boca Raton, Fla., 1980, chap. 2.
8. **Roberts, E. H.**, *Viability of Seeds*, Chapman and Hall, London, 1972, 253.
9. **Roos, E. E.**, Induced genetic changes in seed germplasm during storage, in *The Physiology and Biochemistry of Seed Development, Dormancy and Germination*, Khan, A. A., Ed., Elsevier, Amsterdam, 1982, 426.
10. **Villers, T. A.**, Genetic maintenance of seeds in imbibed storage, in *Crop Genetic Resources for Today and Tomorrow*, Frankel, O. H. and Hawkes, J. G., Eds., Cambridge University Press, Cambridge, 1975, 297.
11. **Bass, L. N. and Stanwood, P. C.**, Long-term preservation of sorghum seed as affected by seed moisture, temperature and atmospheric environment, *Crop Sci.*, 18, 575, 1978.
12. **Harrington, J. F.**, Seed storage and longevity, in *Seed Biology*, Kozlowski, T. T., Ed., Academic Press, New York, 1972, chap. 3.
13. **Stanwood, P. C. and Bass, L. N.**, Seed germplasm preservation using liquid nitrogen, *Seed Sci. Technol.*, 9, 423, 1981.
14. **Stanwood, P. C.**, unpublished data, 1982.
15. **Harrison, J. J. and Carpenter, R.**, Storage of *Allium cepa* seed at low temperatures, *Seed Sci. Technol.*, 5, 699, 1977.
16. **Lipman, C. B. and Lewis, G. N.**, Tolerance of liquid-air temperatures by seeds of higher plants for sixty days, *Plant Physiol.*, 9, 392, 1934.
17. **Styles, E. D., Burgess, J. M., Mason, C., and Huber, B. M.**, Storage of seed in liquid nitrogen, *Cryobiology*, 19, 195, 1982.
18. **Stanwood, P. C. and Roos, E. E.**, Seed storage of several horticultural species in liquid nitrogen ($-196^{\circ}C$), *HortScience*, 14(5), 628, 1979.
19. **Gresshoff, P. M. and Gartner, E.**, Cryopreservation of *Arabidopsis thaliana* and other seeds by storage in liquid nitrogen, *Arabodopsis Inf. Serv.*, 14, 12, 1977.
20. **Rajewsky, B.**, The limits of the target thoery of the biological action of radiation, *Br. J. Radiol.*, 25(298), 550, 1952.
21. **Stanwood, P. C.**, Tolerance of crop seeds to cooling and storage in liquid nitrogen ($-196^{\circ}C$), *J. Seed Technol.*, 5(1), 26, 1980.
22. **Illingworth, J. E.**, Peanut plants from single de-embryonated cotyledons, *HortScience*, 3(4), 275, 1968.
23. **Robbins, M. L. and Whitwood, W. N.**, Deep-cold treatment of seeds: effect on germination and on callus production from excised cotyledons, *Hortic. Res.*, 13, 137, 1973.
24. **Brown, H. T., and Escombe, F.**, Note on the influence of very low temperatures on the germination power of seeds, *R. Soc. Proc.*, 57, 160, 1897.
25. **White, J.**, The ferments and latent life of resting seeds, *R. Soc. Victoria*, 21(New Series), 417, 1908.
26. **Thiselton-Dryer, W.**, On the influence of the temperature of liquid hydrogn on the germanative power of seeds, *R. Soc. Proc.*, 62, 161, 1928.
27. **Adams, J. M. A.**, The effect of very low temperatures on moist seeds (the scientific proceedings), *R. Dublin Soc.*, II, 1, 1908.
28. **Mumford, P. M. and Grout, B. W. W.**, Desiccation and low temperature ($-196^{\circ}C$) tolerance of *Citrus limon* seed, *Seed Sci. Technol.*, 7, 407, 1978.
29. **Fedosenko, V. A.**, Use of superlow temperatures for long term storage of seed (methods and techniques) [Genetic Plant Resources of the All-Union Institute of Plant Industry](RU), *Biul. Uses. Inst. Rastenievod*, 77, 53, 1978.
30. **Fedosenko, V. A. and Yuldashava, L. M.**, The preservation of *Cucumis sativa* seeds at extremely low temperatures (RU), *Biul. Vses. Inst. Rastenievod*, 64, 60, 1976.

31. **Stanwood, P. C. and Bass, L. N.,** Ultracold preservation of seed germplasm, in *Plant Cold Hardiness and Freezing Stress,* Sakai, A. and Li, P., Eds. Academic Press, New York, 1978, 361.

32. **Yeo, R. R. and Thurston, J. R.,** Survival of seed tubers of dwarf spikerust (*Eleochans coloradoensis*) after exposure to extreme temperatures, *Weed Sci.,* 27(4), 434, 1979.

33. **Lipman, C. B.,** Normal viability of seeds and bacterial spores after exposure to temperatures near the absolute zero, *Plant Physiol.,* 11, 201, 1936.

34. **Sakai, A., and Noshiro, M.,** Some factors contributing to the survival of crop seeds cooled to the temperature of liquid nitrogen, in *Crop Genetic Resources for Today and Tomorrow,* International Biology Program,Cambridge University Press, Cambridge, 1975, 317.

35. **Junttila, O. and Stushnoff, C.,** Freezing avoidance by deep supercooling in hydrated lettuce seeds, *Nature (London),* 67, 325, 1977.

36. **Bequerel, P. M.,** La vie latente des graines aux confins du zero absolu., *C. R. Hebd. Sci.* 231(2), 1274, 1950.

37. **Bequerel, P.M.,**Action de l'air liquide sur la vie de la graine, *C. R. Acad. Sci. Tome,* 140, 1652, 1905.

38. **Grout, B. W. W.,** Low temperature storage of imbibed tomato seeds: a model for recalcitrant seed storage, *Cryo-Lett.,* 1(2), 71, 1979.

39. **Mumford, P. M. and Grout, B. W. W.,** Germination and liquid nitrogen storage of Cassava seed, *Ann. Bot.,* 42, 255, 1978.

40. **Busse, W. F.,** Effect of low temperatures on germination of impermeable seeds, *Bot. Gaz.,* 89, 169, 1927.

41. **Ishikawa, M. and Sakai, A.,**Freezing avoidance in Rice and Wheat Seeds in Relation to Water Content, *Low Temp. Sci.,* B36, 39, 1978.

42. **Sun, L. N.,** The survival of excised pea seedling after drying and freezing in liquid nitrogen, *Bot. Gaz.,* 119, 234, 1958.

43. **Jordan, J. L., Jordan, L. S., and Jordan, C. M.,** Effects of freezing to $-196°C$ and thawing on *Setaria lutescens* seeds, *Cryobiology,* 19, 453,1982.

44. **Locket, M. C. and Luyet, B. J.,** Survival of frozen seeds of various water contents, *Biodyamica,* 7(134), 67, 1951.

45. **Roos, E. E. and Stanwood, P. C.,** Effects of low temperature, cooling rate and moisture content on seed germination of lettuce,*J. Am. Soc.Hortic. Sci.,* 106(1), 30, 1981.

46. **Schopmeyer, C. S.,** Seeds of woody plants in the United States, USDA Handbook No. 450, U.S. Department of Agriculture, Washington, D.C., 1974.

47. **Bewley, J. D. and Black, M.,** *Physiology and Biochemistry of Seeds,* Springer-Verlag, Basil, 1978.

48. **Burke,M. J., Gusta,, L. V., Quamme, M. A., Weiser, C. J., and Li, P. H,** Freezing and injury in plants, *Ann. Rev. Plant Physiol.,* 27, 507, 1976.

49. **Brooks, H. J. and Barton, D. W.,** A plan for national fruit and nut germplasm repositories, *HortScience,* 12, 298, 1977.

50. **Chin, H. F. and Roberts, E. H.,** *Recalcitrant Crop Seeds,* Tropical Press SDN, BHD1, Kuala, 1980.

51. **Roberts, E. H.,** Predicting the storage life of seeds, *Seed Sci. Tecnol.,* 1, 499, 1973.

52. **King, M. W. and Roberts, E. H.,** Maintenance of recalcitrant seeds in storage, in *Recalcitrant Crop Seeds,* Chin, H. F. and Roberts, E. H., Ed., Tropical Press SDN, BHD., Kuala Lumpur, India, 53, 1980.

53. **King, M. W. and Roberts, E. H.,** The desiccation response of seeds of *Citrus lemon* L., *Ann. Bot.,* 45, 489, 1980.

54. **Becwar, M., Stanwood, P. C., and Leonhardt, K.,** Dehydration effects of freezing characteristics and survival in liquid nitrogen of desiccation-tolerant and desiccation-sensitive seeds,*J. Am. Soc. Hortic. Sci.,* in press, 1983.

55. **Goldbach, H.,** Imbibed storage of cacao seeds in osmotic solutions, *Plant Genet. Res. Newsl.,*43, 16, 1980.

56. **Merryman, H. T. and Williams, R. J.,** Mechanisms of freezing injury and natural tolerance and the principles of artificial cryoprotection, in *Crop Genetic Resources, the Conservation of Difficult Materials,* Withers, L. A. and Williams, J. T., Eds., International Union of Biological Sciences, B42, 5, 1980.

57. **Withers, L. A.,** *Tissue Culture Storage for Genetic Conservation,* Intl. Board Plant Genetic Resources, Rome, 1980.

58. **Conniff, K. and Stanwood, P. C.,** unpublished data.

59. **Stanwood, P. C., Roos, E. E., Leonhardt, K. W., Anderson, J., Towill, L., and Stushnoff, C.,** *NPGC Working Group on Cryopreservation of Plant Germplasm-Interim Report,* U.S. National Plant Germplasm Committee, Fort Collins, Colo., 1982.

Chapter 11

CRYOPRESERVATION OF EMBRYOS

Y. P. S. Bajaj

TABLE OF CONTENTS

I. INTRODUCTION

The larvae, caterpillars, and sometimes entire insects are able to withstand ultralow temperatures.[1,2] Likewise, cryostorage of the embryos of animals is possible. The technique, especially for mouse embryos, has been very well refined and is being employed for the study of developmental biology and in clinical research.[3] Mouse embryos kept in the frozen state at $-196°C$ or $-269°C$ for various periods and then transplanted into the foster mother develop into normal, full-term fetuses or newborn mice.[4] The possibility of the extension of this work to higher mammals, including human beings, is not ruled out.

Work on the cryopreservation of plant systems is rather recent; however, sufficient progress has been made during the last decade and entire plants can now be regenerated from frozen protoplasts, cells, tissues, and organs of a number of plant species.[5] These observations have far-reaching implications for the conservation of germplasm of rare, elite, and endangered species of plants. For this purpose, the author has urged the establishment of germplasm banks based on the technology of cryopreservation.[6-8] Although not much work has been done regarding plant embryos, the available literature (Table 1) on the survival of embryos in liquid nitrogen (LN), and their subsequent regeneration into plants, points to a number of potential uses,[9-25] the most important being the conservation of recalcitrant species.

The complete organs and large-sized embryos are particularly vulnerable to cryodamage and methods are still to be refined in reviving them after freezing. The main task is the prevention of recrystallization of the ice, and also intracellular freezing which is invariably lethal. Thus, the water content of these embryos needs to be controlled (by desiccation) to bring it to a level at which intracellular freezing is minimal. The cryodamage can be further prevented by adopting a method of thawing which discourages growth of ice crystals. The technology of cryopreservation of embryos is thus a multiple event involving dehydration-freezing-storage-thawing-culture, and a fault at any stage may cause irrepairable damage.

In this chapter, the results of the work on the embryos, protocol for their cryopreservation, various problems encountered, factors influencing their cryoability, potential uses and prospects for the technology of cryopreservation of embryos of various types, such as zygotic, somatic, and nucellar, and also the ones derived from cultured pollen, are summarized.

II. POTENTIAL FOR CRYOPRESERVATION OF EMBRYOS

A. Recalcitrant Seeds

The storage of seeds is the customary means of conservation and international exchange of germplasm. However, in a number of tree species and plantation crops the seeds are recalcitrant.[26] Such seeds are sensitive to humidity and temperature, and thus cannot be preserved under ordinary conditions for long periods due to degeneration of the embryos. Their maximum longevity varies from a few weeks to a few months. Some common examples are cacao (*Theobroma cacao* L.), rubber (*Hevea brasiliensis*), mango (*Mangifera indica*), chestnut (*Castanea crenata*), horse chestnut (*Aesculus hippocastanum*), walnut (*Juglans* sps.), cinnamon (*Cinamommum zeylanicum*), tea (*Thea sinensis*), cardamom (*Elettaria cardamomum*), and avocado (*Persea americana*) (for more examples see Chapter 6).[26] In such cases where the seeds are short-lived, the germplasm could possibly be conserved through the cryopreservation of excised embryos or their segments.

B. Substitute for Seeds and Mass Propagation

The somatic embryos are being looked upon as "seeds" in plants which do not set seeds. In such cases, somatic embryos can be cryostored, and used. Since the somatic embryos can be produced in large numbers from the cell suspensions, or by the in vitro culture of the segments and tissues, it would be rewarding to use them for large-scale multiplication of desirable and elite plants in which seeds are produced in lesser quantities. Moreover, in

Table 1
SUMMARY OF THE WORK DONE ON THE FREEZE-PRESERVATION OF EMBRYOS

Type of embryo	Plant species	Response	Ref.
Zygotic embryos	*Triticum aestivum*	70% Callused/developed plants	9
	Oryza sativa	83% Callused/developed shoots	10
	Triticale	73% Callused/developed plants	11
	Gossypium arboreum	42% Survival of embryo/ovule callus	12
	Brassica napus	15% Callusing of root region	13
	Hordeum vulgare	75—100% Recovery	13
	Cocos nucifera	Embryo segments proliferated	14
Somatic embryos	*Daucus carota*	Up to 80% survival, proembryos underwent normal embryogenesis, callused, or formed plants	15—17
Nucellar embryos (ovules)	*Citrus* spp.	Proliferated to form pseudobulbils and shoots	18
Pollen-embryos (anthers)	*Arachis hypogaea* *A. villosa*	29—38% Survival, callused, shoot formation	19
	Atropa belladonna	Entire haploid plants	20, 21
	Brassica campestris *B. napus*	31—44% Survival, callus, plants	19
	Gossypium arboreum	34% Survival of anther/pollen derived callus	12
	Nicotiana tabacum	Androgenesis, complete plants	21—23
	Oryza sativa	Androgenic anthers callused, plant regeneration	24
	Petunia hybrida	Callusing	21
	Primula obconica	Androgenesis, complete plants	25
	Triticum aestivum	5—19% Survival, callus, shoots	19

trees such as coffee, seeds cannot be stored,[27,28] but asexual embryos can be produced in large number, and they may be used for mass propagation. This is an area which can be commercially exploited by freeze-storage of somatic embryos, especially those of recalcitrant species.

C. Wide-Scale Hybridization

In the incompatible crosses in which the hybrid embryos generally abort at early stages, the young embryos can be excised, cryopreserved, and cultured when the need arises.

D. Haploid Germplasm

The haploid cell cultures are known to be genetically unstable[29] and within a short period of time they revert to a diploid state. In such cases, cryopreservation of pollen embryos or their calluses would be rewarding.

III. PROTOCOL FOR THE CRYOPRESERVATION OF EMBRYOS

For all practical purposes, the technology of cryopreservation of embryos is basically the same as that of the cells or tissues. It involves bringing the embryos to an inactive state of metabolism by subjecting them to ultralow temperatures ($-196°C$) in the presence of suitable cryoprotectants. Four types of embryos have been employed: i.e., zygotic, somatic, nucellar, and the pollen embryos obtained from the androgenic anthers. The young and immature embryos do not need any special manipulations, but the fully differentiated and maturing embryos have to be subjected to partial dehydration for successful preservation.

The small-sized embryos are pooled in 2-mℓ capacity vials containing 1 mℓ of the ap-

Table 2
**EFFECT OF SUDDEN FREEZING (−196°C) ON THE SEEDS, EXCISED
EMBRYOS, AND ENDOSPERM OF RICE PRESERVED FOR 3 WEEKS IN
LIQUID NITROGEN[a]**

| | | Frozen | |
Material	Control	Growth response (survival)	Growth (percentage of control)
Seeds	98% Germination	96% Germination	98
Dehusked seeds	94% Germination	82% Germination	87
Excised embryo with a portion of endosperm	86% Growth	71% Embryos callused and developed shoots	83
Segments of mature endosperm	16% Callused	11% Proliferated to form callus	68

[a] The seeds were germinated on moist filter papers in a Petri dish, whereas dehusked seeds, excised endosperm, and the embryos were cultured on MS + 2,4-D 2 mg/ℓ. Data based on 350 seeds, 92 dehusked seeds, and 360 cultures of embryos and endosperm.

From Bajaj, Y. P. S., *Curr. Sci.*, 50, 947, 1981. With permission.

propriate cryoprotectant. Dimethyl sulfoxide (DMSO 5 to 10%) is commonly employed, and a mixture of DMSO, sucrose, and glycerol at low concentrations (5% each) has given encouraging results. The vials are frozen in LN by various methods. The rate of freezing varies from 0.5°C/min to rapid freezing (about 12,000°C/min). The vials are then stored in LN cylinders. For large-sized embryos and plantlets, dry freezing is recommended.[17] After desiccation and treatment with the cryoprotectant, the material is wrapped in aluminum foil and frozen. The vials are generally thawed by immersing them in a beaker containing water held at 35 to 40°C. The embryos are then rinsed with distilled water or the appropriate liquid medium and cultured to induce growth. The method for the establishment of cultures and the problems involved are discussed by Withers and Bajaj.[30]

For estimation of the survival of the embryos, staining reactions alone, in addition to being insufficient, may even be misleading in determining viability. Sometimes, immediately after freezing and thawing, the cells are in a state of cryoshock and they do not show any sign of life. However, after an extended period of culture (even up to 6 months), they may resume growth.[16] Thus, from a practical point of view the survival data should be based on the resumption of growth, which may be judged by the embryos' ability to (1) increase in size, (2) turn green, (3) proliferate to form callus, and (4) develop shoots and plantlets.

IV. ZYGOTIC EMBRYOS

In a number of fruit and timber trees, and also in plantation crops, the seeds are large-sized and recalcitrant (refractory) and thus cannot be preserved under ordinary conditions. In such cases, the cryopreservation of zygotic embryos would be a logical alternative. Moreover, in wide-scale hybridization programs, especially those dealing with the intergeneric crosses which are incompatible due to degeneration or abortion of the embryos, possibly they can be dissected out at immature stages and cryopreserved.

There is a limited number of reports on the cryobiology of zygotic embryos of rice[10] (Table 2), wheat[11] (Table 3), barley, mustard,[13] and coconut[14] (Table 4). Although entire plants could be regenerated from the retrieved embryos in the first four cases, only callus was obtained with coconut. In all these experiments, quick freezing, followed by thawing at 35 to 40°C, was employed; however, their viability varied considerably (Table 1).

Table 3
SURVIVAL OF IMMATURE EMBRYOS OF WHEAT, RICE, AND TRITICALE FROZEN IN LIQUID NITROGEN IN THE PRESENCE OF DMSO (10%) + SUCROSE (4%)[11]

Species	Experiment no.	Control			Frozen			
		No. of unfrozen embryos cultured	No. of growing cultures	Control (% growing)	No. of embryos frozen	No. of embryos revived	Survival (%)	Revival (% of control)
Triticum aestivum	1	31	28	90.3	37	14	37.8	41.8
(cv. Kalyansona)	2	43	39	90.9	52	23	44.2	48.7
Oryza sativa	1	34	29	85.3	38	18	47.3	55.4
(cv. B370)	2	53	47	88.8	47	22	46.7	52.5
Triticale	1	40	35	87.5	61	39	63.9	73.0
	2	26	22	84.6	45	24	53.3	63.0

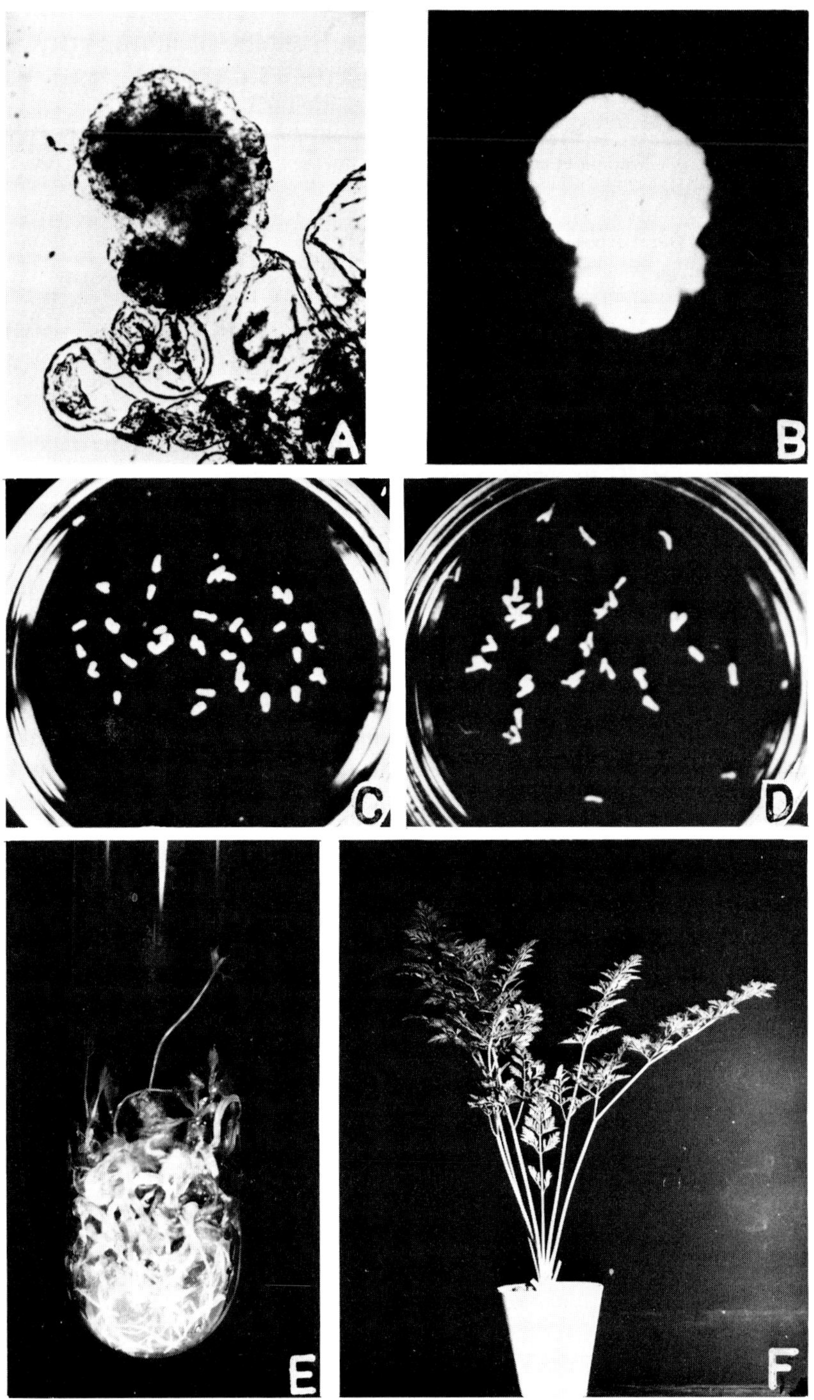

FIGURE 1.   Regeneration of carrot plants from somatic embryos (and proembryonic cell suspensions) frozen at − 196°C. (A) A proembryo along with cell suspension frozen at the rate of 2°C/min and stored for 1 year in liquid nitrogen, photographed in tungsten light (magnification × 75); (B) same specimen photographed in ultraviolet (UV) light after staining with fluorescein diacetate — note the survival of the embryo and death of the callus cells (magnification × 75); (C and D) frozen somatic embryos after 3 and 4 months (lag phase) in culture; (E) formation of plantlets from retrieved embryos and proembryonic callus; (F) same specimens 6 weeks after transfer to a pot (From Bajaj, Y. P. S., *Physiol. Plant.*, 37, 263, 1976, and *Acta Hortic.*, 63, 75, 1976. With permission.)

Table 6
**SURVIVAL OF NUCELLAR EMBRYOS AND OVULES OF *CITRUS*
SPP. FROZEN AT −196°C IN THE PRESENCE OF DMSO (7%) +
SUCROSE (1%)[18]**

Material	No. of cultures frozen	No. of cultures proliferating	Survival (%)	Remarks
Entire ovule	59	17	28.8	Proliferation to form
Micropylar half of the ovule (with nucellar embryos)	37	9	24.3	pseudobulbils and shoots

the number of plants per anther was considerably reduced. Mostly, the anther tissue became soft and spongy and the pollen embryos inside the anther aborted, perhaps due to the toxic substances produced by the degenerating anther.

C. Transversely Cut Halves of Cultured Anthers

Since little success was achieved with the entire anthers, it was considered that if somehow more pollen embryos were exposed, but not removed from the anther, it might improve their chances of survival. By cutting the anther into transverse halves (to expose the pollen embryos) and then freezing them, more pollen embryos survived and the number of plantlets obtained per anther was considerably increased (Table 7).

D. Isolated Pollen Embryos

The frozen-thawed pollen embryos of *Atropa* and tobacco cultured in liquid medium were also able to revive and regenerate plants.[20,22] The percentage of survival was strongly influenced by the stage of embryo development and the cryoprotectant used, in addition to the freezing method employed.

Globular embryos withstood freezing better (Figure 2C and 2D) and showed higher viability than the older embryos. Higher survival of the golublar embryos is attributed to the highly cytoplasmic, thin-walled, and nonvacuolated nature of the cells comparable to the young meristems and actively growing cell suspensions. In culture, the pollen embryos showed a lag period which was linearly proportional to their age and the extent of freeze injury. Younger embryos continued to undergo androgenesis and formed haploid plants, whereas the fully differentiated embryos showed a long lag phase and their survival percentage was low. Some of them survived only partially and had a tendency to callus or produce abnormal and malformed plants and multiple shoots.[21] In recent experiments, however, improvements have been made and maturing embryos were observed to develop into plants.

The survival of pollen embryos, to a great extent, was affected by the nature of the cryoprotectants and their concentration. Dimethyl sulfoxide at 7% gave optimal survival, whereas in the absence of a cryoprotectant the pollen embryos died. An experiment was also conducted to compare the effects of DMSO, sucrose, and glycerol as cryoprotectants. The results obtained with 15% glycerol (survival, 34%) were quite comparable to those obtained with 7% DMSO (survival, 33%). The anther/pollen embryo-derived callus of cotton, however, showed the highest survival (34%) with a mixture of DMSO, sucrose, and glycerol (Figure 3).[12]

Recent experiments[19] on the androgenic anthers and the suspension of pollen embryos of *Arachis*, *Brassica*, and *Triticum*, all subjected to freezing, yielded higher survival, and their response was highly genotypically oriented. The suspension of pollen embryos of *Brassica campestris* obtained from the 5-week-old anther cultures showed survivals of 31 and 44%,

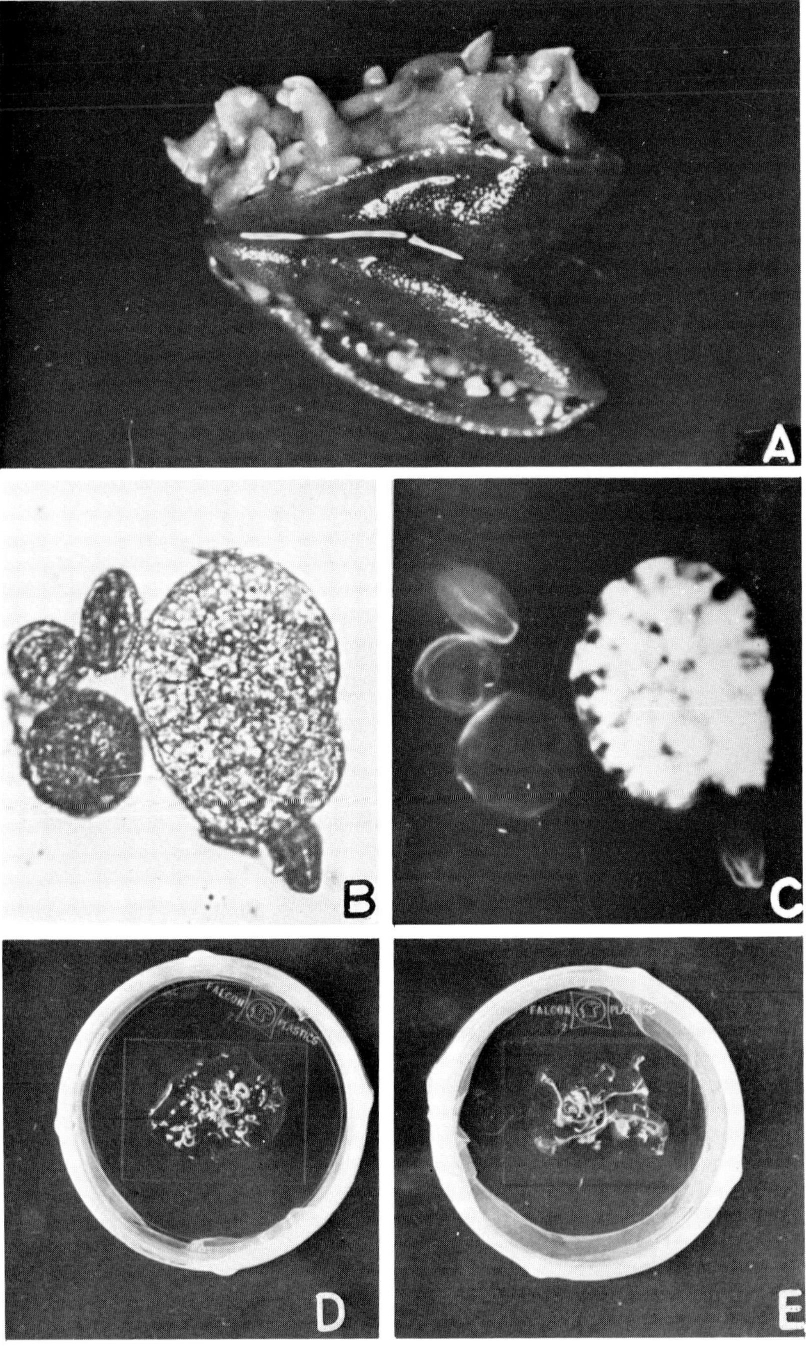

FIGURE 2. Regeneration of plantlets from pollen embryos frozen to −196°C. (A) Four-week-old culture of an excised anther of *Nicotiana tabacum* undergoing androgenesis; (B and C) survival of pollen embryos of *Atropa belladonna* shown in (B) tungsten light and in (C) UV light (magnification × 30.); (D and E) stages in the development of haploid tobacco plantlets obtained from frozen pollen enbryos grown in drop cultures (0.5 mℓ) in a liquid medium containing glutamine (800 mg/ℓ) and serine (100 mg/ℓ). (Figures B and C from Bajaj, Y. P. S., *Acta Hortic.*, 63, 75, 1976; and Figures A, D, and E from Bajaj, Y. P. S., *Phytomorphology*, 28, 171, 1978. With permission.)

Table 7
RESPONSE OF ANDROGENIC *NICOTIANA TABACUM* ANTHERS SUBJECTED TO VARIOUS TREATMENTS AND RECULTURED 4 WEEKS AFTER CULTURE[21]

Treatments	No. of anthers cultured	No. of growing anthers	Survival (%)	Total no. of plantlets	No. of plants per anther
Experiment 1 (whole anthers)					
Controls (untreated)	85	69	81.1	890	12.8
Treated with 7% DMSO	70	54	77.1	530	9.8
Treated with 7% DMSO and warmed at 37°C for 10 min	66	45	68.1	390	8.6
Treated with DMSO, cooled at the rate of 2°C/min, subjected to − 196°C and thawed at 37°C	130	2	1.5	7	3.5
Experiment 2 (transversely cut halves of androgenic anthers)					
Controls (untreated)	45	30	66.6	623	20.7
Frozen at the rate of 2°C/min to − 196°C	60	4	6.6	10	2.5

From Bajaj, Y. P. S., *Phytomorphology*, 28, 171, 1978. With permission.

respectively (Table 8). The pollen embryos underwent a lag phase and did not show any visible sign of growth for the first 5 weeks in cultures; however, thereafter they started to elongate. The embryos resumed growth and either directly developed into plantlets or growth was accompanied by callus formation.

The viability of the anther segments as well as those of pollen embryos was highest in *Arachis villosa* (31 to 38%). Wheat showed the poorest response, as only 5% of the anther segments resumed growth. The retrieved cultures underwent a lag period of 4 to 6 weeks; however, after that the anther segments underwent proliferation and the callus occasionally differentiated malformed shoots and plants.[19] The long lag phase, low viability, and the regeneration of malformed plantlets reflect on the extent of cryodamage caused to the system during pre- and postfreezing stages. Thus, to obtain higher survival values it is highly desirable that various factors affecting the cryoability may be thoroughly worked out.

VIII. FACTORS INFLUENCING THE CRYOABILITY OF EMBRYOS

The survival of the frozen embryos depends on the extent of damage caused during the pre- and postfreezing phase, and is influenced by a number of factors:

1. Genotype
2. Size and stage of development
3. Water content
4. Cryoprotectants
5. Method and rate of freezing
6. Thawing temperature

A. Genotype

The genotype plays an important part in determining the capacity of the plant to withstand low temperatures. Tropical plants in general are presumed to be more sensitive as compared to the plants grown under subzero temperature. Some species are more tolerant than the others (Table 1). The pollen embryos of *Brassica campestris* and *B. napus* showed survival frequencies of 31 and 44%, respectively. Similarly, immature zygotic embryos of *Hordeum*,[13]

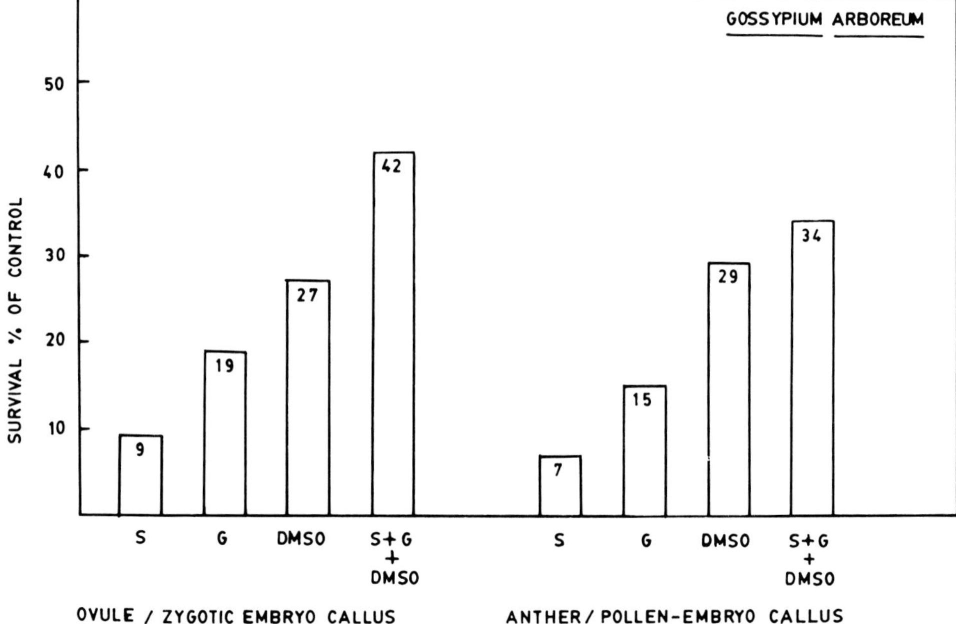

FIGURE 3.   Histogram showing the effect of various cryoprotectants on the survival of anther- and ovule-derived callus of *Gossypium arboreum* frozen in liquid nitrogen. (From Bajaj, Y. P. S., *Curr. Sci.*, 51, 139, 1982. With permission.)

wheat, triticale, and rice (Table 1), when subjected to rapid freezing, survived at rates of 75 to 100, 41 to 48, 63 to 73, and 52 to 55%, respectively.

B. Size and Stage of Development

Young and relatively small embryos are rather easy to manipulate and show higher survival values as compared to the maturing embryos. The pollen embryos at the globular stage and heart-shaped stage easily withstood freezing to −196°C, whereas with advancement in their developmental stage the survival capacity decreased. Moreover, fully differentiated embryos were cryosensitive and had a tendency to partially survive. The same was the case with the zygotic embryos of wheat; the immature embryos survived the usual freezing protocol, but the maturing embryos needed prefreezing dehydration.

C. Water Content and Dehydration

The longevity of the seeds depends on the water content at the time of storage. The orthodox seeds such as those of legumes and cereals which have a water content of 5% or less can easily be stored under ordinary dry conditions. They also survive freezing in the dry state, but are killed if stored after soaking.[34] Nevertheless, vacuum-dried, germinated seeds have been reported to survive at −269°C.[35] A strong correlation between water content and survival has been confirmed in peas.[36,37] Young pea seedlings (7 to 12 mm) survived freezing when desiccated to a water content of 40%. In these seedlings, while the root meristems with normal water content survived, the embryos died. In large, desiccated seedlings, only the stem-tip survived after exposure to liquid nitrogen. The root tips survived exposure to −196°C only when they had a moisture content of 27 to 40%, and they failed to revive when moisture content was less than 14.1%.

From the work done on the excised embryos, it has emerged that the extent of their

Table 8
SURVIVAL OF POLLEN EMBRYOS AND SEGMENTS OF THE ANDROGENIC ANTHERS OF *ARACHIS, BRASSICA,* AND *TRITICUM* FROZEN AND STORED AT − 196°C FOR 12 MONTHS[19]

Plant species	No. of anther segments frozen	No. of anther segments resumed growth	Survival (%)	No. of pollen embryos frozen	No. of pollen embryos surviving	Survival (%)
Arachis hypogaea	107	15	14	73	21	29
A. villosa	120	37	31	121	46	38
Brassica campestris	111	12	11	94	29	31
B. napus	114	21	19	81	35	44
Triticum aestivum	155	8	5	112	21	19

From Bajaj, Y. P. S. *Curr. Sci.*, 52, 484, 1983. With permission.

cryoability depends on their water content. Thus, partial dehydration of the maturing embryos is desirable for obtaining satisfactory results. Withers[17] reported on the successful survival of the somatic embryo-derived carrot plantlets subjected to desiccation. Likewise, zygotic embryos of coconut[14] and citrus,[38] when subjected to partial desiccation, have been observed to withstand freezing. Although desiccation would enable the cryopreservation of embryos of recalcitrant species, it has to be carefully manipulated as subcellular damage would be obvious below certain critical levels of moisture.

D. Nature and Concentration of Cryoprotectants

The long-term freeze-storage of germplasm requires special caution with regard to the choice and concentration of the cryoprotectant. It should not bring about any genetic aberrations in the form of chromosome breakage or mutations. Dimethyl sulfoxide has been extensively used and proved to be an excellent cryoprotectant, for animal as well as for plant cultures. An efficient cryoprotectant should (1) have low molecular weight, (2) be easily miscible with the solvent, (3) be nontoxic even at low concentration, (4) be easily washed from the cells, and (5) permeate rapidly into the system. All these criteria are fulfilled by DMSO.

For the somatic embryos and embryogenic cell suspensions of carrot, DMSO in the range of 5 to 10% has been used.[15-17] In the author's experience, a mixture of various cryoprotectants at low concentrations (5% each of DMSO, sucrose, and glycerol) has proved to be superior to any of these chemicals used alone (see also Chapters 5 and 12) for the ovule- and anther-derived callus of cotton,[12] pollen embryos,[19] or for the immature zygotic embryos of wheat, rice, and triticale where a survival ranging from 42 to 73% was obtained.[11] However, Withers[13] claims 15% DMSO to be optimal for *Hordeum* embryos. It is nevertheless emphasized that perhaps it may be better to use lower concentrations for a longer period or to employ a mixture of different cryoprotectants.

E. Method and Rate of Freezing

The young pollen embryos of *Atropa* and *Nicotiana*, subjected to freezing at the rate 1 to 5°C/min underwent a lag period of 2 to 6 weeks, and the maximum viability obtained was 34%.[21] The pollen embryos of *Arachis, Brassica,* and wheat, on quick freezing, revived and showed 38, 44, and 19% survivals, respectively.[19] Recently, survival of 75 to 100% has been obtained for immature embryos of *Hordeum* frozen by slow, regulated cooling as well as by rapid freezing.[13] Likewise, zygotic embryos of wheat, rice, and triticale subjected to quick freezing showed highest revival values of 48.7, 55.4, and 73%, respectively (Table

Table 2
CELL SUSPENSION CULTURE LINES IN LIQUID NITROGEN STORAGE AT THE FRIEDRICH MIESCHER-INSTITUT, BASEL, SWITZERLAND, AS OF MARCH, 1982

Species	Line designation	Special features	Pregrowth medium	Cryo-protectant
Acer pseudoplatanus	AM	—	S/M	DGP/DGS
Berberis dictyophilla	—	—	M	DGS
Catharanthus roseus	—	—	M	DGS
Corydalis sempervirens	—	—	S/M	DGS
Daucus carota	Wild	—	S	DGP/DGS
	CΔ 6A3	5MT[R]	S/M	DGP/DGS
	A1C5/1	—	S/M	DGP/DGS
	CV8/1	AEC[R]	M	DGP/DGS
Glaucium flavum	—	—	M	DGS
Glycine max	—	—	S/M	DGP/DGS
Hyoscyamus muticus	Wt	Wt	S/M	DGP/DGS
	VA5	His$^-$	S/M	DGP/DGS
	VIIIB9	Trp$^-$	S/M	DGS
	IVH2	Nic$^-$	S/M	DGP
	MA2	NR$^-$	S/M/P	DGS
Nicotiana tabacum	S3	AEC[R]	S/M	DGP/DGS
	TX4	Wt	S/M	DGP/DGS
	MFPr8	MFP[R]	S/M	DGP/DGS
Onobrychis viciifolia	—	—	S/M	DGS
Oryza sativa	Sask.	—	S/M	DGP/DGS
Pennisetum americanum	—	—	S/M	DGP/DGS
Rhazya orientalis	—	—	M	DGS
R. stricta	—	—	M	DGS
Rosa (Paul's Scarlet)	Wt	Wt	S/M	DGP/DGS
	OMT[R]	OMT[R]	S/M	DGP/DGS
	aca[R]	aca[R]	S	DGP/DGS
	aca[R]	aca[R]	S	DGS
Solanum melongena	—	—	S/M	DGS
Sorghum bicolor	—	—	S/M	DGP/DGS
Triticum monococcum	—	—	S/M	DGS
Zea mays	B73	—	S	DGP/DGS
	BMS	Wt	S/M	DGP/DGS
	BMS-C	ADH$^-$	S/M	DGP/DGS

Note: The cells were cryopreserved as follows.[20] They were pregrown in standard medium (indicated by S), or medium supplemented with mannitol at 6% w/v (indicated by M), or proline at 10% w/v (indicated by P); cryoprotected with 0.5 M DMSO plus 0.5 M glycerol plus 1 M proline (indicated by DGP), or 0.5 M DMSO plus 0.5 M glycerol plus 1 M sucrose (indicated by DGS); frozen at 1°C min^{-1} to -35°C, held at -35°C for 40 min, and then plunged into liquid nitrogen; stored in liquid nitrogen or liquid nitrogen vapor; thawed rapidly in water at $+40$°C; recovered on standard semisolid medium. (Modified from Ref. 44 with additional information[45].)

Drawing upon wider knowledge in cryopreservation, it is therefore possible to outline a basic protocol for cell cultures: cryoprotection, slow or stepwise freezing, rapid or slow thawing. Clearly, if success were based upon such simple guidelines then the reproducible cryopreservation of cell cultures would have been achieved long ago. However, we are only now approaching that satisfactory situation; not entirely as a result of "fine tuning" of the cryopreservation protocol *sensu stricto*, but through careful attention to all stages of the procedure including the growth of cultures before harvesting for preservation and recovery procedures after freezing and thawing.

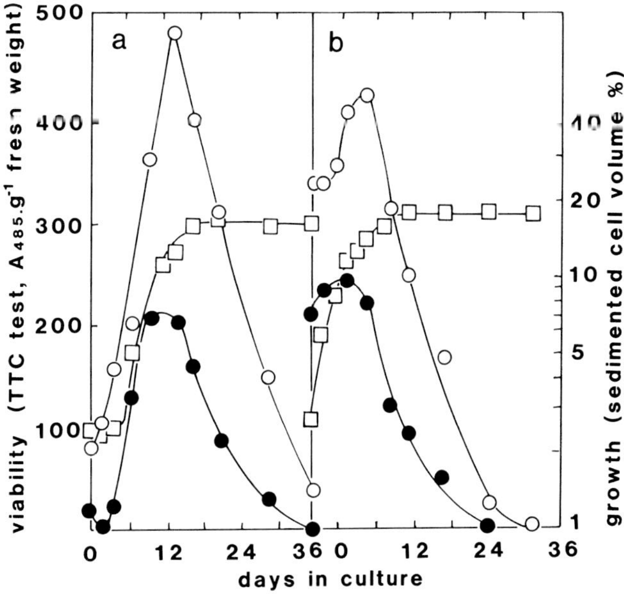

FIGURE 1. Growth (estimated by sedimented cell volume, □) of a cell suspension of *Oryza sativa* inoculated from (a) a 24-day-old culture and (b) a 14-day-old culture. Viability of cells before (○) and after (●) freezing was estimated by the TTC test.[2,31]

B. Pregrowth — Preparing the Culture for Cryopreservation

The stage in the growth cycle at which the cells are harvested can influence survival markedly. A typical growth curve[39] involves: (1) a lag phase during which changes in cellular structure take place, including synthesis of cytoplasmic components and a decrease in proportional vacuolar volume; (2) a period of growth, initially exponential in many cases, but slowing down eventually as nutrients are depleted; (3) a decline in the rate of cell division towards zero; and (4) a stationary phase at which time cells often undergo marked increases in vacuolar and cellular volume, and eventual senescence. Unless fresh nutrients are supplied the cells will die.

Where the subculturing interval is relatively long, cells in early lag phase will closely resemble those in stationary phase. This similarity is reflected in the typical response to cryopreservation. At the beginning of lag phase and in stationary phase post-thaw viability levels are low, often zero. Toward the end of lag phase and in early exponential growth, there is a dramatic increase in freeze tolerance. This phenomenon has been reported widely, but the most careful documentation has been made by Sala and colleagues,[31] who examined the behavior of cells of *Oryza sativa* sampled during a 36-day passage (Figure 1a). Viability was assessed by the TTC test which revealed, interestingly, that a peak in the respiratory performance (the criterion of viability in this test) of the unfrozen cultures occurred some time *after* the peak in survival.

When the experimental passage was initiated from cells harvested at an earlier stage (during exponential growth) in the parent culture, the lag in onset of growth was reduced but the difference in performance in the TTC test before and after freezing was increased (Figure 1b). Clearly, cells which are apparently at a "physiological peak" according to the viability test are not necessarily most amenable to freezing. Nonetheless, it is undoubtedly necessary to have an understanding of the growth pattern of the cells before attempting cryopreservation.

As can be deduced from the data in Figure 1b, it should be possible to eliminate, by

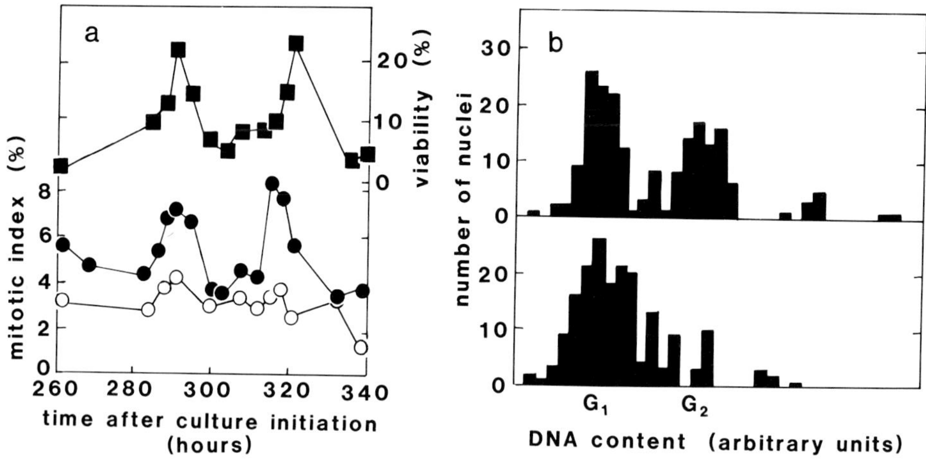

FIGURE 2.   (a) A cell suspension of *Acer pseudoplatanus* in synchronous growth was sampled during the period including the fourth and fifth divisions, frozen, thawed and its percentage of viability estimated by fluorescein diacetate staining (■).[41] Parallel samples were double-stained with Feulgen reagent and Evan's blue[42] to reveal the mitotic index of the total (●) and surviving (○) populations. (b) Microdensitometric measurement of nuclear DNA contents in cells harvested at the first peak of viability (ca. 290 hr) indicate's that the total population consists of cells with the G_1 and G_2 levels of DNA (upper histogram), whereas surviving cells are predominantly in G_1 phase (lower histogram).[19]

timely transfer, unfavorable phases from the growth pattern of the culture and increase the chances of harvesting freeze-tolerant cells. By adopting a 7-day subculturing regime for a cell suspension of *Acer pseudoplatanus*, Withers and Street[9] were able to avoid the most freeze-sensitive phases of growth. Microscopic examination confirmed that the highly vac-uolated cells typical of early lag and stationary phases were absent and that, overall, mean cell volumes were reduced. However, after a number of passages of reduced length, a pattern of oscillating viability emerged. Inspection of the cells suggested that synchrony of cell division might be occurring as a result of the frequent transfers. Testing of the hypothesis that cell cycle may influence freeze tolerance was a relatively easy matter in this species whose cells can be synchronized by nitrate starvation followed by inoculation at low density into complete medium.[40] Accordingly, it was possible to harvest cells over a period covering two tight (in this case the fourth and fifth) synchronous divisions and subject them to cryopreservation and cytological analysis.[19]

TTC testing and staining with fluorescein diacetate[41] (see later section) were used to estimate viability. Staining with Evans' blue[42] (which accumulates in dead cells — again, see later section) and Feulgen reagent were combined to respectively mark lethally damaged cells and indicate the DNA content and mitotic cycle stage of both surviving and lethally damaged cells. Rather surprisingly, survival appeared to follow changes in mitotic index (Figure 2a), suggesting that cells in mitosis have an enhanced tolerance of freezing stresses. However, inspection of the actual values in each case (maximum of 25 and 8%, respectively) refutes this. An explanation can be found by recording the stage in the cell cycle of individual cells. Near to the peak of mitosis and survival, the total population contains cells with nuclear DNA levels corresponding to G_1 and G_2 phases of the cell cycle (Figure 2b). However, surviving cells (unmarked by Evans' blue) had a lower mitotic index than the total population throughout the two cell cycles examined (Figure 2a) and at the point examined in detail; they consisted mostly of those with the G_1 level of DNA.

Thus it appears that cells which have recently passed through mitosis (perhaps in G_0 phase) are more tolerant of freezing than similarly sized or even smaller cells at other phases (see original data[19]). This pattern is different from that found in animal cells where oscillations in survival are less extreme, the highest values in the clearest case coinciding with late G_1 and S phases. Changes in the cell membrane structure and permeability have been suggested as contributory factors in the animal cells.

Since the achievement of tight synchrony is technically demanding[43] and the periods of maximum freeze tolerance may be relatively brief, synchronization and selection of suitable cell-cycle stages are not recommended as a practicable means of achieving consistent success in cryopreservation. Observations of the synchronized culture suggest that a *very* high level of survival may never be attained in an unsynchronized culture since it will surely contain a majority of cells at unfavorable stages. However, recent studies by Withers and King,[20] using an improved cryopreservation procedure, refute this. For example, in the case of cells of *Zea mays*, immediate post-thaw viability levels approaching 80% are recorded and these remained well above 70% during the recovery period. Therefore, in the case of *Acer pseudoplatanus*, we may be observing a situation where, under the prevailing (now known to be suboptimal) cryopreservation conditions, cell-cycle stage was just one factor contributing to overall loss of viability. Perhaps synchronization may even have entrained several of the critical factors.

In summary, it is possible to identify and choose cells which are likely to have a relatively high freeze tolerance, the reasons for which are not entirely clear but which are related to some extent to growth stage, cell volume, and general physiological condition. Other possible contributory factors, which might include detailed aspects of physiology such as levels of metabolite reserves, membrane permeability, etc. remain to be examined.

In the growth cycle of cells of some species, particularly those with a very large mean cell volume, there may never be a completely suitable phase. Consequently, one has to look further for means of enhancing freeze tolerance. Supplementation of the culture medium with osmotically active compounds is a relatively simplistic approach to the enhancement of freeze tolerance, based upon the hypothesis that if forces are applied which inhibit cell expansion, the cell and its vacuole will be reduced in volume and the cell water content will, in consequence, be lowered. This *is* found to be an effective means of enhancing freeze tolerance but, as will become evident, we are seeing a process which more closely resembles stress hardening, with both morphological and physiological aspects, than just a resistance to dimensional growth.

The most widely exploited and generally most effective additive is mannitol. Its use was first reported by Withers and Street in 1977,[9] who demonstrated that levels of 1.1 and 3.3% (w/v) in the culture medium (equivalent in osmosity to 1 and 3 times the major osmotic component in the culture media — sucrose) would both increase the survival potential at its peak and extend the period of the growth cycle during which viability was recorded. The higher level of mannitol brings about a significant reduction in mean cell volume. Cells of the species in this example, *Acer pseudoplatanus*, could be cultured for prolonged periods of time in the supplemented media with no apparent detrimental effect.

In more recent studies, Withers and King[20] have used the higher level of 6% w/v mannitol. A 3- to 4-day-long pregrowth period is highly effective in enhancing freeze tolerance and has now been incorporated into a standard cryopreservation procedure for cell suspensions. In their hands, it has been applied successfully to a wide range of species (Table 2).[44,45] Several other workers have confirmed its efficacy in cryopreservation procedures basically similar to that of Withers and King. A rare case of the failure of mannitol to enhance freeze tolerance adequately has been reported by Maddox and colleagues, working with a cell suspension of *Nicotiana sylvestris*. However, the closely related compound, sorbitol, did prove suitable for this species. The reason for the difference in response is not clear (see below also).

Pritchard and colleagues[46] have attempted to go beyond the superficial observation of a reduction in cell size as an explanation for the efficacy of mannitol as a pregrowth supplement.

Using the cryopreservation procedure of Withers and King,[20] they have noted that a number of critical changes occur in cells of *Acer pseudoplatanus* during a 7-day pregrowth period in medium supplemented with mannitol (and these were matched by similar preliminary data for sorbitol). The proportion of cells with a diameter of 25 μm or less rose from 55% in controls to 79%. This reduction was achieved by a loss of vacuolar volume from 34 to 14% of the total cell volume. Further, a single large vacuole in control cells was replaced by several smaller vesicles in mannitol-pregrown cells. Withers[47] also observed this change in cells of the same species pregrown in the lower level of 3.3% w/v mannitol, and Kartha et al.[22] found that cells of a line of *Catharanthus roseus*, which had a particularly high freeze tolerance, were of the multivacuole type. The latter cells reverted to a single, large vacuole form after freezing, thawing, and recovery. It would be very useful to know whether freeze tolerance was then reduced.

Pregrowth in the presence of 6% w/v mannitol depresses respiration and growth as measured in terms of dry-weight accumulation and rate of cell division.[46,48] A reduction in cell water content is indicated by a marked increase in the 50% plasmolysis point. During freezing, these cells will have a reduced requirement for protective dehydration. Observation of the cells in the cryomicroscope reveals that during freezing the cell wall, which is thinner in the mannitol-pregrown cells as compared with controls, appears to behave "osmotically". This means that separation of the protoplast from the cell wall and consequent damage will be less severe.

Pritchard and colleagues[46,48] suggest that since the control cultures include some cells in the under-25-μm size range which do not show the freeze tolerance of similarly sized mannitol-progrown cells, cell size cannot be the single critical factor. They propose that cytoplasm-to-vacuole ratio, cellular water content, presence of accumulated cytoplasmic solutes, and the osmotic behavior of the cells are all important. Some preliminary experiments in continuation of this work indicate that there is little uptake of mannitol during pregrowth, that total lipid levels increase, and that although total and soluble protein levels increase, the level of free proline decreases.[48] With the exception perhaps of the latter, these changes generally reflect those found in cells undergoing cold or other stress hardening.[49]

In view of the wide efficacy of mannitol, its apparently low toxicity and ease of application, there is little practical purpose in looking more widely for alternatives except in the case of difficult subjects. Two such possible alternative compounds are dimethyl sulfoxide (DMSO) and proline, both of which are, in conventional use, cryoprotectants. Nag and Street[16] first reported the use of DMSO at 5% v/v in the pregrowth of cells of *Atropa belladonna* which showed an increase in post-thaw viability from 30 to 40% (as estimated by the TTC test). Recently, the application of this compound in pregrowth of both shoot-tips[50,51] and cell suspensions[22] has been explored more fully by Kartha and colleagues. Cells of *Catharanthus roseus*, pregrown for 24 hr in the presence of 5% v/v, DMSO showed a similar proportional enhancement in freeze tolerance to the example above. However, it should be noted that the DMSO pregrowth reduced viability by 34%, signaling quite considerable toxicity. Nonetheless, DMSO pregrowth was beneficial on balance. Whether this would be the case for other species remains to be seen. Kartha et al.[22] give a caution in the use of this compound since it is known to have effects on membrane permeability (largely beneficial in freezing and thawing per se, but of unknown consequences in prolonged exposure) and interferes with the synthesis of RNA and proteins.

Problems of toxicity appear to be absent in the use of proline as a pregrowth additive. This compound was chosen by Withers and King[37] for the pregrowth of cells of *Zea mays* because it was known to be accumulated in plant tissues subjected to various environmental stresses. Pregrowth for 3 to 4 days in medium supplemented with 10% w/v proline increased post-thaw viability (as estimated by FDA staining — see later section) from ca. 20 to 60%. No serious loss of viability occurred during pregrowth although some darkening of the cells was noted. This pregrowth treatment has not proved widely successful in others' hands but has been found by Withers and King to be effective for cells of two additional species: *Rosa* Paul's Scarlet and *Hyoscyamus muticus*.[20,44,45]

C. Cryopreservation Proper

1. Cryoprotection

General aspects of cryoprotection are receiving attention elsewhere in this volume (see Chapters 2, 5, and 6) so they will not be examined extensively here. Cryoprotection is essential to the survival of frozen cell-suspension cultures, as indicated earlier. The first success in cryopreservation at the temperature of liquid nitrogen involved the use of DMSO for cells of *Daucus carota*,[3] which are deceptively amenable to freezing. Few other species have been cryoprotected successfully by DMSO alone. Glycerol, so successful in microbial and animal cell systems,[52] is only occasionally suitable when used alone for cell suspensions. Note that there is evidence in a number of biological systems for its having an effect of predisposing cells to deplasmolysis injury.[53] However, glycerol is a useful component of cryoprotectant mixtures. The only other compounds found to be effective when used alone are the amino acids proline[37] and γ-amino butyric acid,[38] which have been used to cryoprotect cells of *Zea mays*. Attempts to utilize proline more widely have failed and γ-amino butyric acid has not been examined further. Thus, the single cryoprotectant has little place in cell-culture methodology.

The combination of DMSO with glycerol dramatically increases the efficacy and utility of either compound. In early studies, Nag and Street[16,17] found a mixture of 5% DMSO and 10% glycerol to be suitable for cells of *Acer pseudoplatanus*. Much more recently, Maddox and colleagues[28] and Hauptmann and Widholm[23] have used similar mixtures (ca. 10% DMSO plus 10 to 12% glycerol) very successfully with the species *Nicotiana tabacum, N. sylvestris*, and *N. plumbaginifolia*, as well as *Daucus carota* and *N. tabacum*. However, in terms of range of application and success in achieving good cryopreservation, no mixtures exceed those adopted routinely by Withers and King:[20] 0.5 M DMSO plus 0.5 M glycerol plus 1 M sucrose, or 0.5 M DMSO plus 0.5 M glycerol plus 1 M proline. Their effective use with all of the species listed in Table 2 indicates the potential scope of their application. Generally, either mixture is effective, but occasionally one is superior (e.g., the proline-containing mixture with one line of *Hyoscyamus muticus*).[45] Several solutes, including sugars, sugar alcohols, and amino acids, can form the third component of such mixtures with good results, but none have been found to be superior to sucrose and proline.[38] Practical matters may influence the choice of components. The sucrose-containing mixture is very viscous and more difficult to filter-sterilize (especially by hand pressure) than the proline-containing mixture. The latter, though, is considerably more costly and must be prepared in culture medium (see below). Various effective cryoprotectant applications are listed in Table 3.

Little consideration has been given to the contribution of the medium components to overall cryoprotective effects. Although the media usually employed for the growth of cell suspensions contain lower levels of solutes than, say, media for protoplast culture (see later sections), the presence of standard medium components and pregrowth additives cannot be ignored. In order to examine this question, Withers compared the survival after freezing of cells of *Zea mays* treated with various cryoprotectants prepared in water or culture medium.[38] In water-based preparations, the concentrations of cryoprotectants were increased to compensate osmotically for the absence of medium constituents. In this study, only one cryoprotectant mixture, DMSO plus glycerol plus sucrose (as detailed above), performed equally well, however prepared. Further examination of other cryoprotectants combined with individual medium components (major and minor salts, hormones, vitamins, coconut milk, and sucrose) failed to reveal clearly which component conferred the greatest cryoprotective effect. However, surprisingly, salts appeared to be more significant in their effect than the

Table 3
CRYOPROTECTANT PREPARATIONS APPLIED SUCCESSFULLY TO CELL SUSPENSION CULTURES AND PROTOPLASTS

DMSO	Glycerol	Proline	Sucrose	Glucose	Ethylene glycol	Polyethylene glycol 6000	Ref.
Cell Suspension Cultures							
3.5							1
5							17
7							24
10							16
	5						17
	10						16
	15						29
		10					37
2.5	2.5						12
5	1						34
5	5						33
(0.5 *M*)	(0.5 *M*)						28
5	10						16, 35
10	10						23
12				5			18
(0.5 *M*)	(0.5 *M*)	(1 *M*)					20
(0.5 *M*)	(0.5 *M*)	(1 *M*)					20
3				4	2.5		10
10				8			13—15
10				8		10	15,36
Protoplasts							
5							7
	(0.7 *M*)						54
5				10			55, 56
10				10			23

Note: Concentrations are expressed as percentages except when otherwise stated. References given are for example only and do not cover all instances of use.

standard level of sucrose in the medium. Absence of medium components affects both initial post-thaw viability and capacity to resume growth.

Sterilization by filtration is recommended for all cryoprotectant preparations and is essential for the more highly concentrated mixtures which will caramellize if heat-sterilized. Further points to note are that the cryoprotectant solution should be prepared at double the final desired concentration and its pH adjusted to that appropriate to the cell culture.

Conventionally, cryoprotectants have been added slowly to cell suspensions, both the cryoprotectants and the cultures being chilled beforehand.[9,17,31] However, in the routine procedures adopted by Withers and King,[20] and in other recent reports,[23] the chilled materials are mixed in one step. Further, Finkle and Ulrich[36] found that a cryoprotectant mixture of polyethylene glycol, DMSO, and glucose was effective when added at 0 or 22°C (Chapter 5). It may be necessary to modify the application procedure in order to adjust the period of exposure to cryoprotectants prior to freezing. This currently stands at ca. 1 hr in most reports. Hauptmann and Widholm[23] have found that loss of viability in cells of *Daucus carota* and *Nicotiana tabacum* during cryoprotection may be reduced by between 10 and 15% by omitting a preincubation stage. They did find, however, that it was necessary to cool the cells slowly to a lower temperature before plunging in liquid nitrogen, indicating a need for additional protective dehydration (see Section II.C.3 below).

2. Preparation for Freezing

In preparation for freezing, cells are dispensed into a suitable container. Commonly used are 2-mℓ volume, presterilized, polypropylene, screw-cap ampules. These usually contain 1 mℓ but occasionally are filled to capacity.[23] Increased compressive forces, possibly deleterious, might be expected under such circumstances. Heat-sealable polypropylene or glass vials are available and the latter have been used successfully for cell suspensions.[22] Care is needed in the use of glass vessels since liquid nitrogen penetration resulting from incomplete sealing may lead to explosion upon warming. At best, this may cause loss or contamination of samples; at worst, injury to the operator.

Takeuchi and colleagues[55] have used heat-sealable, plastic-lined, aluminum-foil packets for protoplasts. Slightly higher post-thaw viabilities are reported in comparison with conventional ampules. Using a method developed for somatic embryos per se, Withers[26] has frozen an embryogenic cell suspension of *Daucus carota* successfully in foil envelopes. A preliminary test in this laboratory of the use of plastic envelopes for cell suspensions suggests that they also may yield higher viabilities than ampules.[57] More uniform thawing rates, more rapid thawing, and reduced compression forces may contribute to the effect.

At this stage, it is important to consider the amount of material being frozen, since cells can be very sensitive to inoculum density.[39] Loss of viability upon cryoprotection and freezing, giving a dual effect of reducing viable cell density and the risk of inhibition of growth by the presence of dead or dying cells, must also be taken into account (see below). Most reports fail to state densities of cells at the time of freezing but available data indicate that although densities as low as 3% packed-cell volume may be suitable for *Oryza sativa*,[31,32] for other species much higher densities (20% for *Catharanthus roseus*,[22] 30% for *Nicotiana* spp.[28]) may be necessary. Studies of packing effects carried out with human erythrocytes[58] have not been paralleled with with plant cells and would warrant investigation. No doubt the local environment of the cell during freezing would vary with the cell density.

3. Freezing

Once concentrated to an adequate cell density and placed within a suitable enclosure, the cell suspension is frozen. Without exception, freezing programs for cells have relied for protection upon extracellular freezing and consequent dehydration of the cytoplasm. Various ways of achieving adequate dehydration are available: slow cooling at a uniform rate, nonuniform slow cooling, or stepwise freezing by exposure to one or more intermediate subzero temperatures.

Some of the most successful attempts at cryopreservation have involved a combination of slow and stepwise freezing in which the specimen is cooled, for example, at 1°C min^{-1} to a holding temperature in the region of -30 to $-40°C$, held at that temperature for up to 1 hr, then plunged into liquid nitrogen.[20,23] Although programable freezers may be used, the improvised freezing apparatus described by Withers and King[20] (Figure 3) achieves such a cooling profile without sophisticated temperature control. Maddox and colleagues[28] employ an even simpler method of freezing (adapted from cryopreservation work with other biological systems) in which the specimen ampules are enclosed in a polystyrene block placed within a deep freeze.

The choice of holding temperature and time in a stepwise freezing program may be critical; 30 to 40 min at $-35°C$ is suitable for cells cryoprotected with 0.5 M DMSO plus 0.5 M glycerol plus 1 M sucrose,[20] but may not be suitable for cells cryoprotected otherwise. Hauptmann and Widholm[23] found that a lower holding temperature ($-40°C$ compared to $-30°C$) was necessary when the preincubation in cryoprotection was eliminated (see above).

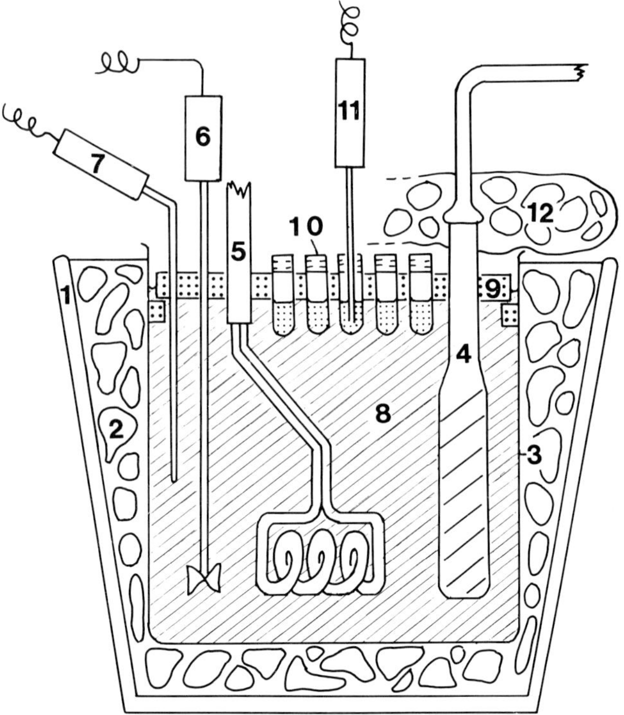

FIGURE 3.   Improvised freezing apparatus. Key: 1 = plastic bin, 2 = polystyrene chips for insulation, 3 = glass beaker — 5-ℓ capacity, 4 = dip cooler, 5 — heating coil to rewarm alcohol, 6 — stirrer (spark-proof), 7 = temperature probe for dip cooler thermostatic control, 8 = alcohol (e.g., industrial methylated spirits), 9 = plastic raft resting on supports attached to inside of beaker, 10 = specimen ampule, 11 = thermometer reading temperature of dummy specimen, 12 = plastic bag loosely filled with polystyrene chips.[20]

In the development of a cryopreservation protocol, the use of stepwise freezing (or ''pre-freezing'') has distinct advantages since it is possible within one experiment to compare a range of holding times (and therefore amounts of cellular dehydration). Over-dehydration is likely to be deleterious, leading to ''solution effect'' damage (see Chapter 2) which is either irreversible or reversible only by the use of slow thawing (see below). Once the point of adequate dehydration is reached, slow freezing is best terminated by quenching the specimens in liquid nitrogen. In fact, very few freezing programs involve slow cooling below a temperature in the region of $-40°C$. Rapid cooling from the intermediate temperature is likely to have the further beneficial effect of minimizing crystal size in ice formed from remaining intracellular water (see Figure 8). Interactions between terminal slow-cooling temperature and thawing rate have been observed, as detailed below under ''Thawing'', Section II.C.5.

4. Storage

For practical, long-term conservation, storage must be at a temperature approaching that of liquid nitrogen. At higher temperatures there is a risk of structural damage to the cell as a result of progressive ice recrystallization. Accordingly, reports of apparently successful cryopreservation at temperatures ranging from -23 to $50°C$ (see Table 1, Section A) must be disregarded as useful approaches to long-term storage. In some cases, the terminal temperature may even be too high for intracellular ice to form.

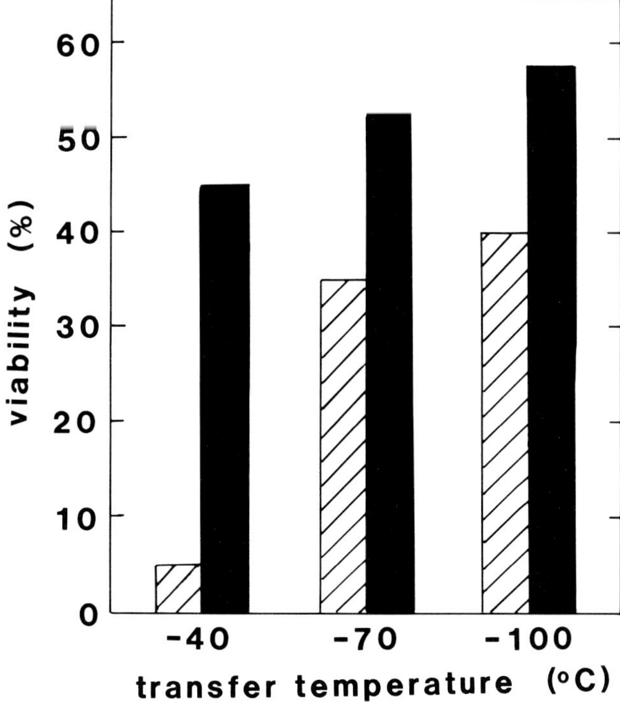

FIGURE 4. Survival (indicated by fluorescein diacetate staining[41]) of cells of *Daucus carota* frozen at a rate of 2°C min^{-1} to different temperatures before plunging into liquid nitrogen and then thawing either slowly (in air at +20°C, cross-hatched blocks) or rapidly (in water at + 40°C, solid blocks).[16]

Due to the relative infancy of cryopreservation studies, no serious work has been carried out on the effect of duration of storage and stability. Nonetheless, reports of successful storage, for example for periods of 12 months as achieved by Maddox and colleagues,[28] reinforce the assumption, based upon theory, of long-term stability at ultralow temperatures.

5. Thawing

Conventionally, rapid thawing has been used for most cell suspensions.[9,16-24] Ampules are retrieved from liquid nitrogen and plunged directly into a water bath at ca. +40°C. With constant agitation, thawing is complete in 1 or 2 min. Precautions to avoid the introduction of microbial contaminants include the use of an inner container of sterile water in the water bath, and wiping the warmed ampule with a sterilant before opening.

The general success of rapid thawing implies the presence of some intracellular ice capable of undergoing damaging recrystallization. However, circumstances may be manipulated so that slow thawing is feasible. A pointer to this is seen in an early report by Nag and Street;[17] they compared the survival of cells of *Daucus carota* frozen to −40, −70, and −100°C before quenching in liquid nitrogen and thawing either rapidly or slowly (in water at +37°C and air at +20°C, respectively). The advantage of rapid thawing was most evident at the highest transfer temperature (Figure 4). However, the level of survival was very adequate after slow thawing in the specimens transferred at −100°C. More recently, Withers[38] has cryopreserved cells of *Zea mays* cryoprotected with 0.5 *M* DMSO plus 0.5 *M* glycerol plus 1 *M* sucrose. Once these cells had been cooled beyond a transfer temperature of ca. −40°C, both rapid and slow thawing gave high post-thaw viabilities and recoveries (Figure 5). Experiences with embryogenic cell suspension of *Daucus carota*[26] endorse the possibility

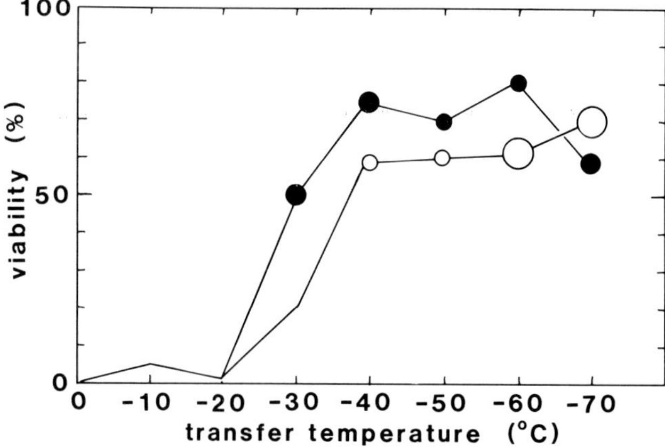

FIGURE 5.   Performance of cells of *Zea mays* frozen slowly at 1°C min⁻¹ to various temperatures before plunging into liquid nitrogen and then thawing either slowly (in air at +20°C, open circles) or rapidly (in water at +40°C, solid circles). Initial post-thaw viability was estimated by fluorescein diacetate staining[41] and is indicated by the lines drawn. Relative recovery in terms of number and size of colonies produced is indicated by circle size.[38]

of using slow thawing. Perhaps, given adequate protective dehydration by use of an appropriate cryoprotectant application and cooling program, many more specimens might yield to slow thawing. Certain practical advantages would then become available: more uniform thawing, a reduction in the (albeit slight) risk of breakage of ampules upon rapid expansion, and a reduction in the risk of rehydration/deplasmolysis injury in highly vacuolated cells which have to be dehydrated severely to avoid ice damage.

D. Post-Thaw Treatments and Recovery Growth

In the earlier attempts to store cell suspensions in liquid nitrogen, the approach to post–thaw treatment was to return the cells to standard culture conditions as quickly as possible. This involved washing to remove cryoprotectants and inoculation into liquid medium for culture on a shaker. Such a procedure has proved satisfactory for the species, *Daucus carota*[9] (which, as must be becoming evident, is particularly flexible in its requirements), and one or two others, e.g., *Oryza sativa*[30-32] and *Catharanthus roseus*.[22] However, studies concentrating upon more recalcitrant species such as *Zea mays, Rosa* Paul's Scarlet and *Acer pseudoplatanus*[20,37] suggest that the approach outlined above may be imposing unnecessary stress upon the cells, leading to an otherwise avoidable loss of viability.

Let us examine the components of this post-thaw phase: washing, mode of culture, and culture medium composition. There is no categorical evidence to suggest that washing is necessary to remove cryoprotectants, although washing has probably been introduced into the procedure on the reasonable assumption of cryoprotectant toxicity by prolonged exposure. Where the cells have been subjected to a range of treatments, washing has been shown to be deleterious. For example, Withers and King,[37] using cells of *Zea mays* cryoprotected with proline, compared (1) washing in distilled water, (2) washing in standard liquid medium, and (3) washing in standard liquid medium supplemented with 10% w/v proline. The cells were then spread onto the surface of semisolid medium. In no case did the cells recover as rapidly as when washing was omitted. Washing in water was particularly damaging in that immediate post–thaw viability was halved as well as recovery being delayed. In terms of recovery rate, all washing treatments had a similar effect to that of removing the suspending liquid and layering the cells "dry" on the surface of the semisolid medium. For other

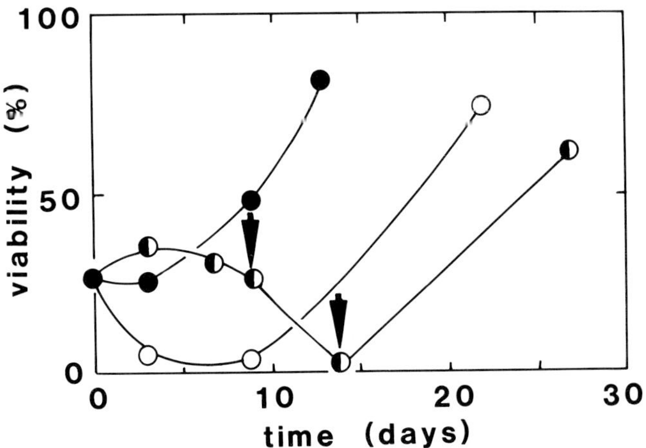

FIGURE 6.   Recovery (estimated by increase in percentage viability, as revealed by fluorescein diacetate staining[41]) in cryopreserved cells of *Zea mays* returned to culture either in liquid medium (○), hanging droplets (◑), or layered over semisolid medium (●). Addition of fresh medium to droplets at the times indicated by arrows prevented senescence. Recovery from the later addition is shown.[37]

cryoprotectant treatments (e.g., 10% w/v glycerol + 10% w/v DMSO) washing was *more* deleterious than just removing the suspending liquid.

Thus, it appears that washing has two effects (their relative importance depending upon the cryoprotectants in use): (1) removal of a factor or factors which are supportive of recovery growth and (2) physical injury, perhaps related to deplasmolysis. Further evidence for a membrane lesion in freshly thawed cells can be found in the observations of Cella, Sala and their colleagues[30,32] that cells of *Oryza sativa* could produce very few stable protoplasts immediately after thawing. Only within 2 to 4 days did the capacity to yield stable protoplasts at the level of control preparations return.

As general advice it would appear best to avoid washing unless absolutely essential. Should it prove to be essential, however, two additional observations suggest ways of minimizing damage. First, Finkle and Ulrich,[36] studying the effect of varying the temperature of cryoprotectant application and removal, found that warm (ca. 22°C) washing was less harmful than cold washing (ca. 0°C, see also Chapter 5). Second, Maddox and colleagues[28] supplemented the post–thaw medium with the mannitol or sorbitol present during pregrowth to minimize deplasmolysis injury (see below).

Adoption of alternative modes of culture may offer ways of removing cryoprotectants which avoid the dangers of washing. Again using cells of *Zea mays*, Withers and King[37] compared their recovery rate (1) in hanging droplets; (2) over semisolid medium, as described earlier; and (3) by inoculation into standard liquid medium. The treatments provide an increasing "dilution" of the original suspending liquid and increasing availability of fresh nutrients. For cells cryoprotected with proline and reintroduced into culture by the first two methods, post-thaw viability levels stabilized and only declined in the droplets as nutrients were depleted (Figure 6). Addition of fresh medium after several days led to rapid recovery. Culture in liquid medium led to a rapid loss of viability and recovery from a very small proportion of the original cell population. Cells cryoprotected with DMSO and glycerol responded even more severely to dilution in liquid medium, lost viability completely after a few days in droplets, and recovered only slowly when layered over semisolid medium. Clearly, the simplest possible recovery method has proved the most satisfactory for these cells of *Zea mays* (and the other species listed in Table 2).

Nonetheless, studies by other workers still support the use of liquid culture in certain cases. Maddox and colleagues[28] have used liquid culture for the recovery of cell suspensions of several *Nicotiana* species in order to avoid having to pass through a post–thaw callus phase. (Although in the experience of the author[57] callusing in cells layered over semisolid medium is rare; usually a "slurry" forms which rapidly regenerates a fine suspension upon return to liquid culture.) The cells of *Nicotiana* spp. were damaged by culture in shaking liquid medium, but static culture was successful.

Little attention has been given to post-thaw medium composition other than in relation to physical factors, as already outlined above. However, since cells returned to culture suffer a drop in inoculum density both by dilution and by the death of a proportion of the cell population, special media may be required to aid recovery from very low viable cell densities.[39] Maddox and colleagues[28] inoculated freshly thawed cells of *Nicotiana* spp. at a high density overall (30% packed-cell volume diluted 1 in 5 or 1 in 10) into the low auxin Medium A of Caboche.[59] This medium is capable of supporting growth from 10 cells per mℓ. Within 2 to 3 weeks a suspension formed which could provide inoculum for shaking liquid culture. During the recovery period, 1 or 2 mℓ of medium was progressively removed and replaced by fresh, unsupplemented medium to replenish nutrients and reduce the levels of osmoticum and carried-over cryoprotectant.

In order to support growth in cryopreserved cell suspensions of *Daucus carota* and *Nicotiana tabacum*, Hauptmann and Widholm[23] used feeder plates. In these, the culture medium is effectively supplemented with conditioning factors produced by unfrozen cells inoculated into the semisolid medium from which the recovering cells were separated by a wire-screen or foam-pad barrier.

Some recent observations by King[45] indicate that even simpler methods than those described above may overcome inoculum density problems and reduce the influence of the suspending liquid present at the time of freezing. Cells often recover more rapidly if drawn together into a mound in the center of the dish, and if the level of liquid medium over the semisolid medium is lowered, either by tilting the dish or cutting a channel in the semisolid medium into which the liquid can drain. Minor modifications to established procedure such as this may be all that is needed to extend current techniques to include more difficult subjects.

E. The Quantitative and Qualitative Examination of Cryopreserved Cells

Over the years, the approaches taken to the investigation of the properties of cryopreserved cells have reflected the state of development of preservation techniques. Thus, at first workers were concerned only with post-thaw viability. Then, as recovery growth became possible, the rate of recovery was monitored. Physiological and electron microscopical studies were carried out on frozen and thawed cells initially, extending to include frozen, thawed, and recovering cells as cryopreservation procedures improved. Most recently of all, as interest has moved towards the implementation of cryopreservation as a routine laboratory technique, the retention of certain traits by frozen and thawed cells has come under investigation.

Looking first at viability tests: the value of these should neither be over- nor underestimated. Those available at present, such as fluorescein diacetate staining,[41] phenosafranine,[41] and Evans' blue[42] dye exclusion, and reduction of 2,3,5–triphenyl tetrazolium chloride (TTC test),[2] can be valuable indicators of immediate post-thaw viability, but should not be taken as categorical signals of capacity to recover growth. Latent injury may not be apparent shortly after thawing and cells may behave very erratically at first. Additionally, anomalous inorganic reactions may mimic the production of formazan from TTC.[33] Thus, a time lag — preferably overnight — before taking a viability test is likely to increase the clarity and meaningfulness of the assay. Although a single estimation may be adequate, monitoring changes in viability over a period of time is even more valuable, revealing accurately the progress of continued cell death or recovery (e.g., as in Withers and King's[20,37] studies on the cryopreservation of cells of *Zea mays*).

Parameters of growth have been employed by several workers to the same end. Changes in cell density, packed or sedimented cell volume, and fresh weight all are useful pointers to recovery (but will not indicate decline). Data so obtained suggest that in many cases growth is resumed rapidly and the rate of recovery growth closely resembles the normal growth rate of unfrozen cells (e.g., in *Oryza sativa*,[31] *Nicotiana tabacum*,[23] and *Daucus carota*[23]). Where suboptimal treatments have been used, lag phases of several days to months, and reduced rates of growth may be experienced (e.g., in *Nicotiana tabacum*[24] and *Daucus carota*[9]).

Observations of ultrastructure and physiological parameters should be interpreted circumspectly. There is a danger that one may be witnessing pathological changes characteristic of general cellular damage exacerbated by the manipulations necessary to the examination procedure, rather than changes unique to cryodamage. Thus, the observations of structural damage in freshly thawed cells must be assumed to include some fixation artifacts. Accepting the limitations which must be placed upon interpretation here, it is still possible to obtain some useful corroborative evidence regarding the refinement of cryopreservation procedures. For example, suboptimal treatments such as the omission of cryoprotectants lead to greater organelle and membrane disruption and some cells, especially more highly cytoplasmic, suffer less damage than other, more highly vacuolated ones.[47,60] However, when we turn to material which is undergoing recovery and more likely to react normally to fixation, more useful observations can be made.

Withers[53,61] has examined cells of *Daucus carota* cryopreserved by a procedure which is now known to be less than ideal but which, at the time, led to reproducible recovery from a proportion of the cell population. Cells which failed to undergo recovery growth declined in various ways. Some died and disintegrated immediately after thawing but others were initially capable of synthesizing wall material or developing cytoplasmic deposits of osmiophilic material. Cells able to undergo recovery and enter mitosis showed signs of damage, particularly to the cell membranes. Dilation of organelles, evident immediately after thawing (and therefore subject to the caution given above), was reversed but some irreversible damage to membrane structures was indicated by the development of masses of spherosomes in the cytoplasm. Within several days, the spherosomes became eroded and cell membrane regeneration ensued. The new membrane at first accumulated in discrete masses but eventually disappeared, presumably becoming incorporated into organelles as the viable population increased by cell division.

The cryopreservation procedure for the cells described above involved post-thaw washing and recovery in liquid medium.[9] We now know this to be less than ideal[20] and parallel observations of cells of *Zea mays* subjected to an improved procedure of layering onto semisolid medium reveal a completely different pattern of recovery.[53,61] Within 2 days of thawing, cells were clearly either dead and disintegrating or healthy and capable of growth and cell division. The latter cells contained no signs of membrane or organelle damage (although some plasmodesma breakage, presumably a result of plasmolysis, was noted). Thus the complex process of accumulation of membrane material in spherosomes and subsequent "recycling" as growth is resumed appears to be avoidable.

Various studies have been carried out on cold- and freeze-damaged plant tissues from a purely physiological stance, but only one serious piece of work has been undertaken to support the development of cryopreservation methods for cell cultures. Using cells of *Oryza sativa*, Cella, Sala and their colleagues[30,32] have been able to gather data on physiological performance which all point to the existence of a distinct recovery phase lasting up to 4 days, during which cells should be handled carefully. (Some of these data relating to the capacity to produce stable protoplasts have already been mentioned above.)

B. Isolation, Cryoprotection, and Preparation for Freezing

Protoplast isolation involves enzymic digestion of the cell wall while at the same time placing the tissue under the influence of a plasmolyzing osmotic stabilizer (e.g., 0.4 to 0.7 *M* mannitol, sorbitol, or sucrose). The isolation procedure, which takes a few hours usually, is selective in that not all of the original cell population will yield protoplasts. Thus, senescing and highly vacuolated cells which are particularly sensitive to plasmolysis may fail to produce stable protoplasts. Since plasmolysis stress during freezing and deplasmolysis stress during thawing and post-thaw treatment are likely contributors to overall cryoinjury, a preparative stage which necessitates survival of severe plasmolysis may well be beneficial.

In all reports but one in which an unspecified "equilibration period"[54] is quoted, cryoprotection follows immediately after isolation and washing. (A lengthy intervening period would permit wall regeneration, turning the protoplasts into cells.) Thus a pregrowth period is not a feature of the cryopreservation procedure for protoplasts, although there is scope for modifying the preculture conditions to which the parent tissue is subjected.

In one study, protoplasts of *Daucus carota* have survived freezing without cryoprotection other than by the protoplast culture medium.[5,61] All others involve the addition of single or mixed cryoprotectants. Most of those listed in Table 3B have been used effectively for the latter species, but in the absence of comparative studies, none can be specifically recommended. The mixture of 5% DMSO and 10% glucose appears to have the widest application, however, being used effectively by Takeuchi and colleagues[56] in the cryopreservation of protoplasts of *Marchantia polymorpha*, *Glycine max* (2 varieties), *Triticum aestivum*, and *Hordeum vulgare*, as well as 2 lines of *D. carota* (habituated and nonhabituated). Most usually, cryoprotectants are chilled and then added slowly to the protoplasts. The protoplasts are transferred to ampules or plastic-lined, aluminum-foil envelopes for freezing (as described earlier). The latter may be superior.[55]

C. Freezing, Storage, and Thawing

Although protoplasts of *Daucus carota* are, characteristically, more flexible in their freezing requirements, being tolerant to some extent of rapid and slow freezing,[5] the overriding pattern is that of a requirement for protective dehydration. Rates of slow freezing in the region of 1 or 2°C min^{-1} are effective and, in the most detailed study carried out to date by Takeuchi and colleagues,[55,56] a transfer to liquid nitrogen at a temperature between -30 and $-40°$ is beneficial (Figure 8). Successful storage in liquid nitrogen for up to 4 months has been carried out in the latter study, indicating that long-term conservation is feasible. Usually, rapid thawing in water at ca. $+40°C$ has been used.

D. Post-Thaw Treatments, Wall Regeneration, and Recovery Growth

Withers[57] has found that protoplasts of *Daucus carota* will recover and undergo embryogenesis when cultured in hanging droplets without post-thaw washing. In the more extensive study by Takeuchi and colleagues,[55,56] involving the liverwort *Marchantia polymorpha* and several higher plant species, the protoplasts were washed and then cultured in a Petri dish containing liquid medium. Viabilities were monitored using fluorescein diacetate. Staining with Calcofluor white indicated the progress of wall regeneration. For the various species, viability levels ranged from 30 to 68%. The deposition of wall material commenced within a few hours of thawing and, in the case of *M. polymorpha*, 80% of the surviving protoplasts were capable of wall regeneration. Cell division commenced within 2 days. All species formed callus and a totipotent response was observed where expected (thallus, rhizoid, and gemma cup formation in *M. plymorpha* and embryogenesis in nonhabituated *D. carota*). Hauptmann and Widholm[23] were able to achieve wall regeneration and recovery growth in protoplasts of *D. carota* by a method based upon that of Takeuchi and colleagues,[55,56] but were unsuccessful with *Datura innoxia* and *Nicotiana tabacum*.

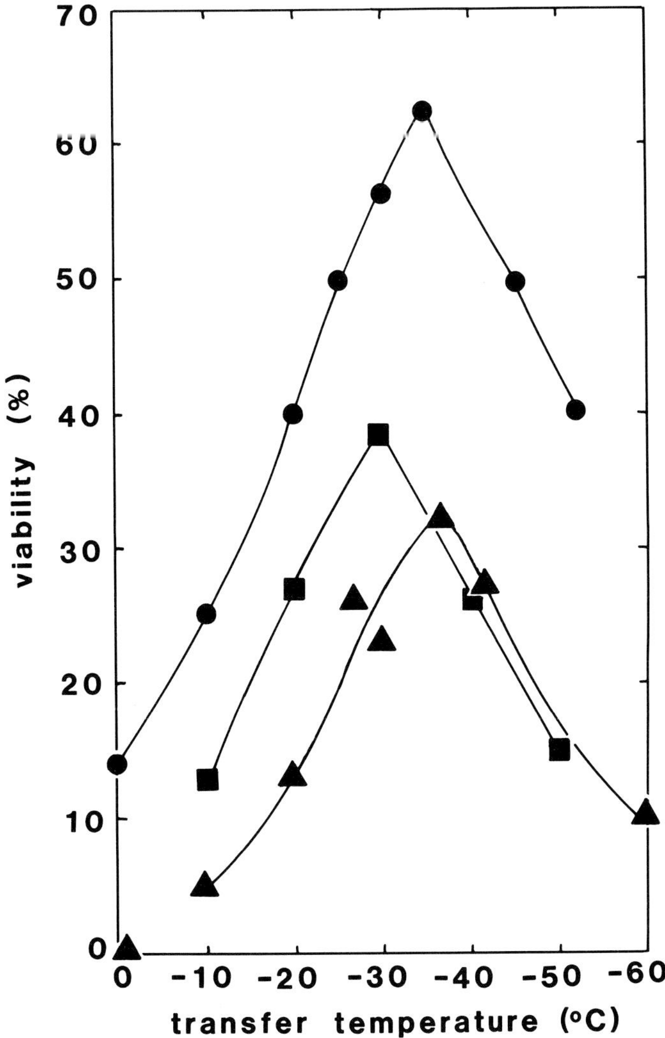

FIGURE 8. Viability (estimated by fluorescein diacetate staining[41]) of protoplasts of *Glycine max* (●), *Marchantia polymorpha* (■), and *Daucus carota* (▲) frozen slowly to various temperatures before plunging into liquid nitrogen and then thawing rapidly.[56]

Clearly, much more work is needed in this area. However, it is evident that protoplasts are capable of surviving cryopreservation in a way similar to cell suspensions and, given appropriate handling, the absence of a cell wall is not an impediment to survival.

IV. CONCLUDING REMARKS—PRESENT LIMITATIONS AND FUTURE PROSPECTS

If the current state of technical expertise in the cryopreservation of cultured cells and protoplasts is compared to that of only a few years ago, it can be seen that considerable progress has been made. Nonetheless, we should not be unaware of the limitations which prevent widespread implementation of storage by cryopreservation in the tissue culture laboratory.

The problems are different when we consider protoplasts as opposed to cells. Taking the former, the main problem with protoplasts is a general lack of work adequate in breadth and depth. The most notable success has been with *Daucus carota* which, as a model system, is so flexible in its requirements that it does not aid technical development. The studies of Takeuchi and colleagues[55,56] give hopeful indications that protoplasts of more species may be amenable to cryopreservation. Yet even in their work, the whole area of pregrowth and post-thaw manipulation to improve survival potential and recovery growth has been largely unexplored.

If cryopreservation is to be useful, particularly to those working in genetic manipulation, it will be necessary for techniques to be available which give good recovery rates in a very wide range of species. Further, the methods must be appropriate for protoplasts which have been the subjects of traumatic treatments such as mutagenesis and fusion that may well affect the response to freezing and thawing.

Turning now to cell suspensions, the problems here have until recently been similar to those with protoplasts, but as the volume of appropriate work is increasing, technical impediments to progress are becoming far fewer in number. Clearly, it remains to be seen whether all species are amenable to freezing and we should give particular attention to tropical and other cold-sensitive species, which are not well represented in successes to date. These will become increasingly important as cryopreservation is utilized more in plant genetic conservation.[5,6] Nonetheless, it can reasonably be anticipated that there should be few species which remain recalcitrant to improving cryopreservation methodology.

Problems which may hinder progress and the widespread, routine implementation of cryopreservation are more of a perceptual than technical nature. First, some studies indicate that small modifications, particularly to freezing rates and post-thaw procedures, yield success where none was reported before. Important as these may be, it is felt that there is a danger of oversophistication to the extent that special "tricks" seemingly will be required in many cases. Better that we should be looking for common denominators (especially in terms of the physiological condition of the specimens prior to freezing) than developing species-by-species modifications which may well be superfluous.

The second perceptual problem relates directly to implementation. It has been possible to cryopreserve some species very successfully for several years. During the last 2 to 3 years, routine methods have emerged and simple freezing units designed,[20] yet the laboratories adopting cryopreservation are very few in number. As long as maintenance in continuous growth satisfies most storage requirements (despite the expense and work load), many workers fail to embrace the convenience and unique security which cryopreservation offers. Perhaps it will take the serious loss of irreplaceable cell lines to illustrate the potential of cryopreservation as a service to the tissue culturist.

Taking the optimistic view that scientists will embrace cryopreservation methodology, we may anticipate that in the future higher plant cultures will be stored, exchanged, and treated as "biological currency" in the way that we presently observe for microbes, sperm, and animal cells.

ACKNOWLEDGMENTS

The author is grateful to several colleagues who made available unpublished and "in press" information. The Agricultural Research Council and Science and Engineering Research Council are thanked for financial support during the period when this chapter was prepared.

REFERENCES

1. **Quatrano, R. S.,** Freeze-preservation of cultured flax cells utilizing DMSO, *Plant Physiol.*, 43, 2057, 1968.
2. **Towill, L. E. and Mazur, P.,** Studies on the reduction of 2,3,5-triphenyl tetrazolium chloride as a viability assay for plant tissue cultures, *Can. J. Bot.*, 1097, 1974.
3. **Nag, K. K. and Street, H. E.,** Carrot embryogenesis from frozen cultured cells, *Nature (London)*, 245, 270, 1973.
4. **Bajaj, Y. P. S. and Reinert, J.,** Cryobiology of plant cell cultures and establishment of gene-banks, in *Applied and Fundamental Aspects of Plant Cell, Tissue and Organ Culture*, Reinert, J. and Bajaj, Y. P. S., Eds., Springer-Verlag, Basel, 1977, 757.
5. **Withers, L. A.,** *Tissue Culture Storage for Genetic Conservation*, No. AGP:IBPGR/80/8, Int. Board Plant Genet. Res., Rome, 1980.
6. **Withers, L. A.,** The storage of plant tissue cultures, in *Crop Genetic Resources — The Conservation of Difficult Material, Series B42*, Withers, L. A. and Williams, J. T., Eds., Int. Union Biol. Sci./Int. Board Plant Genet. Res., Rome, 1982, 49.
7. **Withers, L. A. and Street, H. E.,** Freeze-preservation of plant cell cultures, in *Plant Tissue Culture and Its Bio-technological Application*, Barz, W., Reinhard, E., and Zenk, M.-H., Eds., Springer-Verlag, Basel, 1977, 226.
8. **Towill, L. E. and Mazur, P.,** Effect of cooling and warming rates on survival of frozen *Acer saccharum* tissue cultures, *Plant Physiol.*, 54(Suppl.), 10, 1974.
9. **Withers, L. A. and Street, H. E.,** Freeze-preservation of cultured plant cells. III. The pregrowth phase, *Physiol. Plant.*, 39, 171, 1977.
10. **Finkle, B. J., Sugawara, Y., and Sakai, A.,** Freezing of carrot and tobacco suspension cultures, *Plant Physiol.*, 56 (Suppl.), 80, 1975.
11. **Hollen, L. B. and Blakely, L. M.,** Effect of freezing on cell suspensions of *Haplopappus ravenii*, *Plant Physiol.*, 56 (Suppl.), 39, 1975.
12. **Latta, R.,** Preservation of suspension cultures of plant cells by freezing, *Can. J. Bot.*, 49, 1253, 1971.
13. **Finkle, B. J. and Ulrich, J. M.,** Effect of combination of cryoprotectants on the freezing survival of sugar cane cells, in *Plant Cold Hardiness and Freezing Stress*, Li, P. J. and Sakai, A., Eds., Academic Press, New York, 1978, 373.
14. **Finkle, B. J. and Ulrich, J. M.,** Freezing survival of sugar cane cultures in mixtures of cryoprotective compounds, in *Abstracts: Fourth Int. Congr. Plant Tissue Cell Culture, Calgary*, Int. Assoc. Plant Tissue Culture, No. 1201, 103, 1978.
15. **Finkle, B. J. and Ulrich, J. M.,** Effect of cryoprotectants in combination on the survival of frozen sugar cane cells, *Plant Physiol.*, 63, 589, 1979.
16. **Nag, K. K. and Street, H. E.,** Freeze-preservation of cultured plant cells. I. The pretreatment phase, *Physiol. Plant.*, 34, 254, 1975.
17. **Nag, K. K. and Street, H. E.,** Freeze-preservation of cultured plant cells. II. The freezing and thawing phases, *Physiol. Plant.*, 34, 261, 1975.
18. **Sugawara, Y. and Sakai, A.,** Survival of suspension cultured sycamore cells cooled to the temperature of liquid nitrogen, *Plant Physiol.*, 54, 722, 1974.
19. **Withers, L. A.,** The freeze-preservation of synchronously dividing cultured cells of *Acer pseudoplatanus* L., *Cryobiology*, 15, 89, 1978.
20. **Withers, L. A. and King, P. J.,** A simple freezing unit and cryopreservation method for plant cell suspensions, *Cryo-Lett.*, 1, 213, 1980.
21. **Withers, L. A.,** The development of cryopreservation techniques for plant cell, tissue and organ cultures, in *Proc. Fifth Int. Congr. Plant Tissue Cell Culture*, Fujiwara, A., Ed., Maruzen, Tokyo, 1982, 793.
22. **Kartha, K. K. et al.,** Cryopreservation of periwinkle, *Catharanthus roseus* cells cultured *in vitro*, *Plant Cell Rep.*, 1, 135, 1982.
23. **Hauptmann, R. M. and Widholm, J. M.,** Cryostorage of cloned amino acid analog-resistant carrot and tobacco suspension cultures, *Plant Physiol.*, 70, 30, 1982.
24. **Bajaj, Y. P. S.,** Regeneration of plants from cell suspensions frozen at -20, -70, and $-196°C$, *Physiol. Plant.*, 37, 263, 1976.
25. **Dougall, D. K. and Wetherell, D. F.,** Storage of wild carrot cultures in the frozen state, *Cryobiology*, 11, 410, 1974.
26. **Withers, L. A.,** Freeze-preservation of somatic embryos and clonal plantlets of carrot (*Daucus carota* L.), *Plant Physiol.*, 63, 460, 1979.
27. **Anderson, J. O.,** Cryopreservation of apical meristems and cells of carnation (*Dianthus caryophyllus*), *Cryobiology*, 16, 583, 1979.
28. **Maddox, A., Gonsalves, F., and Shields, R.,** Successful preservation of plant cell cultures at liquid nitrogen temperatures, *Plant Sci. Lett.*, 28, 157, 1982.

29. **Shillito, R. D.,** *Problems Associated wit the Regulation of Auxotrophic Mutants from Plant Cell Tissue Cultures,* Ph.D. thesis, University of Leicester, Leicester, England, 1978.

30. **Cella, R., Sala, F., Nielsen, E., Rollo, F., and Parisi, B.,** Cellular events during the regrowth phase after thawing of freeze-preserved rice cells, in *Abstracts: Meeting of Fed. European Soc. Plant Physiol.,* Edinburgh, 127, 1978.

31. **Sala, F., Cella, R., and Rollo, F.,** Freeze-preservation of rice cells, *Physiol. Plant.,* 45, 170, 1979.

32. **Cella, R., Colombo, R., Galli, M. G., Nielsen, E., Rollo, F., and Sala, F.,** Freeze-preservation of *Oryza sativa* L. cells: a physiological study of freeze-thawed cells, *Physiol. Plant.,* 55, 279, 1982.

33. **Binder, W. and Zaerr, J. B.,** Freeze-preservation of suspension cultured cells of a hardwood poplar, *Cryobiology,* 17(Abstr.), 625, 1980.

34. **Binder, W. and Zaerr, J. B.,** Freeze-preservation of suspension cultured cells of a gymnosperm, Douglas-fir, *Crobiology,* 17(Abstr.), 624, 1980.

35. **Chen, W. H., Cockburn, W., and Street, H. E.,** Preliminary experiments on freeze-preservation of sugar cane cells, *Taiwania,* 24, 70, 1979.

36. **Finkle, B. J. and Ulrich, J. M.,** Cryoprotectant removal temperature as a factor in the survival of frozen rice and sugarcane cells, *Cryobiology,* 19, 329, 1982.

37. **Withers, L. A. and King, P. J.,** Proline — a novel cryoprotectant for the freeze-preservation of cultured cells of *Zea mays* L., *Plant Physiol.,* 64, 675, 1979.

38. **Withers, L. A.,** The cryopreservation of higher plant tissue and cell cultures — an overview with some current observations and future thoughts, *Cryo-Lett.,* 1, 239, 1980.

39. **Street, H. E.,** Cell (suspension) culture techniques, in *Plant Tissue and Cell Culture,* Street, H. E., Ed., Blackwell Scientific, Oxford, 1977, 61.

40. **Gould, A. R. and Street, H. E.,** Kinetic aspects of synchrony in suspension cultures of *Acer pseudoplatanus* L., *J. Cell Sci.,* 17, 337, 1975.

41. **Widholm, J. M.,** The use of fluorescein diacetate and phenosafranine for determining viability of cultured plant cells, *Stain Technol.,* 47, 189, 1972.

42. **Gaff, D. F. and Okongo-Ogola, O.,** The use of non-penetrating pigments for testing survival of cells, *J. Exp. Bot.,* 22, 756, 1971.

43. **King, P. J.,** Plant tissue culture and the cell cycle, in *Advances in Biochemical Engineering, Plant Cell Cultures II,* Vol. 18, Fiechter, A., Ed., Springer-Verlag, Basel, 1980, 1.

44. **Withers, L. A.,** Germplasm storage in plant biotechnology, in *Soc. Exp. Biol. Seminar Ser. 18: Plant Biotechnology,* Mantell, S. H. and Smith, H., Eds., Cambridge University Press, London, 1983. 187.

45. **King, P. J.,** personal communication, 1982.

46. **Pritchard, H. W., Grout, B. W. W., Reid, D. S., and Short, K. C.,** The effects of growth under water stress on the structure, metabolism and cryopreservation of cultured sycamore cells, in *The Biophysics of Water,* Franks, F. and Mathias, S. F., Eds., John Wiley & Sons, New York, 1982, 315.

47. **Withers, L. A.,** A fine-structural study of the freeze-preservation of plant tissue. II. The thawed state, *Protoplasma,* 94, 235, 1978.

48. **Pritchard, H. W.,** personal communication, 1982.

49. **Levitt, J.,** An overview of freezing injury and survival and its interrelationship to other stresses in *Plant Cold Hardiness and Freezing Stress,* Li, P. H. and Sakai, A., Eds., Academic Press, New York, 1978, 3.

50. **Kartha, K. K., Leung, N. L., and Gamborg, O. L.,** Freeze-preservation of pea meristems in liquid nitrogen and subsequent plant regeneration, *Plant Sci. Lett.,* 15, 1, 1979.

51. **Kartha, K. K., Leung, N. L., and Pahl, K.,** Cryopreservation of strawberry meristems and mass propagation of plantlets, *J. Am. Soc. Hortic. Sci.,* 105, 481, 1980.

52. **Ashwood-Smith, M. J. and Farrant, J.,** *Low Temperature Preservation in Medicine and Biology,* Pitman Medical, Tunbridge Wells, 1980.

53. **Withers, L. A.,** Low temperature storage of plant tissue cultures, in *Advances in Biochemical Engineering, Plant Cell Cultures II,* Vol. 18, Fiechter, A., Ed., Springer-Verlag, Basel, 1080, 101.

54. **Mazur, R. A. and Hartmann, J. X.,** Freezing of plant protoplasts., in *Plant Cell and Tissue Culture: Principles and Applications,* Sharp, W. R., Larsen, P. O., Paddock, E. F., and Raghavan, V., Eds., Ohio State University Press, Columbus, 1979, 876.

55. **Takeuchi, M., Matsushima, H., and Sugawara, Y.,** Long-term freeze-preservation of protoplasts of carrot and *Marchantia, Cryo-Lett.,* 1, 519, 1980.

56. **Takeuchi, M., Matsushima, H., and Sugawara, Y.,** Totipotency and viability of protoplasts after long-term freeze-preservation, in *Proc. 5th Int. Congr. Plant Tissue Cell Cult.,* Fujiwara, A., Ed., Maruzen, Tokyo, Japan, 1982, 797.

57. **Withers, A.,** unpublished data, 1982.

58. **Pegg, D. E. and Diaper, M. P.,** Cell packing and the recovery of human erythrocytes frozen and thawed in the presence of glycerol, *Cryobiology,* 17(Abstr.), 609, 1980.

59. **Caboche, M.,** Nutritional requirements of protoplast derived haploid tobacco *Nicotiana tabacum* cells grown at low cell densities in liquid medium, *Planta,* 149, 7, 1980.
60. **Withers, L. A. and Davey, M. R.,** A fine-structural study of the freeze-preservation of plant tissue cultures. I. The frozen state, *Protoplasma,* 94, 207, 1978.
61. **Withers, A.,** Preservation of germplasm, in *Int. Rev. Cytol. Perspectives in Plant Cell and Tissue Culture,* (Suppl. 11B), Vasil, I. K., Ed., Academic Press, New York, 1980, 101.

INDEX